NONTHERMAL FOOD PROCESSING TECHNOLOGIES

Impact on Color Profile

Postharvest Biology and Technology Series

NONTHERMAL FOOD PROCESSING TECHNOLOGIES

Impact on Color Profile

Edited by

Kaavya Rathnakumar, PhD
R. Pandiselvam, PhD

AAP APPLE ACADEMIC PRESS

First edition published 2025

Apple Academic Press Inc.
1265 Goldenrod Circle, NE,
Palm Bay, FL 32905 USA

760 Laurentian Drive, Unit 19,
Burlington, ON L7N 0A4, CANADA

CRC Press
2385 NW Executive Center Drive,
Suite 320, Boca Raton FL 33431

4 Park Square, Milton Park,
Abingdon, Oxon, OX14 4RN UK

Library and Archives Canada Cataloguing in Publication

CIP data on file with Canada Library and Archives

Library of Congress Cataloging-in-Publication Data

CIP data on file with US Library of Congress

ISBN: 978-1-77491-636-0 (hbk)
ISBN: 978-1-77491-637-7 (pbk)
ISBN: 978-1-00348-691-6 (ebk)

ABOUT THE BOOK SERIES: POSTHARVEST BIOLOGY AND TECHNOLOGY

Editor-in-Chief

Mohammed Wasim Siddiqui, PhD

Scientist-cum-Assistant Professor, Department of Food Science and Technology, Bihar Agricultural University, Sabour, Bhagalpur, Bihar, India

AAP Acquisitions Editor | Horticultural Science

Founding/Managing Editor | *Journal of Postharvest Technology*

As we know, preserving the quality of fresh produce has long been a challenging task. In the past, several approaches were in use for the postharvest management of fresh produce, but due to continuous advancement in technology, the increased health consciousness of consumers, and environmental concerns, these approaches have been modified and enhanced to address these issues and concerns.

The Postharvest Biology and Technology series features edited books that address several aspects related to postharvest technology of fresh produce. The series presents existing and novel management systems that are in use today or that have great potential to maintain the postharvest quality of fresh produce in terms of microbiological safety, nutrition, and sensory quality.

The books are aimed at professionals, postharvest scientists, academicians researching postharvest problems, and graduate-level students. This series is intended to be a comprehensive venture that provides up-to-date scientific and technical information focusing on postharvest management for fresh produce.

Books in the series address the following themes

- Nutritional composition and antioxidant properties of fresh produce
- Postharvest physiology and biochemistry
- Biotic and abiotic factors affecting maturity and quality
- Preharvest treatments affecting postharvest quality

- Maturity and harvesting issues
- Nondestructive quality assessment
- Physiological and biochemical changes during ripening
- Postharvest treatments and their effects on shelf life and quality
- Postharvest operations such as sorting, grading, ripening, de-greening, curing etc.
- Storage and shelf-life studies
- Packaging, transportation, and marketingVase life improvement of flowers and foliage
- Postharvest management of spice, medicinal, and plantation crops
- Fruit and vegetable processing waste/byproducts: management and utilization
- Postharvest diseases and physiological disorders
- Minimal processing of fruits and vegetables
- Quarantine and phytosanitory treatments for fresh produce
- Conventional and modern breeding approaches to improve the postharvest quality
- Biotechnological approaches to improve postharvest quality of horticultural crops

BOOKS IN THE SERIES

Postharvest Biology and Technology of Horticultural Crops: Principles and Practices for Quality Maintenance
Editor: Mohammed Wasim Siddiqui, PhD

Insect Pests of Stored Grain: Biology, Behavior, and Management Strategies
Ranjeet Kumar, PhD

Postharvest Management of Horticultural Crops: Practices for Quality Preservation
Editors: Mohammed Wasim Siddiqui, PhD, and Asgar Ali, PhD

Innovative Packaging of Fruits and Vegetables: Strategies for Safety and Quality Maintenance
Editors: Mohammed Wasim Siddiqui, PhD,
Mohammad Shafiur Rahman, PhD, and Ali Abas Wani, PhD

Advances in Postharvest Technologies of Vegetable Crops
Editors: Bijendra Singh, PhD, Sudhir Singh, PhD, and Tanmay K. Koley, PhD

Plant Food By-Products: Industrial Relevance for Food Additives and Nutraceuticals
Editors: J. Fernando Ayala-Zavala, PhD, Gustavo González-Aguilar, PhD, and Mohammed Wasim Siddiqui, PhD

Emerging Technologies for Shelf-Life Enhancement of Fruits
Editors: Basharat Nabi Dar, PhD, and Shabir Ahmad Mir, PhD

Sensor-Based Quality Assessment Systems for Fruits and Vegetables
Editors: Bambang Kuswandi, PhD, and Mohammed Wasim Siddiqui, PhD

Emerging Postharvest Treatment of Fruits and Vegetables
Editors: Kalyan Barman, PhD, Swati Sharma, PhD, and Mohammed Wasim Siddiqui, PhD

Packaging and Storage of Fruits and Vegetables: Emerging Trends
Editor: Tanweer Alam, PhD

Postharvest Physiological Disorders of Fruits
Editors: Ghulam Khaliq, PhD, and Sajid Ali, PhD

Bioengineered Fruit and Vegetables: Trends in Postharvest Quality Improvement
Editors: Mohammad Irfan, PhD, and Pankaj Kumar, PhD

Novel Packaging Systems for Fruits and Vegetables
Editors: Kirtiraj K. Gaikwad, PhD, and Suman Singh, PhD

Ozone Applications in the Food Industry: Trends and Future Prospects
Editors: Ravi Pandiselvam, PhD, Anjineyulu Kothakota, PhD, and Kaavya Rathnakumar, PhD

Nonthermal Food Processing Technologies: Impact on Color Profile
Editors: Kaavya Rathnakumar, PhD, and R. Pandiselvam, PhD

Ready-to-Eat Snacks: Emerging Technologies for Production and Safety
Editors: R. Pandiselvam, PhD and Gülşah Çalışkan Koç, PhD

ABOUT THE SERIES EDITOR

Mohammed Wasim Siddiqui, PhD Scientist-cum-Assistant Professor, Department of Food Science and Technology, Bihar Agricultural University, Sabour, Bhagalpur, Bihar, India, and author or co-author of peer-reviewed research articles, book chapters, two manuals, and conference papers. He has over 20 edited and an authored books to his credit, published by Elsevier, CRC Press, Springer, and Apple Academic Press. Dr. Siddiqui has established the international peer-reviewed *Journal of Postharvest Technology.* He is Editor-in-Chief of two book series (*Postharvest Biology and Technology and Innovations in Horticultural Science*), published by Apple Academic Press. Dr. Siddiqui is also a Senior Acquisitions Editor for Apple Academic Press, USA, for Horticultural Science. He has been serving as an editorial board member and active reviewer of several international journals.

Dr. Siddiqui has received several grants and respected awards for his research work by a number of organizations. He is an active member of the organizing committees of several national and international seminars, conferences, and summits. He is one of key members in establishing the World Food Preservation Center (WFPC), LLC, USA, and is currently an active associate and supporter.

Dr. Siddiqui acquired a BSc (Agriculture) degree from Jawaharlal Nehru Krishi Vishwa Vidyalaya, Jabalpur, India. He received his MSc (Horticulture) and PhD (Horticulture) degrees from Bidhan Chandra Krishi Viswavidyalaya, Mohanpur, Nadia, India, with specialization in Postharvest Technology. He is a member of Core Research Group at the Bihar Agricultural University (BAU), which provides appropriate direction and assistance in prioritizing research.

ABOUT THE EDITORS

Kaavya Rathnakumar, PhD
Senior Food Scientist, Balchem Human Nutrition and Health, St. Louis, Missouri, USA

Kaavya Rathnakumar, PhD, is a Senior Food Scientist at Balchem Human Health and Nutrition in St. Louis, Missouri, USA. Previously, she was an innovation scientist at Beyond Meat, Los Angeles, California, and a postdoctoral research associate/director at Frozen Dessert Center, Department of Food Science at the University of Wisconsin–Madison, USA. Her expertise includes waste by-product utilization using nonthermal and sustainable techniques, product and process development, ingredient formulations for dairy and non-dairy frozen desserts and beverages, plant-based meat analogs, strategic long-term innovation in dairy and plant-based proteins, milk lipids and phospholipids, structured fats, and emulsions through novel process engineering techniques. She is a member and plays various leadership roles in the Institute of Food Technologists (IFT), the American Association of Agricultural and Biological Engineers (ASABE), the American Dairy Science Association (ADSA), and the American Oil and Chemists Society (AOCS). She has received nine international research awards from the above-mentioned organizations. In 2020, she was honored with the prestigious John Brandt Award from Land O'Lakes, Midwest Dairy, USA. Additionally, she was awarded the 2022 Women in Engineering Award from the American Association of Agricultural and Biological Engineers (ASABE). She has published 22 peer-reviewed articles, six book chapters, and three books, and has presented over 30 conference presentations.

Dr. Rathnakumar earned her PhD in Dairy and Food Science at South Dakota State University, USA. She earned a Master of Engineering in Food Engineering and Bioprocess Technology at the Asian Institute of Technology (AIT, Thailand) and her bachelor's in Food Processing Engineering at the Indian Institute of Food Processing Technology (IIFPT, India).

R. Pandiselvam, PhD
Scientist, Physiology, Biochemistry, and Post-Harvest Technology Division, ICAR-Central Plantation Crops Research Institute, Kasargod, India

R. Pandiselvam, PhD, is a young academician actively engaged in food engineering research. He is currently a Scientist in the Physiology, Biochemistry and Post-Harvest Technology Division at the ICAR-Central Plantation Crops Research Institute, Kasaragod, India. During the last five years, he has contributed in the design and development of a minimal processing machine for tender coconut, a continuous type coconut testa removing machine, tender coconut punching and cutting machine, and a preservation protocol for trimmed tender coconut. He was also a developer/co-developer of many value-added food products (Kalpa Krunch, Frozen Coconut Delicacy, matured coconut water-based beverages), which are commercialized by many entrepreneurs. Dr. Pandiselvam has authored over 185 articles published in national and international journals of repute. He has been the author or coauthor of five books and 60 book chapters. Dr. Pandiselvam is the Associate Editor of *Measurement: Food* (Elsevier) and *Frontiers in Nutrition*. Dr. Pandiselvam also features as an editorial board member of *Food & Humanity, Ozone Science and Engineering, Journal of Texture Studies,* and *Quality Assurance, and Safety of Crops & Foods Journal*. He was conferred with a National Academy of Agricultural Sciences (NAAS) Young Scientist Award, NAAS Associate and Indian Society of Agricultural Engineers (ISAE)-Distinguished Service Award.

CONTENTS

CONTRIBUTORS

Amirhossein Abedini
Department of Environmental Health, Food Safety Division, Faculty of Public Health, Tehran University of Medical Sciences, Tehran, Iran; Students' Scientific Research Center, Tehran University of Medical Sciences, Tehran, Iran

C. Ashokkumar
Department of Food Technology, College of Food and Dairy Technology, TANUVAS, Chennai, Tamil Nadu, India

Nancy Awasti
Sensory Scientist, Lactalis American Group, Buffalo, New York, USA

Karthik Sajith Babu
Department of Animal Sciences and Industry, Food Science Institute, Kansas State University, Manhattan, Kansas, USA

Gayathri Balakrishnan
Department of Food Science and Human Nutrition, University of Florida, Gainesville, Florida; Noblesville, Indiana, USA

Khilendra Bhanduriya
Sr. Product Development Scientist, Milk Specialties Global, Eden Prairie, Minnesota, USA

Anto Pradeep Raja Charles
Doctoral Research Assistant, Food Ingredients and Biopolymer Laboratory, Department of Plant Sciences, North Dakota State University, Fargo, North Dakota, USA

U. Gnana Moorthy Eswaran
PhD Research Scholar, Department of Agricultural and Food Engineering, Indian Institute of Technology, Kharagpur, West Bengal, India

Proshanta Guha
Professor, Department of Agricultural and Food Engineering, Indian Institute of Technology, Kharagpur, West Bengal, India

Osorio-Arias Juan
Corporación Universitaria Minuto de Dios – Uniminuto, Research Direction, Bogotá DC, Colombia

Sangeetha Karunanithi
PhD Research Scholar, Department of Agricultural and Food Engineering, Indian Institute of Technology, Kharagpur, West Bengal, India

Anandu Chandra Khanashyam
Department of Food Science and Technology, Kasetsart University, Ladyao, Chatuchak, Bangkok, Thailand

Chatrapati Veera Raghava Kumar
Scientist, Idaho Milk Products, Idaho, USA

Pulivarthi Manoj Kumar
PhD Scholar, Department of Grain Science and Industry, Kansas State University, Manhattan, Kansas, USA

Gokulprasanth Murugan
PG Research Scholar, Department of Fish Processing Technology, Dr. M.G.R. Fisheries College and Research Institute, Tamil Nadu Dr. J. Jayalalithaa Fisheries University, Ponneri, Tamil Nadu, India

Pankaj Parmar
Parul Institute of Technology, Parul University, Faculty of Agriculture, Dairy Technology Department, Gujarat, India

Akshay R. Patil
National Institute of Food Technology, Entrepreneurship, and Management, Thanjavur, Tamil Nadu, India

Rajib Lochan Poudyal
Japan Ginger Company, Koch-Shi, Kochi, Japan

Guru Yeshwanth Subbiah Prabhakaran
Graduate Research Scholar, Department of Food Science, University of Wisconsin, Madison, Twin Falls, Idaho, USA

Bharathipriya Rajasekaran
PhD Research Scholar, International Center of Excellence in Seafood Science and Innovation, Faculty of Agro-Industry, Prince of Songkla University, Hat Yai, Songkhla, Thailand

Dakshayani Rajendran
PhD Research Scholar, Food Science and Technology, Department of Food Product Development, National Institute of Food Technology, Entrepreneurship, and Management, Thanjavur (NIFTEM-T), Tamil Nadu, India

Ramnarain Ramakrishna
Postdoctoral Scientist, Food Ingredients and Biopolymer Laboratory, Department of Plant Sciences, North Dakota State University, Fargo, North Dakota, USA

Jaganmohan Rangarajan
Professor and Head, School of Sensory Science, Department of Food Product Development, National Institute of Food Technology, Entrepreneurship, and Management, Thanjavur (NIFTEM-T), Tamil Nadu, India

Jiajia Rao
Associate Professor, Food Ingredients and Biopolymer Laboratory, Department of Plant Sciences, North Dakota State University, Fargo, North Dakota, USA

Kaavya Rathnakumar
Department of Food Science, University of Wisconsin, Madison, USA

Kartik Shah
Technology Principal, Sargento Foods, Elkhart Lake, Wisconsin, USA

Sachin Sharma
Department of Animal, Veterinary, and Food Sciences, University of Idaho, Moscow

Pratiksha Shrestha
Department of Food Technology and Quality Control (DFTQC), Kathmandu, Nepal

Arshdeep Singh
Division of Food, Nutrition, and Exercise Sciences, University of Missouri, Columbia, Missouri

Lovepreet Singh
Department of Animal, Veterinary, and Food Sciences, University of Idaho, Moscow

Prem Prakash Srivastav
Professor, Department of Agricultural and Food Engineering, Indian Institute of Technology, Kharagpur, West Bengal, India

Venkateswarlu Sunkesula
Department of Food and Dairy Science, Midwest Dairy Foods Research Center, South Dakota State University, Brookings, South Dakota, USA; VP Research and Product Development, Idaho Milk Products, Twin Falls, Idaho, USA

Priyamvada Thorakkattu
Department of Animal Sciences and Industry, Food Science Institute, Kansas State University, Manhattan, Kansas, USA

Pranav Vashisht
Food Biosciences and Technology Program, Department of Agricultural and Environmental Sciences, Tennessee State University, Nashville, Tennessee, USA

Karthikeyan Venkatachalam
Associate Professor, Faculty of Innovative Agriculture and Fishery Establishment Project, Prince of Songkla University, Surat Thani Campus, Makham Tia, Muang, Surat Thani, Thailand

Pratishtha Verma
Department of Dairy and Food Science, South Dakota State University, Brookings, South Dakota, USA; Graduate Research Scholar, Department of Food Science, South Dakota State University, Twin Falls, Idaho, USA

Mohana Yoganandan
Graduate Research Scholar, Department of Food Science, Kansas State University, Findlay, Ohio, USA; Scientist, The Mennel Milling Company, Fostoria, Ohio, USA

Duarte-Correa Yudy
BIOALI Research Group, Food Department, Faculty of Pharmaceutical and Food Sciences, Universidad de Antioquia, Medellín, Colombia

PREFACE

Food is the most basic commodity in daily life, and food technology keeps evolving to meet the needs of humans. In food products, the consumer often assesses the initial quality of the product by its color and appearance. The color of fresh produce and processed food products is the primary indicator to judge its freshness and quality. The functional properties of foods, such as pigments, antioxidants, and bioactive compounds, are affected during processing and preservation. Many processes involved in the preservation and shelf-life extension of food cause detrimental effects on the color and appearance. Retention of the color properties in food products requires innovative nonthermal approaches for process design because of their sensitivity to a variety of physical and chemical factors. Alternative methods for thermal processing of food are gaining importance due to increased consumer demand for minimally processed fresh-like food products with high sensory, appearance, and nutritional qualities.

This book gives an overview of the science today, especially nonthermal treatment, and covers all aspects of this multidisciplinary field. It offers comprehensive coverage of the application of emerging technologies on the color profile of different food products such as fruits, vegetables, beverages, dairy, and meat. In this context, there is a growing interest in nonthermal processes in food processing. This book is a collection of the most important emerging nonthermal technologies such as cold plasma, ozone, pulsed light, UV light, pulsed electric field, high-pressure processing, etc. The content of the book offers the influence and impact of these emerging technologies on the color and appearance of foods and beverages, along with the challenges and prospects. This book provides a great context on the application of emerging food processing technologies to solve food safety issues and enhance shelf-life extension, which is the current need. The book provides details about the latest advancements in the design and efficiency of emerging food technologies and their impact on color.

In brief, the readers will explore in-depth information regarding emerging technology and its role in food processing and the quality of food reaching the plate of the consumer. It will also provide a comprehensive overview of future aspects. This book serves as a one-stop reference for food technologists, processing industries, and nutritionists looking forward to understanding the implications of emerging food processing technologies on the color profile of foods.

We expect this book to be a great resource for all the food engineers, processors, and industries who are looking for a future in nonthermal techniques. We are indebted to the valuable contributions made by the most knowledgeable authors who are experts in this field.

We take this opportunity to thank the Almighty God for strength and blessings; we would also like to thank family, friends, and mentors who were very supportive in every way, and last but not least, once again, to all the contributors for making valuable contributions as book chapters. We also extend our sincere thanks to the editorial office of Apple Academic Press, Inc. for their help and support throughout the assignment.

—Kaavya Rathnakumar, PhD,
R. Pandiselvam, PhD

CHAPTER 1

FOOD PIGMENTS AND COLOR MEASUREMENT TECHNIQUES

GURU YESHWANTH SUBBIAH PRABHAKARAN,[1] MOHANA YOGANANDAN,[2] and PRATISHTHA VERMA[3]

[1]*Department of Food Science, University of Wisconsin, Madison, Twin Falls, Idaho, USA*

[2]*Department of Food Science, Kansas State University, Findlay, Ohio, USA*

[3]*Department of Food Science, South Dakota State University, Twin Falls, Idaho, USA*

ABSTRACT

Pigments play a vital role in making this world colorful and beautiful. For centuries, these colors have helped humans distinguish foods and ensure safe food consumption. In plant systems, they play a crucial role by attracting pollinators, aiding photosynthesis, and offering protection against environmental and physiological stressors. Since the early 19th century, artificial colors have dominated due to their ease of production, application, and wider range of color hues compared to natural pigments. However, the detrimental effects of artificial pigments on human health, coupled with the numerous health benefits of natural food pigments, have led scientists and food manufacturers to gain insights into the extraction and applications of the latter in food products. This chapter delineates the structure, functionality, stability, degradation pathways, extraction, and analysis of four major natural food pigments – chlorophylls, anthocyanins

Nonthermal Food Processing Technologies: Impact on Color Profile.
Kaavya Rathnakumar and R. Pandiselvam (Eds.)

(ACNs), carotenoids, and betalains. The presence of a chromophore structure in all these pigments helps them function as light-absorbing components and antioxidants. The same chromophore structure also makes them susceptible to degradation via several agents. Understanding the details of how their structure is affected by environmental changes would give us immense understanding of their extraction and viability in food applications. In the past five decades, the potential of these pigments as antioxidants and the numerous health benefits they impart to humans have garnered considerable interest. Several *in vitro* studies have evaluated the benefits of these pigments as food and nutraceuticals. The functionality of the pigments in plant systems and human health has been briefly discussed in this chapter, in addition to safe and efficient extraction techniques needed to retain the chemical structure and antioxidant properties.

1.1 INTRODUCTION – NATURAL FOOD PIGMENTS

Since the dawn of mankind, fruits and vegetables have been consumed for survival and well-being. Humans have learned and evolved by judging the color and features of fruits and vegetables and associating their appearance with their functions. With the rise of civilizations, humans started cultivating fruits and vegetables that were healthy and nutritious for consumption. The appearance of these cultivated produce has played a huge role in their acceptance from then until today. Fresh-looking produce with vibrant colors is usually sorted and sold at premium costs, while faded produce is sold at economical costs. In addition to the impact of the products' color on their direct consumption, their impact on food products made from them is also significantly important for their acceptance. The chemical compounds that produce vibrant colors in fruits and vegetables are called pigments. Understanding the nature of these pigments and their role in food coloration and human health goes a long way in designing both nutritious and aesthetically appealing foods.

Food pigments are broadly classified into two categories: natural pigments and artificial (synthetic) food colorants. Natural food pigments are those that occur naturally in food produce. Most foods, especially fruits and vegetables, are colored by four natural food pigments, namely chlorophylls, carotenoids, ACNs, and betanins [1]. Given the poor stability of natural pigments in processing lines, artificial colorants revolutionized the processed food industry. With the advancement of science, a wide range of artificial or synthetic colorants has been synthesized in the past couple of centuries to be incorporated into manufactured food products, improving their appearance and appeal [2]. Although artificial food colorants are very versatile and capable of achieving hues that natural pigments cannot achieve, their negative impact on the health of consumers has brought a lot of attention to their use in food products. This has led several researchers to explore the benefits of natural pigments, their extraction, and their application in food products. Understanding the role these pigments play in plant systems, their stability, and degradation pathways helps us understand their functionality and behavior, thereby helping us design processing steps to safely extract these pigments without losing their functionality.

This chapter discusses the structure, functionality, stability, degradation, extraction, and analysis of the four major natural pigments.

1.2 CHLOROPHYLLS

1.2.1 STRUCTURE AND FUNCTIONALITY

Chlorophylls provide green pigmentation in fruits and vegetables. They are natural pigments primarily identified in photosynthetic organisms such as plants and certain bacteria. They play a crucial role in photosynthesis and act as indicators of ripeness and quality. Chlorophylls are highly unstable and prone to degradation in senescing leaves and ripening fruits. In plants, chlorophylls are primarily found in subcellular organelles called chloroplasts. Here, they are arranged along with other polypeptides to trap incoming light particles, which are in turn used to generate carbohydrates and amino acids using various chemical pathways. Chlorophyll molecules are relatively large complexes consisting of a porphyrin ring (1.5 nm^2) and a phytyl side chain [2 nm (nanometer)] (Figure 1.1). The porphyrin chromophore, with its Mg bonding, forms an extensive ring structure and is responsible for trapping energy from photons [3]. The major types of chlorophylls are chlorophyll α, chlorophyll β, chlorophyll χ, and chlorophyll δ. Chlorophyll α and chlorophyll β are predominantly found in higher photosynthetic plants, with chlorophyll α usually being two or three times more predominant than chlorophyll β, while chlorophyll χ and d are found in algal and diatomic species [4]. The different chlorophylls vary depending on the side chains in their porphyrin ring. The slight structural changes dictate which wavelengths of light can be absorbed and affect the pigmentation they produce [3]. In a living cell, chlorophylls are found within the chloroplast membranes, also known as thylakoids.

Here, they exist as complexes between chlorophylls and polypeptide complexes called light-harvesting complexes (LHCs). Depending on the organism, chlorophylls are often associated with carotenes, xanthophylls, and tocopherols. While chlorophylls are responsible for generating excitation energy through photosynthesis, carotenoids and tocopherols act as scavengers to absorb excess excitation energy and act as traps for singlet oxygen and other peroxides.

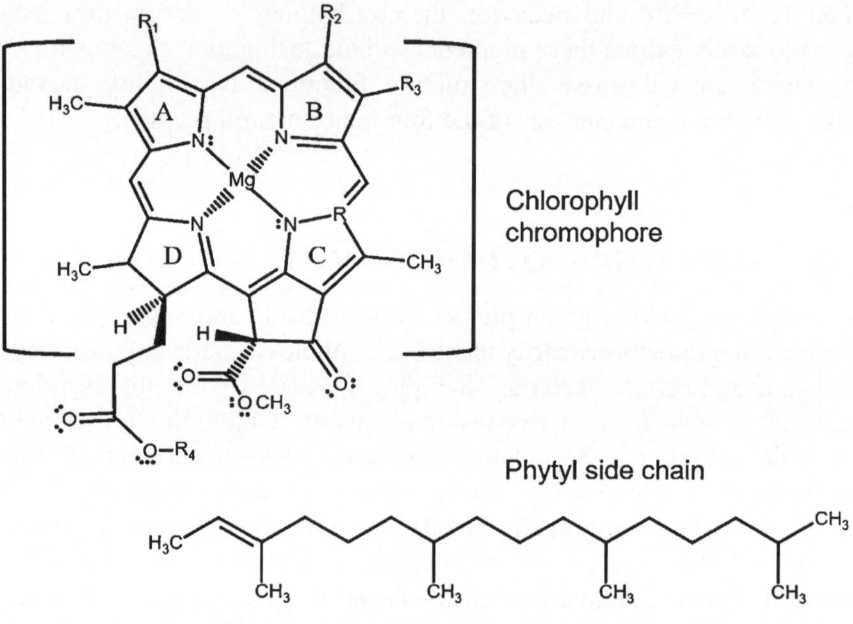

Pigment	R_1 Substitution	R_2 Substitution	R_3 Substitution	R_4 Substitution
Chlorophyll α	$-CH=CH_2$	$-CH_3$	$-CH_2-CH_3$	Phytyl chain
Chlorophyll β	$-CH=CH_2$	$-CHO$	$-CH_2-CH_3$	Phytyl chain
Chlorophyll χ	$-CH=CH_2$	$-CH_3$	$-CH_2-CH_3$	H
Chlorophyll δ	$-CHO$	$-CH_3$	$-CH_2-CH_3$	Phytyl chain

FIGURE 1.1 Structural differences between four major types of chlorophylls.

1.2.2 STABILITY

The application of chlorophyll pigments in food products as food colorants has been extensively studied, and understanding their stability in food processing unit operations is crucial. While magnesium (Mg) is essential for the functionality of chlorophyll, it also has a negative impact on its stability. Heating chlorophylls at 70°C for 5 minutes under mildly acidic conditions is sufficient to generate more than six colorless degradative products [5]. This loss of color occurs due to the replacement of Mg by two hydrogen ions, resulting in the formation of pheophytin derivatives. Chlorophyll molecules containing Mg are highly susceptible to oxidation, especially when exposed to light. Although this process also takes place in living organisms, the destabilization caused by electron excitation is immediately balanced by electron rearrangement during photosynthesis. However, in laboratory settings, this destabilization leads to the formation of colorless end products. Severe heating or acidification conditions, such as those used in canning spinach, result in the formation of pyropheophorbides [6]. To mitigate degradation and discoloration, researchers have explored methods such as high-temperature short-time (HTST) processing and pH adjustment. While these methods have shown some success in maintaining green coloration during processing, significant discoloration still occurs during storage. Industrial manufacturers of chlorophyll food colorants overcome this challenge by replacing Mg in chlorophyll pheophytins formed during heat processing with copper (Cu) or zinc (Zn). Copper or zinc substituted chlorophyll molecules, also known as pheophytins, are very stable when exposed to light and do not readily dissolve in mineral acids. This technique, termed Re-Greening, is done by adding zinc or copper salts to foods like spinach or broccoli before canning. The pheophytins and pheophorbides, formed during heating, complex with the salts and provide stability against further degradation and discoloration [7]. Another way to improve chlorophyll stability is to reduce water activity. Since the degradation of chlorophyll molecules to pheophytins under thermal conditions is mediated by water, reducing water activity causes chlorophyll molecules to be bound in non-reactive compartments, thus reducing degradation [8].

1.2.3 DEGRADATION

Degradation and simultaneous synthesis of chlorophyll are naturally occurring phenomena that constantly occur throughout the organism's lifespan. The rate of degradation is highest during the seedling phase, leaf senescence, and fruit maturation [9]. During degradation phases such as senescence and ripening, chlorophylls undergo irreversible degradation along with physical changes to the pigments. Various by-products are formed depending on the cause of degradation – natural vs. unnatural. Understanding the breakdown mechanisms and by-products would provide us with a good understanding of the stability of chlorophylls and the effect of processing on them. One natural pathway for chlorophyll degradation is caused by lipid-protein chlorophyllase which cleaves the phytol ester to form pheophorbide at Mg II or chlorophyllide. This enzymatic degradation also occurs during blanching, resulting in a significant color change from green to olive-brown due to the activation of the chlorophyllase enzyme under this mild thermal treatment [10]. Another degradation pathway involves the loss of Mg induced by acids. Acids are well-known for their ability to dissolve metallic salts. Under *in-vitro* conditions and in the presence of acid, the chlorophyll molecule loses its Mg and forms pheophytins, while chlorophyllide molecules form pheophorbides. Demetallized pheophorbide molecules are oil-soluble chlorophyll molecules without copper, which serve as the starting material for various chlorophyll-derived food colorants. Water-soluble pheophorbides are other derivatives obtained through controlled cleavage of the pentanone ring. However, the complex structure of porphyrins and their unpredictable aggregation behavior in solutions make it challenging to obtain reproducible results from this hydrolysis. Pheophorbides, in turn, break down into different products depending on the source of degradation, such as decarboxyphytoporphyrin (heat + anoxia + time), phytoporphyrin (gut digestion), phyllophorbide (alkali solvation effect + heat), didehydro rhodochrosite (alkali solvation), and pheophorbideatoCuII (gut acid). Regardless of the source of degradation, a significant portion of the chlorophyll molecules ultimately decompose into colorless products, with geo- or petro-porphyrins found in oils and coals being the most stable end form [9].

1.2.4 EXTRACTION AND ANALYSIS

Chlorophylls and their derivatives vary widely in polarity, making it challenging to establish an effective method for their analysis. Extraction and sample handling are crucial factors that can significantly impact the results. While chlorophylls, pheophytins, and pyro pheophytins are relatively nonpolar, chlorophyllides and pheophorbides are polar. Therefore, the choice of solvent plays a major role in achieving successful extraction and analysis. Various solvents such as methanol, ethanol, acetone, and petroleum ether have been reported in many studies for the extraction of chlorophyll and its derivatives [11]. The Association of Official Agricultural Chemists (AOAC) recommends maceration of the substrate containing chlorophylls, followed by the addition of extraction solvents and filtering the solvent along with solubilized chlorophylls. Due to the high photosensitivity of chlorophylls, they should be analyzed immediately or dried and stored at freezing temperatures under non-reactive atmospheres such as argon or nitrogen [4].

The porphyrin structure of the chlorophyll molecule is responsible for its absorption spectra and can be observed as four main spectral bands in the range of 500 to 700 nm, as well as a substantial band at approximately 400 nm. The large band observed at around 400 nm in the near-violet region is also known as the Soret band, which specifically corresponds to the porphyrin ring structure of the chlorophyll molecule and its derivatives. The absence of this band indicates a broken porphyrin structure. Chlorophyll derivatives exhibit different absorption behavior due to structural changes, resulting in slightly different shapes and band positions in absorption spectra. The absorption spectra are also influenced by the dielectric properties of the extraction solvent. Polar solvents such as methanol, acetone, ethanol, and water tend to shift the absorption spectra towards longer wavelengths, while reducing light absorption and specificity [12]. To facilitate analysis and transferability, many researchers have determined the coefficient of specific absorption for pure pigments in different solvent systems and derived equations to calculate their respective concentrations. Some commonly used solvents for developing these equations include diethyl ether (for chlorophylls a and b) [13, 14], 80% ethanol (for pheophytins) [15], and 80% acetone [16, 17].

In addition to spectrophotometric analysis, other methods such as open-column chromatography [18], thin-layer chromatography (TLC), and reverse-phase high-performance liquid chromatography (RP-HPLC) have been successfully employed to analyze chlorophyll and its derivatives. RP-HPLC has predominated over open-column chromatography and TLC since it is rapid and reliable and C-18 columns are the most popular columns for chlorophyll analysis in food samples. Since chlorophylls and their derivatives vary in their degree of hydrophobicity, complex elution gradients with highly polar solvents like methanol with acetone are typically employed for analysis. Photodiode array detection (PAD) in combination with fluorescence detection has been most successful in achieving greater sensitivity and specificity in analyzing chlorophyll and its derivative pigments [19]. Optimal resolution of low pH pigments such as pheophorbides and chlorophylls, which has a $pK_a$of 2 to 4, have been obtained using techniques like ion-suppression and ion-pairing. Buffers such as ammonium acetate provide proton equilibrium and help in separating high polarity, ionizable chlorophyll derivatives such as pheophorbides and chlorophyllides [20]. Metal chlorophyll derivatives such as zinc and copper pheophytins are also detected utilizing an RPC-18 column via both gradient [21] and isocratic elution methods [22].

1.3 CAROTENOIDS

1.3.1 STRUCTURE AND FUNCTIONALITY

Carotenoids are widespread natural food pigments found in plants, insects, birds, and animals. They provide yellow, orange, or red colors and are highly sought after as natural sources for food colorants. Carotenoid synthesis has been a part of early evolution, and while bacteria and algae still retain this ability, higher animals have lost this ability. Hence, most of the carotenoids seem to come from their diet. Major subgroups of carotenoids include carotenes (e.g., lycopene and β-carotene), which are hydrocarbon carotenoids, and oxycarotenoids or xanthophylls (e.g., lutein and zeaxanthin), which are derivatives of carotenes with oxygenated functional groups like hydroxy, keto, epoxy, methoxy, or carboxylic groups [3]. Carotenoids have a tetraterpenoid structure containing eight isoprene units connected by conjugated double bonds, and they primarily exist in the *E*-isomeric form (all-*trans*) in nature. Figure 1.2 shows the most common carotenoids found in nature. The most basic carotene is acyclic lycopene. Other carotenoids are slightly modified with the presence of a six- or five-carbon ring structure on one or both ends of the lycopene molecule. The carotenoid molecules with the end groups removed are known as *apocarotenoids*. The most abundant natural carotenoid is fucoxanthin, found in the photosynthetic membrane of marine brown algae, while the most common carotenoids found in green leaves include β-carotene, lutein, violaxanthin, and neoxanthin [23]. While the carotenoid composition in leaves of most species is usually similar, the total carotenoid content changes between different species and varieties. Some examples of unique carotenoid patterns include those found in tomato (lycopene and its hydroxyl derivatives), peach (β-carotene and its hydroxyl derivatives), and red pepper (capsanthin).

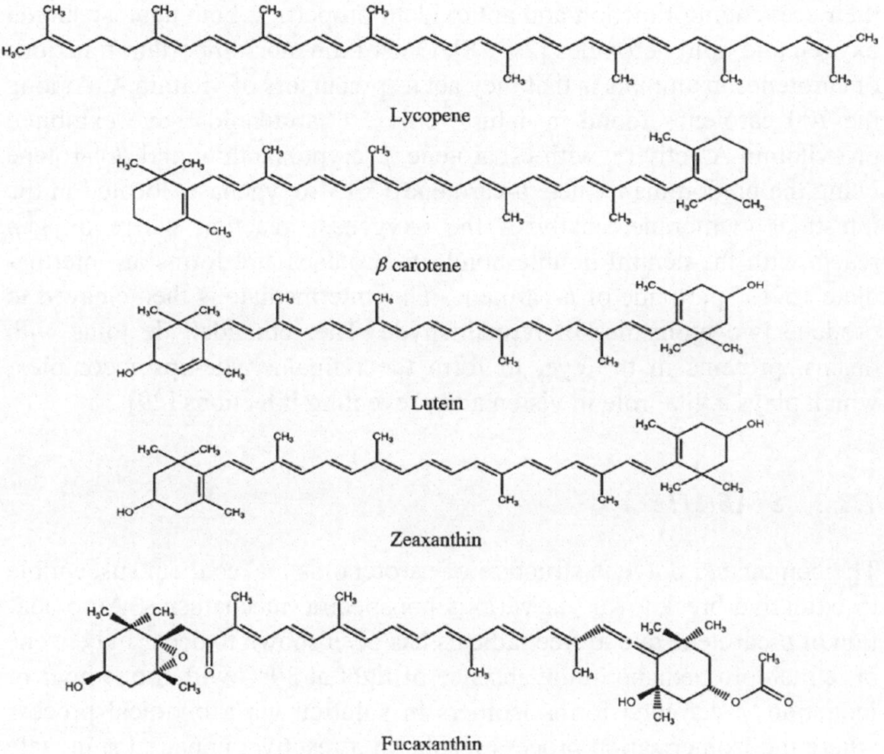

FIGURE 1.2 Structures of commonly found carotenoids.

Carotenoids' light-absorbing properties and their susceptibility to oxidative degradation can be attributed to their long chromophore structure and the presence of 3 to 13 conjugated double bonds. Although their role in primary photochemistry and electron transfer is unclear, their role in absorbing excess electron excitation energy and thereby protecting photosynthetic systems is well established. Within chloroplast thylakoid membranes, carotenoids are found in photosystems 1 and 2 [24]. Here, carotenoids (mainly xanthophylls) function as scavengers of triplet chlorophyll and singlet oxygen generated due to excess absorption of light energy. Carotenoids have also exhibited a positive impact on human health by providing protection against cancer, heart diseases, certain skin conditions, and even providing anti-aging properties [25, 26]. Like their function of protecting photosynthetic systems, carotenoids are also known to provide protection against porphyrin molecules in the skin through their scavenging function and antioxidant properties, both against singlet oxygen and lipid peroxides [27, 28]. One of the most important functions of carotenes in animals is that they act as precursors of vitamin A. Among the 700 carotenes found in nature, only 50 carotenoids have exhibited pro-vitamin A activity, with α-carotene, *β*-cryptoxanthin, and *β*-carotene being the predominant ones. *β*-carotene 15,15'-oxygenase, located in the intestinal membrane, catalyzes the oxygenase reaction where oxygen reacts with the central double bond of carotenes and forms an intermediate 15-15'-peroxide of *β*-carotene. This intermediate is then cleaved to produce two molecules of retinaldehyde. The retinaldehyde joins with opsins, proteins in the eye, to form the retinaldehyde-opsin complex, which plays a vital role in vision and preventing infections [29].

1.3.2 STABILITY

The conjugated polyene structure of carotenoids makes them susceptible to oxidative breakdown via various nonspecific mechanisms. Autoxidation of *β*-carotene due to free radicals has been shown to occur in benzene or tetrachloromethane in the absence of light at 30°C within one hour of induction. *β*-carotene forms isomers in solution via a biradical process where the isomerization process results in a reactive, unpaired spin state that reacts with O_2, resulting in the formation of carbon-peroxyl triplet biradicals [30]. High-temperature-induced degradation of carotenoids produces volatile and non-volatile by-products which are highly dependent on temperature, availability of oxygen, light, etc. Heating *β*-carotene at 30°C resulted in the formation of dihydroactinidiole (DHA), which is most frequently cited as the first volatile compound produced due to heat degradation. Mutatochrome is reported to be the precursor of DHA. Upon heating at 97°C for 3 hours in the presence of air, there was a considerable increase in the quantity of DHA produced, which indicated that other non-volatile compounds such as autochrome and luteochrome could act as precursors, in addition to mutatochrome. When *β*-carotene is heated in the presence of oxygen, 5,6-epoxy-beta-carotene is produced, which is further converted to mutatochrome, 5,6,5,'6'-diepoxy-beta-carotene or luteochrome, which in turn could convert to autochrome and eventually to DHA [31]. Prolonged heating at higher temperatures (180°C for 2 hours) in the presence of oxygen causes pure trans-β-carotene to oxidize, resulting in oxidation products and several *cis* isomers. This reaction is triggered by the oxidation of β-carotene's double bonds in the presence of oxygen [32]. When β-carotene is heated at 60°C in toluene, several oxidation products, including epoxides, are produced. These products also include peroxide radicals, which react with other β-carotene molecules and induce further oxidation reactions [33]. Other common degradation agents of β-carotene include light, which produces photodegradation byproducts such as carotenoid radical cations [34]. When exposed to acid, carotenoids undergo *trans-cis* isomerization through a carotenoid carbocation intermediate [35]. Interaction with iron (ferric chloride) also triggers an oxidation reaction, forming ferrous iron and cation radicals of carotenoids. This oxidation reaction is not affected by light [36].

1.3.3 DEGRADATION

Carotenoids give plants their beautiful color that we perceive. Even after degradation, they produce several volatile and non-volatile compounds that provide plants with physiochemical benefits and have economic value. The degradation of neutral carotenoids and the resulting by-products are highly influenced by the agent responsible for degradation. For instance, photochemical oxidation results in random cleavage and the generation of various oxidation products, including aldehydes, ketones, endoperoxides, and lactones. Volatile derivatives of photooxidation, such as β-cyclocitral, have been shown to induce gene expression, resulting in acclimation to stress conditions in plants [37]. Non-specific enzymatic oxidation by lipoxygenase and peroxidases (PODs), through random attacks on carotenoid molecules, results in the production of unspecified cleavage products. Specific enzymatic breakdown of carotenes is catalyzed by nonheme iron-containing enzymes called carotenoid cleavage dioxygenases (CCDs) or carotenoid cleavage oxygenases (COOs), resulting in the formation of biologically important apocarotenoids in living organisms. Other CCDs that are known to attack double bonds specifically include

9-cis-epoxycarotenoid dioxygenases (NCEDs) from the Arabidopsis family, BoLCD from *Bixa Orellana*, and CsZCD from *Crocus sativus*. One of the best-known compounds derived from carotenoid cleavage is the hormone abscisic acid (ABA), and NCEDs are the primary enzymes responsible for ABA biosynthesis [38]. ABA synthesis increases as a stress response and during seed dehydration [39]. Other functionally important cleavage byproducts include 'yellow pigment' (mycorradicin) bluemenin, which has anti-fungal properties, volatile products such as β-ionone, which have flavor and aroma compounds, and apocarotenoid pigments like bixin (annatto) and crocin (saffron), which have economic value as food colorants [40].

1.3.4 EXTRACTION AND ANALYSIS

Carotenoids are available in a wide range of plants and animals, which makes the extraction and analysis of these compounds challenging. The extraction process varies depending on the source of carotenoids, but generally involves disrupting tissues to release the desired components and removing unwanted components [41]. The choice of solvent for carotenoid extraction depends largely on the polarity of the target carotenoid. Non-polar carotenoids, including esterified carotenoids, are extracted using hydrophobic solvents such as hexane, while polar carotenoids like xanthophylls are extracted using polar solvents [42]. Commonly used extraction solvents for carotenoids in food samples include hexane, acetone, ethanol, and tetrahydrofuran. In addition to individual solvents, solvent mixtures have also been used, with a 4:3 ratio of ethanol to hexane being the most commonly applied for carotenoid extraction [43, 44]. It is important to consider the susceptibility of carotenoids to oxidation from light, heat, and acid when developing an extraction method. Antioxidants such as pyrogallol, butyl hydroxyanisol, butylated hydroxytoluene (BHT), ascorbic acid (AA), and butylhydroquinone are commonly used to protect carotenoids from oxidation in solution. Saponification is another method used to extract non-polar xanthophylls in their fatty-acid esterified form. In this method, a low concentration of alkali is used to remove chlorophyll and lipid compounds, resulting in an isolated sample of carotenoids for chromatographic analysis. After saponification, liquid extraction is employed to extract carotenoids. Since xanthophylls are sensitive to alkali treatments, they are generally not purified using saponification [45, 46]. Supercritical fluid extraction (SFE) is an alternative method commonly used to extract carotenoids as it is faster than liquid extraction and more cost-effective. Additionally, it utilizes carbon dioxide, which is non-toxic, non-flammable, and environmentally friendly [47].

Since carotenoids exist in a wide variety of forms due to isomerization and esterification, their analysis and quantification is very difficult. Liquid chromatography-mass spectrometry (LC-MS) is utilized to establish carotenoid identity, while coupling LC-MS with a photodiode array detector (PAD) or UV-Vis detector with LC-MS/MS helps achieve higher resolution and accurate data [42, 48]. Reversed-phase HPLC is the popular choice for carotenoid separation from extracts isolated from biological matrices, although normal-phase methods have also been studied and applied [42]. Various methods have been established using C_8 [49], C_{18} [50], and C_{30} columns [43]. While C_8 and C_{18} columns are the preferred columns for formative and non-isomeric carotenoids, C_{30} columns showed better separation and higher resolution for less polar carotenes like lycopene and *β*-carotene. Mobile phase solvents can include a gradient with various combinations of acetonitrile, butanol, methanol, methylene chloride, and ethyl acetate or an isocratic elution with acetonitrile, methanol, and isopropyl alcohol in the ratio of 39:43:18, v/v/v [51]. Triethylamine (TEA) is a common antioxidant added to solvents to prevent the degradation of carotenoids during analysis.

1.4 ANTHOCYANINS (ACNS)

1.4.1 STRUCTURE AND FUNCTIONALITY

Anthocyanins (ACNs) are natural pigments that belong to the flavonoid group, serving as a subclass. They contribute to the orange to blue-violet coloration observed in plants. The fundamental structure of ACN is an aglycon known as anthocyanidin, which consists of an aromatic ring (I) connected to an oxygen-containing heterocyclic ring (II), which in turn is attached to another aromatic ring (III) via a C-C bond (Figure 1.3). The benzopyrylium core, composed of the aromatic ring (I) and heterocyclic ring (II), forms the chromophore portion of ACNs. Anthocyanidins are highly unstable and are often found as glycosides called ACNs, with one or more sugar moieties attached to them. To date, more than 700 structurally distinct ACN molecules have been identified in nature, all of which are derivatives of 27 cyanidins [52]. There are six predominant anthocyanidin structures, accounting for approximately 92% of all identified ACNs, each with a unique arrangement and number of hydroxyl and methoxyl subgroups (Figure 1.3). The various ACNs found in nature differ in terms of the basic anthocyanidin structure, the type and number of attached sugars, and the degree of sugar acylation and acylating agent. Common sugar substituents include glucose, galactose, rhamnose, and arabinose, while typical acylating agents include cinnamic acid and aliphatic acids [53]. The most widespread ACNs are those derived from non-methylated anthocyanidins, which are identified in 80% of pigmented leaves, 69% of fruits, and 50% of flowers [54]. The ability of ACNs to produce color can be attributed to their benzopyrilium chromophore structure, which has a strong electron-withdrawing property. This property causes ACNs to behave as weak diacids. At higher pH levels, the deprotonation of these diacids increases electron polarization, resulting in absorption maxima towards higher wavelengths. Therefore, the color produced by ACNs depends on the equilibrium between the acid substituent groups attached to the chromophore at any specific pH. For example, the flower color of Japanese morning glory is impacted by vacuolar pH. The buds produce a reddish-brown color (pH 6.6), which changes to a brilliant blue color as the petals unfold (pH 7.7) [55]. In addition to pH and acid substituents, co-pigmentation of ACNs with co-pigments (such as polysaccharides, nucleotides, alkaloids, metals, amino acids, organic acids, flavonoids, or other ACNs) improves the stability of the ACN structure [56].

In plants, ACNs have several functionalities such as photoprotection, metal chelation, delaying leaf senescence, and responding to stress inducers. ACNs can inhibit and/or reduce the effect of excess excitation energy produced due to photosynthesis, which can damage the photosynthetic apparatus. The specific group of ACNs responsible for photoprotection is the light-responsive flavonoids, which contain a catechol group attached to ring III of the flavonoid molecule. These substituents provide photoprotection by both preventing the synthesis of reactive oxygen species (ROS) and by reducing them once formed [57]. Vacuolar ACNs also contribute to scavenging H_2O_2 (hydrogen peroxide) and thus preventing the simplistic movement of reactive oxygen species between cells. This regulation of H_2O_2 results in the alteration of H_2O_2 levels and affects H_2O_2 signaling, like mitogen-activated protein kinase cascades. ACNs' affinity to proteins

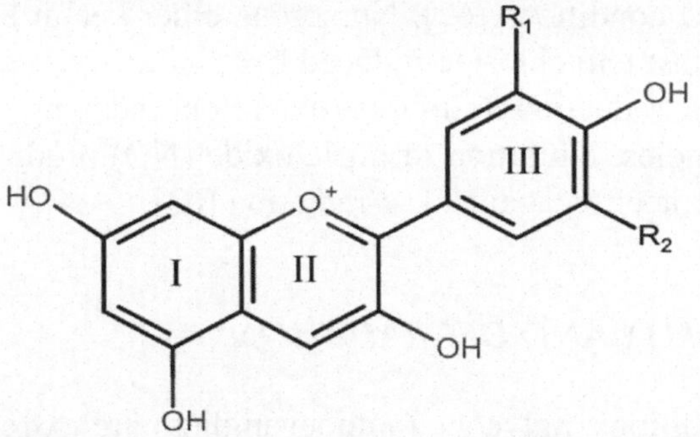

Anthocyanidin	R_1	R_2
Pelargonidin	H	H
Cyanidin	OH	H
Peonidin	O–CH_3	H
Delphinidin	OH	OH
Petunidin	O–CH_3	OH
Malvidin	O–CH_3	O–CH_3

FIGURE 1.3 Structural differences between the six predominant anthocyanidin structures.

that control signaling cascades also gives them the potential to regulate developmental processes. Metal chelation is another functional role that ACNs perform in plants, through which metal ions are chelated in the cytosol or the vacuole to protect plants under metal stress [58]. ACNs also function as alternative sinks for excess sugars during senescence, thereby diminishing sugar-induced leaf senescence in plants with insufficient nutrition [59]. In cellular vacuoles, ACNs and AA form a vacuolar system to reduce peroxides and free radicals, thereby mitigating their impact on programmed cell death [60]. Additionally, ACNs are also proposed to play a definite role in providing resistance to plants against insect attack and act as antibacterial agents.

ACNs are also widely studied for their effect on human nutrition and health. Crude ACN extracts have been shown to reduce vascular permeability and frailty. ACNs found in red wine have been found to be potent in absorbing ROS and hindering oxidative reaction of lipoproteins and platelet aggregation [61]. ACNs and other flavonoids have been shown to suppress HCT-15 cells (sourced from cancer cells in the human large intestine) or AGS (adenocarcinoma gastric cell line) cells (sourced from cancer cells in the human stomach) significantly better than non-anthocyanic cells in *in vitro* conditions [62]. Numerous other health benefits include protection against tetrachloride-induced liver injury, reduction in swelling of the pancreas, reduction in lipid peroxidation, reducing oxidative stress during pregnancies, inhibition of nitric oxide (NO) production, antiulcer activity, and protection against UV radiation [63].

1.4.2 STABILITY AND DEGRADATION

In neutral conditions, aglycons (anthocyanidins) are extremely unstable, and hence, they are present as glycosylated ACNs. The instability of anthocyanidins is mainly due to the lack of methylation on the B ring. Glycosylation and acylation of these aglycons provide improved stability against degrading into phenol and aldehyde acid molecules. Even in their glycosylated forms, ACNs are prone to degradation by light irradiation, heat, pH, organic acids, metal ions, and sugars. The ionic nature of ACNs makes them extremely susceptible to changes in pH. Depending on the pH, ACNs exist in four forms: at pH 1, they exist as flavylium cations (red color); at pH between 2 and 4, they exist predominantly as quinoidal blue species; at pH between 5 and 6, they form colorless compounds like carbinol pseudobase, quinoidal base, or chalcone; at pH higher than 7, they degrade to substituent groups. All four forms can co-exist in the pH 4–6 range, through the balance provided by the flavylium cation [56].

Another major determinant that influences the stability of ACNs is heat. During heat treatment (HT), ACNs undergo various degenerative reactions such as deglucosylation, water-nucleophilic attack, cleavage, and polymerization. Grape extracts heated at 35°C resulted in having less than half the quantity of ACNs compared to when heated at 25°C. These compounds undergo degeneration during heating and eventually produce browning products in the presence of oxygen. Recent studies have established that the heat stability of ACNs is closely associated with their structure, particularly their sugar fraction [64]. Furthermore, ACNs in acylated form showed higher stability at high temperatures than the non-acylated form. Acylated forms of ACNs found in purple yam, sweet potato, and red cabbage exhibited considerable stability compared to their non-acylated counterparts [65]. Heat processing techniques such as blanching (95°C/3 mins) and pasteurization at 90°C to produce blueberry puree have shown a 43% loss of monomeric ACNs. However, it has also increased the quantities of polymer colors from 1% to 12%, indicating that endogenous enzymes play a crucial role in polyphenol degradation [66]. Hence, mild heat treatment such as blanching before pasteurization could help inactivate the degradation enzyme. Certain steps that could mitigate the effect of heat processing on ACN degradation include reducing pH and oxygen concentration.

Oxygen content is another factor that affects the stability of ACNs. Oxygen can increase the rate of ACN degradation through direct oxidation (due to the unsaturated nature of ACNs) and/or through oxidation enzymes such as polyphenol oxidase [67]. Consequently, their stability increases when stored under vacuum or an atmosphere of nitrogen or argon, compared to oxygen. Furthermore, the thermal degradation of pigments is significantly reduced by removing oxygen [64]. In the presence of oxygen, AA is known to degrade ACNs, resulting in the whitening of the polyphenol pigment. Although the exact mechanism is not yet known, the theorized mechanism involves AA condensation with ACNs or the formation of H_2O_2, which in turn oxidizes the pyrylium ring [64]. Light also affects the stability of these compounds. While light is essential for the biosynthesis and accumulation of ACNs, it also accelerates their breakdown and affects their antioxidant activity. This effect is particularly severe in the presence of oxygen or when exposed to fluorescent lighting [68]. Sulfites and sulfates negatively affect the stability of ACNs by forming colorless sulfur derivatives. When these ACNs are heated, sulfur dioxide is released, providing numerous benefits in winemaking such as microbial protection, antioxidant properties, and acting as a solvent. Finally, degradative enzymes called anthocyanases are highly detrimental to the stability of ACNs. The most common anthocyanases include glycosidases, PODs, and phenolases, which may be produced by plants in the tissues or as a result of microbial contamination [64].

1.4.3 EXTRACTION AND ANALYSIS

Anthocyanins (ACNs) are well known for their wide range of colors and antioxidant properties. Therefore, extraction methods should focus on preserving these properties. The basic steps involved in analyzing ACNs include extraction from the food source, concentration, purification, and analysis. Since ACNs are polar in nature, various acidified polar solvents such as acidified ethanol, methanol, acetonitrile, and organic solvents have been used for extraction, with organic solvents being significantly more effective. While the commonly accepted method for extracting the pigments is using a sample of 1% HCl (hydrochloric acid) in ethanol, acetic acid has been shown to provide better extraction than HCl [69]. The presence of acid provides stability to ACNs through the formation of red flavylium cations and prevents degradation of non-acylated pigments. However, care should be taken since the evaporation of HCl and ACNs during the concentration step could result in fractional or total hydrolysis of acyl compounds of the acylated pigments [70]. Traditional extraction processes include soaking, stirring, and enzyme-mediated extraction. With the increased concern about protecting antioxidant properties and color, novel methods such as pressurized liquid extraction (PLE), pulsed electric field (PEF) assisted extraction, and SFE have been applied but with moderate success due to the heat instability and polar nature of ACNs [69]. Aqueous two-phase extraction (ATPE) is another common extraction technique used to separate macromolecules. ATPE with ethanol-ammonium sulfate has been utilized to successfully extract ACNs with a 90% yield [71]. Extraction using ultrasound is another method that has garnered interest due to its short extraction time, less solvent, and higher yield [72]. Concentration is usually done under vacuum, and high temperatures (≥ 30°C) should be avoided to protect heat-sensitive ACNs.

The extraction steps mentioned above are generally non-selective and can result in extracts containing impurities such as sugars and proteins, which can negatively impact the purity of ACNs. Common methods for purifying ACNs include solid phase extraction (SPE), paper, thin-layer, and column chromatography, as well as high-speed counter-current chromatography (HSCCC) [73]. In SPE, the ACN extracted in the mobile phase flows through a solid stationary phase, allowing different components of the extract solution to adsorb based on their polarity. Typical solid phases used in SPE include C-18 and Sephadex for initial purification, and silica gel, Amberlite IRC 80, Amberlite XAD-7HP, and DOWEX 50WX8 for more extensive purification. HSCCC is used in selected applications to purify specific ACN components. Some advantages of HSCCC include low pressures and higher flow rates, a wide range of solvents, and low-cost stationary phase [72]. The choice of solvent plays a significant role in SPE, and methanol with a 0.01% HCl solution [74] or 3% methanoic acid [75] is commonly used for ACN purification. However, caution should be exercised to use only water when purifying ACNs for human consumption.

The chromophore structure of ACNs makes it possible to analyze and quantify these compounds using spectrophotometric methods. The absorption maxima of ACNs are in the wavelength regions of 510–550 nm. Absorbance at different wavelengths can be obtained to quantify individual components such as phenolic content (280 nm), flavanols (360 nm), and anthocyanins (520 nm) [76]. Corresponding standards include 10% ethyl alcohol containing gallic acid and (+)-catechin for total phenolics, 95% ethyl alcohol containing quercetin for flavanols, and malvidin 3-glucoside in 10% ethyl alcohol for ACNs [76]. The most popular format of spectrometric method used for the quantification of monomeric ACNs is the pH-differentiation method. Monomeric ACNs change color from reddish blue at pH 4.1 to almost achromatic at pH 4.5. By measuring the difference in absorbances at these two pH values, monomeric ACNs can be quantified [77]. Care should be taken since this method is prone to several variabilities including, but not limited to, temperature, acylation of ACNs, quantity of disaccharide substituents, lack of consistency in absorptivity value from different studies, and the acid concentration of the solvents used [78]. Reverse-phase liquid chromatography has been the preferred method for analyzing ACNs and flavonoids due to its simplicity and reliability. Common solvents used include acetic acid, formic acid solutions in organic solvents such as methanol and acetonitrile. Elution is typically performed at room temperature. LC coupled with mass spectrometry has been used to quantify ACN pigments. High performance liquid chromatography has also been employed for ACN analysis due to its improved resolution and shorter run time. Capillary zone electrophoresis (CZE) is a novel analysis method that combines electrophoresis and chromatography to separate compounds based on their electrophoretic mobility. CZE has been applied to separate and quantify charged natural compounds like flavonoids. With advancements in analytical technology, various ionization methods such as matrix-assisted laser desorption/ionization (MALDI) and electrospray ionization (ESI) coupled with detectors like ion trap, orbitrap, quadrupole time-of-flight (QToF), and triple QToF are increasingly used in ACN studies. Other innovative methods for determining ACNs include nuclear magnetic resonance and ion mobility spectrometry [72].

1.5 BETALAINS

1.5.1 STRUCTURE AND FUNCTIONALITY

Betalains are hydrophilic pigments with nitrogenous substituents that produce the red-violet to yellow-orange colors in plants and vegetables. These vacuolar pigments are derivatives of the amino acid named tyrosine and are broadly differentiated into two groups: betacyanins and betaxanthins. Betalamic acid forms the chromophore part of betalains owing to its conjugated double bonds, while the residue attached to it determines the type and color of the final betalain. Betacyanins consist of a cyclo-3,4-dihydroxyphenylalanine (cyclo-Dopa) residue, while betaxanthin contains an amino acid or amine side chain (Figure 1.4). Cyclo-Dopa plays a vital part in the absorption behavior of betacyanin by extending the electronic resonance to the diphenolic ring and thereby shifting the absorption maxima to 540 nm (violet coloration) from 480 nm (yellow coloration) [79]. The aglycone structure of most betacyanins is betanidin, and the substitution groups on the aglycone structure, either via glycosylation or acylation, determine the final betacyanin color [80]. The most widespread betacyanin is betanidin 5-*O*-*β*-glucoside found in red beets (*Beta vulgaris* ssp.). Betacyanins exhibit two major absorption maxima depending on the residue attached to it – one occurs in the UV range of 270 to 280 nm, owing to its cyclo Dopa structure, and the other in the visible range of 535 to 538 nm. Betanidin glycosylation usually causes a hypsochromic shift of ~6 nm and further esterification with hydroxycinnamic acids causes a bathochromic shift of ≤12 nm [81, 82]. Betaxanthins are considered to be the yellow equivalent of betacyanins and are formed by betalamic acid condensation with amine or amino acid. Betaxanthins with an amine residue exhibit a lower absorption maximum compared to the ones with an amino acid residue [83]. Indica xanthin was the first isolated plant betaxanthin (from *Opuntia ficus-indica* L.) and it consists of a proline residue [82]. Vulgaxanthin is one of the major betaxanthins and was first isolated from yellow beetroots (*Beta vulgaris* L.). Depending on the amino acid residue, they are classified as vulgaxanthin I (glutamine), vulgaxanthin II (glutamic acid), vulgaxanthin III (asparagine), and vulgaxanthin IV (leucine) [82].

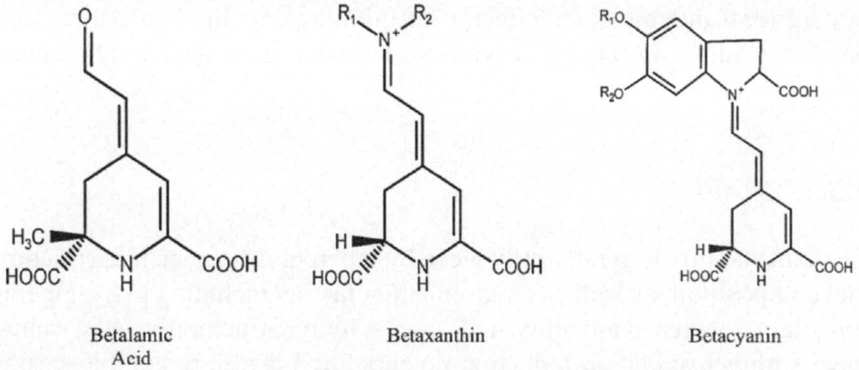

FIGURE 1.4 Basic structural differences between betalamic acid, betaxanthin, and betacyanin.

Betalains are present in various plant parts, and some consumable sources of betalains include beetroot, colored Swiss chard, and cactus fruits [80]. While their primary role is to provide color and attract pollinators, they also provide numerous other functional benefits. The antioxidant property of betalains has gained increased popularity in the past couple of decades. Several *in vitro* studies have successfully shown the antiradical scavenging ability of betalains [84–88]. Betalains have also been shown to have much higher antiradical scavenging activity than vitamin E, vitamin C, rutin, catechin, β-carotene, and α-tocopherol [86]. The antioxidant activity of betalains is due to its betalamic acid, which can donate two electrons to an oxidizing agent, and this reaction is pH dependent with the highest activity in the pH range of 2–4. Betaxanthins are used as food supplements, owing to their amino acid residues, to fortify processed foods with essential amino acids. Betalains have also been shown to exhibit inhibitory effects towards cancer development in the lungs and skin [89], in addition to demonstrating long-term suppression effects towards skin and liver tumors in mice [90]. They also provide protective action against paraquat-induced toxicity in the liver, kidney, and lung of rats. Indica xanthin has been proven to possess anti-inflammatory characteristics against acute inflammation in rats. Further, betalains have been extensively studied and shown to have several health promoting characteristics such as reducing blood pressure, antiglycatic effect, reduction of non-high density lipoprotein cholesterol, low density lipoprotein and cholesterol. Betalains have also been shown to provide beneficial effects in the cardiovascular system (anti-inflammatory effect, protection against lipid oxidation, and heme deposition in blood), nervous system, and gastrointestinal systems [80, 91].

1.5.2 STABILITY

Betalain stability is greatly influenced by inherent factors such as structure and composition, as well as environmental factors including pH, temperature, light, oxygen availability, and water activity. Structurally, betacyanins have a higher oxidation-reduction potential and therefore exhibit greater stability compared to betaxanthins at room temperature and during heating [92]. Esterification with aliphatic acids and aromatic acids enhances the stability of betacyanins [93]. Betalains are susceptible to degradation when they come into contact with enzymes such as β-peroxidases, polyphenol oxidases, and PODs [80]. Betacyanins demonstrate higher stability against enzymatic degradation compared to betaxanthins, especially in the presence of catalase [86]. Betalains exhibit good stability between pH 3 and pH 7, with maximum stability observed at pH 5.5 to 5.8. However, this optimal pH range may vary depending on the structure, temperature, and oxygen levels. Upon heating, betalains show maximum stability at a pH of 6.0. Under anaerobic conditions, betalains exhibit higher stability within a lower pH range of 4.0–5.0. The absorption maxima of betalains shift towards lower wavelengths at pH below 3.5 and towards higher wavelengths at pH above 7.0 [80]. Water activity (a_w) is another factor that greatly affects the stability of betanins, especially the aldimine bond of betaxanthins. The degradation reaction of betanins follows first-order kinetics, and the stability of betanins increases as water activity decreases. This can be attributed to limited reactant mobility and reduced oxygen solubility [94]. The maximum degradation of betalains occurs at an a_w of 0.64, and their stability significantly improves below 0.63 [80]. Oxygen is required for the degradation of betalains during storage, and this degradation increases with increasing atmospheric oxygen levels [95]. Storage stability can be improved by using antioxidants such as citric acid (CA) and isocitric acid [94], and by storing them under a nitrogen environment [96]. The presence of a chromophore structure in betalains makes them prone to degradation under light. In the presence of ultraviolet (UV) and visible light, the π-electrons of the chromophore get excited to a higher energy state, thereby enhancing reactivity or reducing the molecule's activation energy [97]. Furthermore, light-induced degradation of betalains is influenced by the availability of oxygen, with negligible degradation occurring in anaerobic conditions [80]. Betalain degradation has also been reported to be affected by metal cations such as iron, copper, tin, and aluminum. CA and ethylenediaminetetraacetic acid (EDTA) can be used to prevent metal-induced degradation due to their ability to chelate metals. In addition to the above-mentioned factors, temperature is another major factor that influences the stability of betalains, especially during processing and storage. Betalain stability is directly influenced by temperature and the duration of heating, and they are most stable at temperatures below 50°C and 70°C for processing and below 60°C and 80°C for storage, respectively [98].

1.5.3 DEGRADATION

Considering that betalains are not very stable compounds and are prone to degradation due to various agents, understanding the degradation pathways and resulting byproducts would help us comprehend the loss of these compounds during application and processing steps. One significant effect of betalain degradation is the loss of chromatic stability, resulting in colorless products. This loss in color can be measured by spectrophotometry or HPLC [99]. Betanins usually coexist with their isomer isobetanin, formed by C_{15} isomerization. While both isomers exhibit the same color and properties, the ratio of betanin to isobetanin changes due to thermal treatment and the influence of acidic and alkaline conditions [100]. Deglycosylation is another degradation reaction where the glucose moiety is cleaved, leaving behind the aglycone molecule and resulting in a bathochromatic shift of 4 nm. This reaction occurs at very low pH, high temperatures, and under the action of β-glucosidase [97]. Hydrolysis of betalains occurs at the aldimine bond (RCH=NR) during thermal treatment, yielding yellow betalamic acid and achromatic cyclo-Dopa. This hydrolysis is reversible during short-term treatment, via Schiff base condensation of betalamic acid and cyclo-Dopa, resulting in betacyanin. During prolonged heating, betalamic acid is prone to aldol condensation and Maillard browning, and cyclo-Dopa undergoes degradation by deglycosylation and ring opening, resulting in dopamine products. The presence of amino acids would interfere with the betacyanin regeneration reaction, resulting in the formation of yellow-colored betaxanthins. Degradation of betalains also occurs via decarboxylation, leading to the cleavage of the carboxylic group and a smaller absorption maximum. Theoretical decarboxylation can occur at positions 2, 15, and 17 of betacyanins and at positions 11 and 13 in betaxanthins. However, *in vitro* studies have shown that decarboxylation depends on the susceptibility of the carboxylic group and the solvent medium. Unlike other degradation mechanisms, decarboxylation may be beneficial for the chromatic integrity of betalains, as the betanidin chromophore remains unaffected by decarboxylation, and mono-decarboxylated betanins are considerably more stable than non-decarboxylated betanins [101]. Dehydrogenation of betanins is another form of degradation, where two hydrogen atoms are lost to form a double bond in the betanidin chromophore. The resulting compound is a yellow-colored 14,15-dehydrobetanin or neobetanin [102]. The change in color from red betanins to yellow neobetanins has led researchers to believe that dehydrogenation is the main degradation pathway during thermal treatment of betanins [99, 101].

1.5.4 EXTRACTION AND ANALYSIS

Betalains are hydrophilic and hence are commonly extracted from ground or dry plant sources by maceration using a cooled polar solvent such as 20 to 50% v/v methanol/ethanol or water. Considering that the pigments are prone to enzymatic degradation during extraction, pretreatment by heat processing at 70°C for 20 minutes is beneficial in inactivating the enzymes. Furthermore, slight acidification of the water used for extraction with citric acid or hydrochloric acid would enhance the stability of pigments against polyphenol oxidases. Pigment sources such as beet extract are rich in sugars, which decrease the tinctorial power of the pigments and hence need to be removed during the extraction process. Yeast fermentation is usually employed to remove these free sugars. Another method commonly employed to extract betalains includes Soxhlet extraction. While these methods have been successfully applied to extract betalains, the resulting low pigment yield requires a concentration step, which results in browning of the pigments [103]. This has given rise to several novel extraction techniques that improve cellular permeability of pigments, provide improved yield, and are eco-friendly. These novel techniques include PEF, microwave (MW) and ultrasonic (US) assisted extraction, diffusion extraction, membrane separation such as reverse osmosis (RO) and loose RO, and ATPE [104]. The extraction of betalains from red beets was found to yield higher levels of betacyanins than betaxanthins [105]. The use of MW radiation causes the plant cell structure to expand and eventually rupture, facilitating the migration of pigments. MW-assisted extraction, when combined with an acid like AA, could potentially provide higher yields and improve the stability of betacyanins. However, this method is not effective for betaxanthins due to their susceptibility to acid denaturation [104]. Another successful technique for extracting betalains from red beets is PEF treatment, which has been shown to yield over 90% extraction efficiency. Membrane techniques such as RO and loose RO have also been reported as highly efficient methods for producing betalain-rich extracts with minimal quantities of salts and nitrates on a pilot scale [106]. After extraction, the betalains need to be isolated and purified before analysis. Chromatography techniques such as HPLC, ion-exchange chromatography (IEC), TLC, and electrophoresis can be used for separation and purification [107].

Double bonds in the chromophore structure of betalains exhibit absorption characteristics in the visible region–460–475 nm for betaxanthins and 480–540 nm for betacyanins. The electrical resonance of the 1,7-diazaheptamethine moiety of betaxanthin also provides the pigment with the ability to exhibit fluorescence. On the other hand, the resonance extension to the indole ring moiety is the reason for betacyanins exhibiting little to no fluorescence properties. Due to these properties, HPLC with UV-Vis spectrophotometry or fluorescence detection is the predominant separation and quantitation technique [103]. Betalain isolation using HPLC is primarily affected by stereochemistry and overall polarity. C_8 or C_{18} reverse phase silica columns are common stationary phases used with mobile solutions such as acetic acid or phosphate buffer with organic modifiers like acetonitrile or methanol to ensure that pH remains constant at pH 4 during elution [107]. Gradient elution is typically employed to achieve complete separation of different components. HPLC, coupled with UV-Vis spectrophotometry, can detect betacyanins (535–540 nm), betaxanthins (470 nm), betalamic acid (406 nm), and colorless phenolics (280 nm). The photodiode array detector is also commonly used for betalain analysis. The amphoteric property of betalains, resulting from the presence of negatively charged COOH group and weak positively charged nitrogen atom, causes these pigments to exhibit electrophoretic mobility. CZE with a fused-silica capillary maintained at –22 kV (kilovolt) and 15°C has been employed to separate betanins, isobetanins, and their analogous aglycons [107]. Other novel techniques used for analyzing betalains include ESI coupled with mass spectrometry (ESI-MS) in positive ion mode, LC-MS coupled with ESI, and analyzed using quadrupole Time-of-Flight (QToF) analyzer or Fourier Transform ion cyclotron resonance mass spectrometry (FT-ICR-MS) [109].

1.6 SUMMARY

Natural food pigments are extremely complex structures that are sensitive to even minor changes in their environments. Understanding their structures and how they relate to various conditions and treatments would immensely help us design processing parameters to preserve their structure and functionality. This chapter delineates structural differences between major natural food pigments – chlorophylls, carotenoids, ACNs, and betalains – and identifies key causes for structural degradation due to various agents. It also discusses various color measurement techniques that are employed for each pigment, along with their advantages and limitations.

KEYWORDS

- **anthocyanins**
- **betalains**
- **carotenoids**
- **chlorophylls**
- **color measurement**
- **natural pigments**
- **stability**
- **structure**

REFERENCES

1. Rodriguez-Amaya, D. B., (2019). Natural food pigments and colorants. In: *Bioactive Molecules in Food* (pp. 867–901). Springer.
2. Coultate, T., & Blackburn, R. S., (2018). Food colorants: Their past, present and future. *Coloration Technology, 134*(3), 165–186.
3. Hendry, G., (1996). *Natural Food Colorants*. Springer-Science Business Media, B.V.
4. Wrolstad, R. E., Ferruzzi, M. G., & Schwartz, S. J., (2005). *Chlorophylls*. Hoboken, NJ: John Wiley & Sons.
5. Matile, P., Schellenberg, M., & Peisker, C., (1992). Production and release of a chlorophyll catabolite in isolated senescent chloroplasts. *Planta, 187*, 230–235.
6. Schwartz, S. J., Woo, S. L., & Elbe, J. H. V., (1981). *High-Performance Liquid Chromatography of Chlorophylls and Their Derivatives in Fresh and Processed Spinach* (Vol. 29, No. 3, pp. 533–535). American Chemical Society.
7. Jones, I. D., White, R. C., Gibbs, E., & Butler, L. S., (1977). Estimation of zinc pheophytins, chlorophylls, and pheophytins in mixtures in diethyl ether or 80% acetone by spectrophotometry and fluorometry. *Journal of Agricultural and Food Chemistry, 24*(1), 146–149.
8. Dutton, H. J., Bailey, G. F., & Kohake, E., (1943). Dehydrated spinach. *Industrial & Engineering Chemistry, 35*(11), 1173–1177.
9. Hendry, G. A., Houghton, J. D., & Brown, S., (1987). The degradation of chlorophyll – a biological enigma. *The New Phytologist, 107*(2), 255–302.
10. Canjura, F. L., Schwartz, S. J., & Nunes, R. V., (1991). Degradation kinetics of chlorophylls and chlorophyllides. *Journal of Food Science, 56*(6), 1639–1643.
11. Schwartz, S. J., & Lorenzo, T. V., (1990). Chlorophylls in foods. *Critical Reviews in Food Science and Nutrition, 29*(1), 1–17.
12. Strain, H. H., & Svec, W. A., (1966). Extraction, separation, estimation, and isolation of the chlorophylls. In: *The Chlorophylls* (pp. 21–66). Academic Press.
13. Comar, C. L., & Zscheile, F., (1942). Analysis of plant extracts for chlorophylls a and b by a photoelectric spectrophotometric method. *Plant Physiol., 17*(2), 198–209.
14. White, R., & Gibbs, I. J. E., (1963). Determination of chlorophylls, chlorophyllides, pheaopytins, and pheophorbides in plant material. *Journal of Food Science, 28*(4), 431–436.
15. Wickliff, J., & Aranoff, S., (1962). Quantitative measurement of leaf chlorophylls by spectrophotometry of their pheophytins in aqueous alcoholic extracts. *Plant Physiol., 37*(5), 584.
16. Brunisma, J., (1961). A comment on the spectrophotometric determination of chlorophyll. *Biochimica et Biophysica Acta, 52*(3), 576–578.
17. Dietrich, W., (1958). Determination of the conversion of chlorophyll to pheophytin. *Food Technology, 12*, 428.
18. Tswett, M., (1906). Physikalisch-chemischestudienüber das chlorophyll. Die adsorption en. *Berichte der Deutschen Botanischen Gesellschaft, 24*(20), 316–323.
19. Shioi, Y., Fukae, R., & Sasa, T., (1983). Chlorophyll analysis by high-performance liquid chromatography. *Biochimica et Biophysica Acta (BBA) – Bioenergetics, 722*(1), 72–79.

20. Luis, A., Fernandez-Lopez, J. A., & Marıa, J. R., (2000). *Journal of Chromatography A, 870*(1, 2), 483–489.
21. Schwartz, S., (2006). High performance liquid chromatography of zinc and copper pheophytins. *Journal of Liquid Chromotography, 7*(8), 1679–7683.
22. Inoue, H., Yamashita, H., Furuya, K., & Nonomura, Y., (1994). Determination of copper(I1) chlorophyll in by reversed-phase high-performance liquid chromatography. *Journal of Chromatography A., 679*, 99–104.
23. Isler, O., (2013). *Carotenoids*. Verlag: Springer.
24. Siefermann-Harms, D., (1985). Carotenoids in photosynthesis. I. Location in photosynthetic membranes and light-harvesting function. *Biochimica et Biophysica Acta, 811*, 325–355.
25. Peto, R., Doll, R., Buckley, J., & Sporn, M., (1981). Can dietary beta-carotene materially reduce human cancer rates? *Nature, 290*, 201–208.
26. Santamaria, L., & Bianchi, A., (1989). Cancer chemoprevention by supplemental carotenoids in animals and humans. *Preventive Medicine, 18*(5), 603–623.
27. Britton, G., & Goodwin, T. W., (1982). *Carotenoid Chemistry and Biochemistry*. Liverpool: Pergamon Press.
28. Stewart, G. F. F., Schweigert, B. S., Hawthorn, J., & Bauernfeind, J. C., (2012). *Carotenoids as Colorants and Vitamin A Precursors: Technological and Nutritional Applications*. Elsevier.
29. Pitt, G., (1971). Vitamin A. In: *Carotenoids* (pp. 717–742). Basel, Birkhäuser.
30. Mordi, R. C., Walton, J. C., Burton, G. W., Hughes, L., David, I. U., Douglas, L. A., & Douglas M. J., (1993). Oxidative degradation of β-carotene and β-apo-8′-carotenal. *Tetrahedron, 49*(4), 911–928.
31. Kanasawud, P., & Crouzet, J. C., (1990). Mechanism of formation of volatile compounds by thermal degradation of carotenoids in aqueous medium. 1. beta-carotene degradation. *Journal of Agricultural and Food Chemistry, 38*(1), 237–243.
32. Marty, C., & Berset, C., (1990). Factors affecting the thermal degradation of all-trans-beta-carotene. *Journal of Agricultural and Food Chemistry, 38*(4), 1063–1067.
33. Handelman, G. J., Kuijk, F. J. G. M. V., Chatterjee, A., & Krinsky, N. I., (1991). Characterization of products formed during the autoxidation of beta-carotene. *Free Radical Biology and Medicine, 10*, 427–437.
34. Konovalova, T. A., Gao, Y., Schad, R., Kispert, L. D., Saylor, C. A., & Brunel, L. C., (2001). Photooxidation of carotenoids in mesoporous MCM-41, Ni-MCM-41 and Al-MCM-41 molecular sieves. *The Journal of Physical Chemistry B, 105*(31), 7459–7464.
35. Konovalov, V. V., & Kispert, L. D., (1999). AM1, INDO/S and optical studies of carbocations of carotenoid molecules. Acid induced isomerization. *Journal of Chemical Society, Perkin Transactions, 2*(4), 901–910.
36. Boon, C. S., McClements, D. J., Weiss, J., & Decker, E. A., (2010). Factors influencing the chemical stability of carotenoids in foods. *Critical Reviews in Food Science and Nutrition, 50*, 515–532.
37. Havaux, M., (2013). Carotenoid oxidation products as stress signals in plants. *The Plant Journal, 79*, 597–606.

38. Sun, T., Tadmor, Y., & Li, L., (2020). Pathways for carotenoid biosynthesis, degradation, and storage. In: *Plant and Food Carotenoids* (pp. 3–26). Barcelona, Humana Press.
39. Cutler, A. J., & Krochko, J. E., (1999). Formation and breakdown of ABA. *Trends in Plant Science, 4*(12), 472–478.
40. Giuliano, G., Al-Babili, S., & Lintig, J. V., (2003). Carotenoid oxygenases: Cleave it or leave it. *Trends in Plant Science, 8*(4), 145–149.
41. Ishida, B. K., & Chapman, M. H., (2012). Effects of a hydrodynamic process on extraction of carotenoids from tomato. *Food Chemistry, 132*(3), 1156–1160.
42. Amorim-Carrilho, K., Cepeda, A., Fente, C., & Regal, P., (2014). Review of methods for analysis of carotenoids. *Trends in Analytical Chemistry, 56*, 49–73.
43. Lin, C., & Chen, B., (2003). Determination of carotenoids in tomato juice by liquid chromotography. *Journal of Chromotography A, 1012*, 103–109.
44. Taungbodhitham, A. K., Jones, G. P., Wahlqvist, M. L., & Briggs, D. R., (1998). Evaluation of extraction method for the analysis of carotenoids in fruits and vegetables. *Food Chemistry, 63*(4), 577–584.
45. Quiros, A. R. B. D., & Costa, H. S., (2006). Analysis of carotenoids in vegetable and plasma samples: A review. *Journal of Food Composition and Analysis, 19*, 91–111.
46. Irakli, M. N., Samanidou, V. F., & Papadoyannis, I. N., (2011). Development and validation of an HPLC method for the simultaneous determination of tocopherols, tocotrienols and carotenoids in cereals after solid-phase extraction. *Journal of Separation Science, 34*(1, 2), 1375–1382.
47. Marsili, R., & Callahan, D., (1993). Comparison of a liquid solvent extraction technique and supercritical fluid extraction for the determination of α- and β-carotene in vegetables. *Journal of Chromotographic Science, 31*(10), 422–428.
48. Bramley, P. M., (1992). Analysis of carotenoids by high performance liquid chromatography and diode-array detection. *Photochemical Analysis, 3*, 97–104.
49. Gokmen, V., Bahçeci, S., & Acar, J., (2002). Liquid chromatographic method for the determination of chlorophylls, carotenoids, and their derivatives in fresh and processed vegetables. *Journal of Liquid Chromatography & Related Technologies, 25*(8), 1201–1213.
50. Meléndez-Martínez, A. J., Vicario, I. M., & Heredia, F. J., (2003). A routine high-performance liquid chromatography method for carotenoid determination in ultrafrozen orange juices. *Journal of Agricultural and Food Chemistry, 51*(15), 4219–4224.
51. Vági, E., Simándi, B., Daood, H. G., Deák, A., & Sawinsky, J., (2002). Recovery of pigments from *Origanum majorana* L. by extraction with supercritical carbon dioxide. *Journal of Agricultural and Food Chemistry, 50*(8), 2297–2301.
52. Wallace, T., & Giusti, M. M., (2015). Anthocyanins. *Advances in Nutrition, 6*(5), 620–622.
53. Clifford, M. N., (2000). Anthocyanins – nature, occurrence, and dietary burden. *Journal of the Science of Food and Agriculture, 80*, 1063–1072.
54. Dey, P., & Harborne, J. B., (1989). *Methods in Plant Biochemistry: Plant Phenolics.* London: Academic Press Limited.
55. Houghton, A., Appelhagen, I., & Martin, C., (2021). Natural blues: Structure meets function in anthocyanins. *Plants, 10*(7), 26.

56. Castañeda-Ovando, A., Pacheco-Hernández, M. D. L., Páez-Hernández, M. E., Rodríguez, J. A., & Galán-Vidal, C. A., (2009). Chemical studies of anthocyanins: A review. *Food Chemistry, 113*, 859–871.
57. Agati, G., Brunetti, C., Ferdinando, M. D., Ferrini, F., Pollastri, S., & Tattini, M., (2013). Functional roles of flavonoids in photoprotection: New evidence, lessons from the past. *Plant Physiology and Biochemistry, 72*, 35–45.
58. Landi, M., Tattini, M., & Gould, K. S., (2015). Multiple functional roles of anthocyanins in plant-environment interactions. *Environmental and Experimental Botany, 119*, 4–17.
59. Murakami, P. F., Schaberg, P. G., & Shane, J. B., (2008). Stem girdling manipulates leaf sugar concentrations and anthocyanin expression in sugar maple trees during autumn. *Tree Physiology, 28*(10), 1467–1473.
60. Ferreres, F., Figueiredo, R., Bettencourt, S., Carqueijeiro, I., Oliveira, J., Gil-Izquierdo, A., Pereira, D. M., et al., (2011). Identification of phenolic compounds in isolated vacuoles of the medicinal plant *Catharanthus roseus* and their interaction with vacuolar class III peroxidase: An H_2O_2 affair? *Journal of Experimental Botany, 62*(8), 2841–2854.
61. Ghiselli, A., Nardini, M., Baldi, A., & Scaccini, C., (1998). Antioxidant activity of different phenolic fractions separated from an Italian red wine. *Journal of Agricultural and Food Chemistry, 46*(2), 361–367.
62. Kamei, H., Hashimotol, Y., Koide, T., Kojima, T., & Hasegawa, M., (1998). Anti-tumor effect of methanol extracts from red and white wines. *Cancer Biotherapy and Radiopharmaceuticals, 13*(6), 447–452.
63. Kong, J. M., Chia, L. S., Goh, N. K., Chia, T. F., & Brouillard, R., (2003). Analysis and biological activities of anthocyanins. *Phytochemistry, 64*(5), 923–933.
64. Enaru, B., Dretcanu, G., Pop, T. D., Stanila, A., & Zorit, A. D., (2021). Anthocyanins: Factors affecting their stability and degradation. *Antioxidants, 10*(12).
65. Hayashi, K., Ohara, N., & Tsukui, A., (1996). Stability of anthocyanins in various vegetables and fruits. *Food Science and Technology, 2*(1), 30–33.
66. Brownmiller, C., Howard, L., & Prior, R., (2008). Processing and storage effects on monomeric anthocyanins, percent polymeric color, and antioxidant capacity of processed blueberry products. *Journal of Food Science, 5*, 72–79.
67. Patras, A., Brunton, N. P., O'Donnell, C., & Tiwari, B., (2010). Effect of thermal processing on anthocyanin stability in foods; mechanisms and kinetics of degradation. *Trends in Food Science & Technology, 21*, 3–11.
68. Cavalcanti, R. N., Santos, D. T., & Meireles, M. A. A., (2011). Non-thermal stabilization mechanisms of anthocyanins in model and food systems—An overview. *Food Research International, 44*(2), 499–509.
69. Oancea, S., Stoia, M., & Coman, D., (2012). Effects of extraction conditions on bioactive anthocyanin content of *Vaccinium corymbosum* in the perspective of food applications. *Procedia Engineering, 42*, 489–495.
70. Revilla, E., Ryan, J. M., & Martín-Ortega, G., (1998). Comparison of several procedures used for the extraction of anthocyanins from red grapes. *Journal of Agricultural and Food Chemistry, 46*(11), 4592–4597.

71. Liu, X., Mu, T., Sun, H., Zhang, M., & Chen, J., (2013). Optimization of aqueous two-phase extraction of anthocyanins from purple sweet potatoes by response surface methodology. *Food Chemistry, 141*(3), 3034–3041.
72. Ongkowijoyo, P., Luna-Vital, D. A., & Mejia, E. D., (2018). Extraction techniques and analysis of anthocyanins from food sources by mass spectrometry: An update. *Food Chemistry, 250*, 113–126.
73. Harbone, A., (1998). *Phytochemical Methods a Guide to Modern Techniques of Plant Analysis*. Springer Science and Business Media.
74. Hong, V., & Wrolstad, R. E., (1990). Use of HPLC separation/photodiode array detection for characterization of anthocyanins. *Journal of Agricultural and Food Chemistry, 38*(3), 708–715.
75. Garcia-Viguera, C., Zafrilla, P., & Tomás-Barberán, F. A., (1998). The use of acetone as an extraction solvent for anthocyanins from strawberry fruit. *Phytochemical Analysis, 9*(6), 274–277.
76. Mazza, G., Fukumoto, L., Delaquis, P., Girard, B., & Ewert, B., (1999). Anthocyanins, phenolics, and color of cabernet franc, merlot, and pinot noir wines from British Columbia. *Journal of Agricultural and Food Chemistry, 47*(10), 4009–4017.
77. Fuleki, T., & Francis, F., (1968). Quantitative methods for anthocyanins. determination of total anthocyanin and degradation index for cranberry juice. *Journal of Food Science, 33*(1), 78–83.
78. Mazza, G., Cacace, J. E., & Kay, C. D., (2004). Methods of analysis for anthocyanins in plants and biological fluids. *Journal of AOAC International, 87*(1), 129–145.
79. Strack, D., Vogt, T., & Schliemann, W., (2003). Recent advances in betalain research. *Phytochemistry, 62*, 247–269.
80. Azeredo, H. M., (2009). Betalains: Properties, sources, applications, and stability – a review. *International Journal of Food Science and Technology, 44*, 2365–2376.
81. Kobayashi, N., Schmidt, J., Nimtz, M., Wray, V., & Schliemann, W., (2000). Betalains from Christmas cactus. *Phytochemistry, 54*, 419–426.
82. Khan, M. I., & Giridhar, P., (2015). Plant betalains: Chemistry and biochemistry. *Phytochemistry, 117*, 267–295.
83. Stintzing, F. C., & Carle, R., (2004). Functional properties of anthocyanins and betalains in plants, food, and in human nutrition. *Trends in Food Science & Technology, 15*, 19–38.
84. Swarna, J., Lokeswari, T. S., Smita, M., & Ravindhran, R., (2013). Characterization and determination of in vitro antioxidant potential of betalains from *Talinum triangulare* (Jacq.) Willd. *Food Chemistry, 141*(4), 4382–4390.
85. Escribano, J., Pedreño, M. A., García-Carmona, F., & Muñoz, R., (1998). Characterization of the antiradical activity of betalains from *Beta vulgaris* L. roots. *Phytochemical Analysis, 9*(3), 124–127.
86. Slimen, I. B., Najar, T., & Abderrabba, M., (2019). Chemical and antioxidant properties of betalains. *Journal of Agricultural and Food Chemistry, 65*, 675–689.
87. Gliszczyńska-Świgło, A., Szymusiak, H., & Malinowska, P., (2006). Betanin, the main pigment of red beet: Molecular origin of its exceptionally high free radical-scavenging activity. *Food Additives & Contaminants, 23*(11), 1079–1087.

88. Cai, Y. Z., Sun, M., & Corke, H., (2005). Characterization and application of betalain pigments from plants of the *Amaranthaceae*. *Trends in Food Science & Technology, 16*(9), 370–376.
89. Kapadia, G. J., Tokuda, H., Konoshima, T., & Nishino, H., (1996). Chemoprevention of lung and skin cancer by *Beta vulgaris* (beet) root extract. *Cancer Letters, 100*(1, 2), 211–214.
90. Kapadia, G. J., Azuine, M. A., Sridhar, R., Okuda, Y., Tsuruta, A., Ichiishi, E., Mukainake, T., et al., (2003). Chemoprevention of DMBA-induced UV-B promoted, NOR-1-induced TPA promoted skin carcinogenesis, and DEN-induced phenobarbital promoted liver tumors in mice by extract of beetroot. *Pharmacological Research, 47*(2), 141–148.
91. Rahimi, P., Abedimanesh, S., Mesbah-Namin, S. A., & Ostadrahimi, A., (2019). Betalains, the nature-inspired pigments, in health and diseases. *Critical Reviews in Food Science and Nutrition, 59*(18), 2949–2978.
92. Attoe, E., & Von, E. J. H., (1985). Oxygen involvement in betanine degradation—Measurement of active oxygen species and oxidation reduction potentials. *Food Chemistry, 16*(1), 49–67.
93. Reynoso, R., Garcia, F. A., Morales, D., & Mejia, E. G. D., (1997). Stability of betalain pigments from a cactacea fruit. *Journal of Agricultural and Food Chemistry, 45*(8), 2884–2889.
94. Delgado-Vargas, F., Jiménez, A. R., & Paredes-López, O., (2000). Natural pigments: Carotenoids, anthocyanins, and betalains — Characteristics, biosynthesis, processing, and stability. *Critical Reviews in Food Science and Nutrition, 40*(3), 173–289.
95. Attoe, E., & Von, E. J. H., (1985). Oxygen involvement in betanine degradation: Effect of antioxidants. *Journal of Food Science, 50*(1), 106–110.
96. Carrilo-Lopez, A., & Yahia, E. M., (2018). Betalains: Chemistry and biological functions. In: *Fruit and Vegetable Phytochemicals: Chemistry and Human Health* (Vol. 2, pp. 383–392). John Wiley and Sons.
97. Jackman, R., & Smith, J., (1996). Anthocyanins and betalains. In: *Natural Food Colorants* (pp. 244–309). Boston, MA, Springer.
98. Belhadj, S. I., Najar, T., & Abderrabba, M., (2017). Chemical and antioxidant properties of betalains. *Journal of Agricultural and Food Chemistry, 65*(4), 675–689.
99. Herbach, K. M., Strintzing, F. C., & Carle, R., (2006). Betalain stability and degradation — Structural and chromatic aspects. *Journal of Food Science, 71*(4), 41–50.
100. Wilcox, M. E., Wyler, H., & Dreiding, A. S., (1965). Stereochemistry of betanidin and isobetanidin. VIII. Structure of the bark pigment betanin. *Helvetica Chimica Acta, 48*(5), 1134–1147.
101. Herbach, K. M., Stintzing, F. C., & Carle, R., (2005). Identification of heat-induced degradation products from purified betanin, phyllocactin and hylocerenin by high-performance liquid chromatography/electrospray ionization mass spectrometry. *Rapid Communications in Mass Spectrometry, 19*(18), 2603–2616.
102. Alard, D., Wray, V., Grotjahn, L., Reznik, H., & Strack, D., (1985). Neobetanin: Isolation and identification from *Beta vulgaris*. *Phytochemistry, 24*(10), 2383–2385.
103. Coy-Barrera, E., (2020). Analysis of betalains (betacyanins and betaxanthins). In: *Recent Advances in Natural Products Analysis* (pp. 593–619). Elsevier.

104. Ngamwonglumlert, L., Devahastin, S., & Chiewchan, N., (2017). Natural colorants: Pigment stability and extraction yield enhancement via utilization of appropriate pretreatment and extraction methods. *Critical Reviews in Food Science and Nutrition, 57*(15), 3243–3259.
105. Da Silva, H. R. P., Da Silva, C., & Bolanho, B. C., (2018). Ultrasonic-assisted extraction of betalains from red beet (*Beta vulgaris* L.). *Journal of Food Process Engineering, 41*(6), e12833.
106. Mereddy, R., Chan, A., Fanning, K., Nirmal, N., & Sultanbawa, Y., (2017). Betalain rich functional extract with reduced salts and nitrate content from red beetroot (*Beta vulgaris* L.) using membrane separation technology. *Food Chemistry, 215*(15), 311–317.
107. Akbar, H. E., Sadiq, Z., & Zia-Ul-Haq, M., (2018). Analysis of betalains. In: *Betalains: Biomolecular Aspects* (pp. 139–164). Cham, Springer.
108. Singer, J. W., & Von Elbe, J. H., (1980). Degradation rates of vulgaxanthine I. *Journal of Food Science, 45*(3), 489–491.
109. Miguel, M. G., (2018). Betalains in some species of the *Amaranthaceae* family: A review. *Antioxidants, 7*(4), 53.
110. Lodish, H., Berk, A., Zipursky, S. L., Matsudaira, P., Baltimore, D., & Darnell, J. (2000). Oxidation of glucose and fatty acids to CO_2. *Molecular Cell Biology* (4th edn.).
111. Turker, N., Aksay, S., & Ekiz, H. İ., (2004). Effect of storage temperature on the stability of anthocyanins of a fermented black carrot (*Daucus carota* var. L.) beverage: Shalgam. *Journal of Agricultural and Food Chemistry, 52*(12), 3807–3813.

CHAPTER 2

IMPACT OF HIGH-PRESSURE PROCESSING ON FOOD COLORS

NANCY AWASTI,[1] KARTIK SHAH,[2] KHILENDRA BHANDURIYA,[3] AND VENKATESWARLU SUNKESULA[4]

[1]*Lactalis American Group, Buffalo, New York, USA*

[2]*Sargento Foods, Elkhart Lake, Wisconsin, USA*

[3]*Milk Specialties Global, Eden Prairie, Minnesota, USA*

[4]*Research and Product Development, Idaho Milk Products, Twin Falls, Idaho, USA*

ABSTRACT

Food color plays a significant role in identifying the age and suitability of food for human consumption. Consumer acceptability is greatly influenced by the perception of food color. Traditional high-temperature techniques used in the beverage and food manufacturing industry can lead to changes in color components, loss of vitamins, and other nutritional values. However, with the growing demand for fresh flavor, color, and clean label food, it is essential to have a deep understanding of nonthermal technologies that use minimal heat and additives to preserve the freshness and extend the shelf life of the product. Over the past three decades, nonthermal techniques such as high-pressure processing (HPP) on pre-packed food have proven to be less aggressive, ensuring improved safety and quality attributes, as well as an extended shelf life for finished food products. The implementation of pressure treatment, either alone or in combination with thermal and

Nonthermal Food Processing Technologies: Impact on Color Profile.
Kaavya Rathnakumar and R. Pandiselvam (Eds.)

nonthermal methods, may affect the natural attributes of finished products. Therefore, to comprehend the changes in color or appearance of different foods (dairy, seafood, juice, beverages), this chapter focuses on reviewing the impact of HPP on food colors.

2.1 INTRODUCTION

The increase in demand for better quality, good nutrition, functional, and sensory characteristics of food among consumers is one of the major driving factors that has led to the growth of nonthermal practices such as high-pressure processing (HPP). Currently, the application of high thermal techniques is the preferred approach for food preservation, but these treatments result in the degradation of quality, freshness, non-enzymatic browning, and changes in sensory aspects of the food. To meet the demand without compromising safety, quality, and functionality of the finished food product, it is important to utilize the latest techniques in the food manufacturing sector. Over the past three decades, nonthermal methods like HPP have proven to be less aggressive, with improved food safety and quality parameters, as well as extended shelf life of end products [1]. HPP has become a significant part of food manufacturing, especially in the beverage, dairy, and meat sectors. Therefore, for the benefit of consumers and the food manufacturing industry, there is a need to implement innovative and improved mild processing and preservation methodologies that allow the production of safe, stable, and fresh food.

The first food products produced using high pressure were launched in the early 1990s in the Japanese market, including sauces, jams, and jellies [2]. Following this trend, other food products such as fruit juices in Portugal and France, oysters, and guacamole in the United States, were introduced using this technology [3]. After the initial success of HPP treatment with fruit jams and juices, the application of HPP technique has been expanded to a wide range of food products, including wheat and wheat products, seafood, smoothies, dairy and dairy products, as well as other beverages such as ready-to-eat food and protein drinks. In terms of operating cost, a review [4] indicated that high pressure processing (HPP) techniques are much more cost-effective compared to thermal treatment.

Food color plays an essential role in identifying the age of food and its suitability for human consumption [5]. Therefore, color perception greatly influences the sensory experience of customers when deciding and selecting food products from the shelf. To understand consumer acceptability and market requirements, it is crucial to comprehend the impact of any process on food matrices. Visual attributes should be carefully studied when formulating new products. High heat treatments (HTs) are known to cause the degradation of enzymes that affect food color [6]. The use of traditional high temperature technology in the beverage and food manufacturing and processing industry is associated with changes in carotenoids, chlorophyll, and other important components that contribute to the color of food. It also leads to the loss of vitamins and other nutritional values. Conventional HT has been found to have an adverse impact on color-producing enzymes such as anthocyanins (ACNs). With the increasing demand for fresh flavor, color, and clean label food, it is necessary to have an in-depth understanding of nonthermal technologies that utilize minimal heat and additives to retain fresh-like characteristics while extending the shelf life of the product. In comparison to thermal treatments, HPP technology has an insignificant impact on the sensory qualities of the product and hence retains the "fresh-like" qualities of aroma, flavor, color, appearance, and texture [7]. However, a growing number of reviews and recent scientific research studies [8–11] have focused on the impact of HPP technology on the color, texture, and nutritional value of finished foods. The implementation of high-pressure techniques alone or in association with other thermal and nonthermal techniques may influence the natural attributes of finished products. Therefore, to understand the changes in color or appearance of different foods (dairy, seafood, juice, beverages), the current chapter focuses on reviewing the effect of HPP on food colors in different foods.

2.2 PRINCIPAL OF HPP TECHNOLOGY

Unlike thermal methods, the preservation effect after pressure treatment is uniform and independent of the dimensions, size, and geometry of the food. Another term for HPP is high hydrostatic pressure (HHP) technology, which involves applying pressures of about 100–1,000 MPa (mega pascal) to pre-packed food for a short duration, typically ranging from a few seconds to approximately 20 minutes. HPP can be described based on two principles – (i) Le Chatelier's principle; and (ii) the isostatic principle or

Pascal's law. Unlike thermal treatments that utilize temperature gradients, HPP treatment exposes each atom and molecule of food to constant pressure simultaneously [12]. HPP effectively inactivates microbial cells, yeasts, and molds while minimally affecting the appearance and other sensory aspects of the processed food. Additionally, the application of high pressure inactivates the enzyme oxidases that can reduce the quality of food. As an example, polyphenol oxidase (PPO) catalyzes the colorless component in fruit pulp, which is o-diphenol, and therefore results in the development of o-quinones responsible for producing dark-colored melanin after polymerization [13]. Thus, the discoloration rate in fruit and fruit products such as pulp and beverages is directly associated with the level of PPO activity. HPP technology includes the implementation of high pressure to inactivate PPO enzymes and pasteurize the treated food or beverage by inactivating spoilage-causing vegetative microorganisms without impacting the nutrient, color, and flavor [14]. Therefore, the implementation of HPP at an industrial scale is mandatory to preserve and prevent the stability of color, flavor, and nutrients of treated food [10].

2.3 SIGNIFICANCE OF FOOD COLOR TO CONSUMER

Humans perceive the existence of food through touch, sight, smell, taste, and sound. The experience of indulging in food involves these fundamental senses, each of which encompasses a wide range of the spectrum to discern differences. Sight utilizes the spectrum within the visible light zone, covering all visible colors from Red to Violet. We rely on color and shape to distinguish food varieties and determine their suitability for consumption, including assessing their age. Color perception also influences sensory experience, as food colors are associated with flavors that give distinctiveness to fruits and vegetables. For example, we associate orange color with oranges and red color with strawberries. The intensity of color can also be linked to the concentration of associated flavor, such as red versus hot red. However, the correlation between color and flavor is not consistently observed, leading to some degree of uncertainty [15, 16]. Historically, to preserve food, it has always been processed and converted into a different product form [17]. The conversion process changes the primary color to different shades (dried orange appears dark orange in color) or completely different colors (white milk to pale-yellow butter).

A tremendous amount of food is being processed in North America and other countries. Food has evolved in the last century to offer convenience to meet busy lifestyles, and so have food colors. Currently, food manufacturers use food colors to provide an identity to formulated processed food that correlates with flavor. Colors can also be used to correct color loss during processing and offset seasonal/natural color changes [18].

Colors are usually classified as natural and artificial, although there are plenty of ways to categorize colors [19]. The most familiar natural colors are chlorophyll, ACN, and carotenoids. Carotenoids are phytonutrients that are responsible for characteristic colors such as red, yellow, and orange in vegetables and fruits [19]. Generally, carotenoids are hydrophobic and soluble in fat. Carrots, kale, pumpkin, cantaloupe, dairy, etc., are common examples of foods rich in carotenoids. Carotenoids also possess antioxidant properties and reduce the harmful impact of free radicals on tissues. Carotenoids act as a precursor to vitamin A, which helps improve vision, growth, and support immunity [19].

Chlorophyll is a well-known pigment found in plants that plays a significant role in photosynthesis and is responsible for providing green color to the leaves [20]. Chlorophyll exists in two different variants, *a* and *b*, and it is sandwiched between lipid and protein present in leaves. The differences are at position 7, where a contains -CH_3 and *b* contains -CHO, resulting in a blue-green color for *a* and *a* more yellow-green color for *b* [21]. The intensity and brightness of the green color represent the quality and health of the plant. The color can be affected by extrinsic factors such as metal ions, heat, pH, age, enzymes, bleach, and fermentation [22]. These factors can also assist in the isolation, purification, and preservation of chlorophyll [19].

ACN is a phenolic compound that plays an essential role in plant colors. It has been found in flowers, fruits, leaves, stems, and seed coats, contributing to colors ranging from purple to red to blue. There are six major ACNs (cyanidin, delphinidin, petunidin, peonidin, malvidin, and pelargonidin) that contribute to these colors and protect the plant from UV lights [23, 24]. ACNs are phenolic compounds with water-soluble abilities and are formed due to various physiological and biochemical changes in plants [24–26]. The degradation of ACN has been linked to various factors such as increased glycosidic substitution, enzymes (oxidase and glycosidase), ascorbic acid (AA), light exposure, and pH [27].

2.4 THERMAL AND NONTHERMAL TREATMENTS IMPACT ON COLOR

Preservation of food has been the essence of human evolution, allowing for the saving and storing of food for extended periods. This helps to stabilize the food supply, especially when faced with seasonal limitations. Preservation also enables the transportation of food to areas where its availability would otherwise be limited or nonexistent. There are two main methods of food preservation. The first involves treating the food to eliminate pathogens, spoilage bacteria, and their enzymes. The second method involves transforming the food into a different state, where it can be preserved through the addition of chemicals, salt, or by reducing water activity. In both cases, the original structure of the food undergoes various physical, chemical, or microbiological changes that significantly impact its rheology (shape, size, texture, and color) and sensory characteristics.

Thermal treatment has been widely used as a preservation method for liquid and semisolid food. Hot water or steam, whether through indirect or direct contact, are the primary sources of heat energy used in thermal processing. Pasteurization, with batch temperatures of 63°C for 30 minutes or continuous temperatures of 72°C for 15 seconds, is a common method for eliminating pathogens in milk and ensuring its safety for consumption. Thermalization, with temperatures ranging from 57–68°C for 10–20 seconds, is also employed to reduce bacterial load in raw milk when refrigeration capacity is limited for storage and transportation. It is important to note that thermized milk should still be pasteurized before consumption. These processes are generally applied to other liquid products such as juice, wine, and beer, with minimal modifications, to achieve the same objectives. Pasteurization of packaged goods, like beer and fruit juices, is not uncommon. These treatments are considered the most gentle thermal methods for food processing, ensuring that the product remains free from spoilage when stored under required refrigeration conditions. The impact of pasteurization on food color has been extensively studied. For instance, thermal pasteurization of grapefruits has shown significant differences in color before and after treatment. The major pigments, β-carotene and lycopene, did not exhibit any differences after treatment. The observed changes are attributed to the precipitation of suspended particles caused by the applied heat load [28]. These changes appear to increase L^*, a^*, b^*, and ΔE values. For highly pigmented fruit juices, the use of color enhancers is recommended to compensate for any color loss during processing and storage. A previous study [29] demonstrated that the phenolic compounds in pomegranate juices are sensitive to thermal treatment and tend to bind with protein and pectin, resulting in turbidity. The turbidity and color changes accelerate with thermal processing; therefore, fining of phenolic compounds is highly recommended to stabilize color changes in pigmented juices. The need for shelf-stable storage products has led to the widespread use of ultra-heat treatment (UHT) and retort in the food and beverages sector. These processes are designed to achieve "Commercial sterility," which involves the complete destruction of all microorganisms and spores. UHT processing (>135°C/2.0–5.0 seconds) and subsequent storage were reported to have a deteriorating impact on the colors during the heat processing and post-treatment storage [30]. Similarly, Crinó & coworkers [31] evaluated the stability of red/pink natural color during UHT treatment and storage. However, they reported a negative impact of high temperature (HT) on the stability of natural food colors during food processing. Even a temperature increase of 5°C can significantly affect the color [31]. Sugarcane juices become redder after undergoing ultra-high temperature (UHT) treatment, which is attributed to the Maillard reaction in the juice, resulting in browning and loss of chlorophyll [32]. Similarly, liquid eggs lose color during UHT processing and storage [33]. Heat sterilization at 121°C for 15 minutes of green vegetables leads to a color transformation from green to an olive-brown color due to the conversion of chlorophyll to pheophytin [30, 34]. Strawberry canning results in approximately 70% loss of total monomeric anthocyanins (ACNs) and 23.5% loss of total phenolic content [35]. This loss continues during storage and is accelerated at high temperatures. Overall, thermal treatment causes the loss of colors primarily derived from chlorophyll and ACNs, reducing brightness and increasing redness due to Maillard browning. The extent of these changes depends on the heat load, which can be expressed as a function of temperature x time.

To address issues resulting from conventional thermal processing, numerous nonthermal approaches have been developed and applied to food and food systems. These techniques are typically referred to as nonthermal methods. The influence of nonthermal techniques on color preservation has been widely researched and experimented. The results are widely dependent on the techniques' mechanism and the product's intrinsic factors (composition, aw, pH, etc.) and extrinsic factors (temperature, humidity, etc.). Microwave (MW) and radiofrequency use electromagnetic (EM) waves to heat food. The wave frequencies of 915 and 2,450 MHz (megahertz) were applied using MW, and approximately 13 to 40 MHz for radiofrequency [36]. MW's dielectric heating of polarizable molecules has been applied in food preservation. The rapid rate of temperature increase offers some advantages to the phytochemical component of foods and vegetables. The effect of MWs depends on the power, time of treatment, and the type of food it is being applied to. Microwaving was able to maintain the color of both frozen and fresh vegetables when compared with boiling and steaming [37]. The treatment has been shown to be more destructive to the color of pears [38] and carrots [39]. On the other hand, fresh citrus peel of Thompson navel and mandarin did not show any significant loss of color when subjected to the MW treatment [40]. The Maillard browning of potato fries occurs faster when frying compared to microwaving [41]. The ability of radiofrequency to heat the product from core to surface offers a unique moisture migration that reduces over-dehydration on the product surface. Radiofrequency can preserve the color of the walnut kernel for up to two years [36]. Similarly, Ohmic heating (OH) has a significant impact on the phytochemical compounds of vegetable tissues by manipulating the energy levels [42]. The principle of OH works on the transformation of electrical energy into thermal energy via the resistance produced by the media/food. The amount of energy produced is a function of the treatment time, voltage, and type of food/media. Milk treatment with OH has been shown to increase whiteness for the first 15 minutes of treatment, then it starts getting darker as the treatment time passes, but the concentrated milk showed less browning compared to conventional heating [43]. In acerola pulp, frequencies of OH could contribute to pigment and AA degradation, and the rate of degradation increases with a reduction in electric field frequencies [44]. Ultrasound, acoustic sound waves >20 KHz (kilo hertz), produces energy with compression and depression modes in the medium. This technology is effective in preserving color during the thawing of frozen products, seaweed, meat, fruits, and vegetables, cereal products, and dairy products [45] compared to traditional thermal heating. Exposure to medium to higher power for a longer time could still cause the chemical decomposition of ACNs. Controlled variation of amplitude could help extract the bound ACNs from the suspended juice [46]. Pulsed electric field (PEF) is another example of a nonthermal technique that employs short-duration pulses with high electric field (10–80 kV/cm). These pulses are applied for a brief duration (micro- to milliseconds). In terms of color, the pasteurized juices showed increased yellowness (b*) and less redness (a*) with higher color variation than high-intensity PEF-treated juice [47]. Similarly, no color change was observed in fresh and high-intensity pulsed electric field (HIPEF)-treated juice. Ultraviolet (UV) light (200–400 nm) is widely used as a disinfectant for surfaces, water, and air. The effect of UV on food has also been explored in the last few decades, indicating a low to no negative effect on food colors. UV can be classified as Short (200–280 nm), Medium (280–320 nm), and High (320–400 nm) wave UVs [48]. UV increases brightness and reduces *Salmonella typhimurium* load by 61% in chicken legs compared to untreated chicken legs [49], and no color change was observed in lettuce or strawberries after up to 45 minutes of treatment [50].

With the listed work above, it is apparent that nonthermal techniques offer superior and nondestructive solutions to food processing. The effect is not limited to only food safety of the product, but it has been extended to preserving appearance (color and texture), sensory, and nutritional attributes of food. Thus, additional investigation and innovation are needed to translate these novel techniques to a commercial scale.

2.5 IMPACT ON COLORS OF DIFFERENT FOOD

2.5.1 VEGETABLES AND FRUITS

The minimal impact of high-pressure processing on low-molecular-mass chemicals consisting of covalent bonds, such as color, makes it a viable substitute for vegetable and fruit processing. However, incomplete inactivation of the enzymes or microorganisms responsible for the undesirable chemical reaction can also cause alterations in the color components of HPP processed food. These non-enzymatic and enzymatic reactions can impact the color of the HPP-treated foods. As discussed in this chapter, chlorophyll, carotenoids, and ACNs are some of the major color-contributing components. Chlorophyll is attributed to the green color in food products, and chlorophyll degradation to pheophytins can degrade the green color to gray-brown. Carotenoids result in red and orange-yellow colors in vegetables and fruits, whereas vacuoles containing water-soluble flavonoid pigments called ACNs cause red to blue colors. Additionally, some of the color changes are also a result of enzymatic browning, such as browning induced by the polyphenol oxidase enzyme (PPO). Inactivation of this enzyme using HPP can be beneficial to induce irreversible inactivation of PPO.

The impact of high pressure and temperature on the color of vegetables and fruits has also been widely studied. Research by Krebbers and coworkers [51] demonstrated that ambient or moderate temperature high-pressure processing (HPP) resulted in minimal color change in green vegetables [51]. Similarly, a previous study observed that increasing the pressure from 200 to 800 MPa at temperatures above 50°C caused a degradation of chlorophyll (a and b) components by 19.4% and 68.4%, respectively [52]. However, several studies have indicated improved color in the post-treatment food matrices as well. A previous study [51] reported improved green color in treated green beans compared to untreated green beans after the high-pressure treatment. The authors treated the green beans at ambient temperature for 1 minute using 500 MPa pressure. A similar study on HPP of spinach and tomato puree also resulted in improved color compared to the heat-pasteurized version of the product [53]. Although a detailed study may be required to further understand these differences, it is hypothesized that the leakage of chlorophyll into the intracellular space during HPP treatment may have resulted in improved color in the treated samples [54].

The carotenoid pigments are the main cause of the yellow-orange and red color in vegetables and fruits, making them less susceptible to pressure. A study by McInerney et al. [55] found no impact on the level of carotenoids in HPP treatment at 400 and 600 MPa in carrots, green beans, and broccoli [55]. Another study by Rodrigo, Van Loey, & Hendrickx [56], showed no change in major carotenoids such as β-carotene and lutein in broccoli after HPP up to 700 MPa at 65°C for up to 1-hour HPP pre-treatment of 400 MPa at 40°C for 1 minute of peeled "navel" and "caracara" oranges produced orange juice with increased carotenoid concentration than in 200 MPa processed juice [57].

The blue and red colors of vegetables and fruits are associated with water-soluble vacuolar flavonoid pigments – ACNs. Several authors have studied the impact of HPP on ACN stability in various food products, including corn and certain varieties of fruits. An increase in ACNs was reported when purple waxy corn was treated at 700 MPa compared to 500 MPa for 30 to 45 minutes [58]. However, a study conducted by Kouniaki and coworkers [59] examined the influence of 600 MPa pressure on ACNs in blackcurrants and reported the best stability in fruits stored at 5°C [59]. Browning of fruits and vegetables due to polyphenol oxidase (PPO) is very common and often results in a loss of flavor and color. HPP can be used as an alternative treatment to heating methods to inactivate PPO. Enzyme inactivation using HPP depends on the type of food, treatment intensity, treatment duration, initial concentration, and composition of the food, among other factors. Fruits such as grapes, strawberries, and apricots require lower pressures, such as 600, 400, and 100, respectively. On the other hand, vegetables like mushrooms and potatoes require higher pressures (800–900 MPa) for PPO inactivation [60].

2.5.2 JUICES/SMOOTHIE

The loss of functional and sensory qualities that occur during the thermal pressurization of juice is one of the industry's major challenges. Therefore, nonthermal techniques such as HPP technology are seen as potential treatments to stabilize sensory as well as functional attributes during the storage of these products at refrigerated conditions [1]. Thermal techniques, especially pasteurization, thermization, or sterilization, are often applied to inactivate enzymes in fruits and fruit products. These enzymes can deteriorate the food quality nutritionally, organoleptically, or aesthetically if not inactivated. HPP treatment is used as an alternative to destabilize enzymes and microorganisms causing spoilage in food, without altering the nutritional and sensory qualities of the treated products [61].

Cloud destabilization, non-enzymatic browning, and consistency loss are common quality defects in products such as citrus fruits like orange juices and tomato purees. These defects are linked to the enzyme pectin methyl esterase (PME) [7], which is usually inactivated after thermal treatment. To prevent discoloration caused by high temperature treatment of juices, nonthermal technologies like hydrodynamic cavitation and HPP [62] have gained significant interest. Thermal treatments are often used to inactivate PME and prevent quality defects. However, the impact of HPP treatment on PME inactivation and resultant quality has also been extensively studied [7]. Several studies have shown that heat-stable fractions of PME remain active even after HPP treatment [63]. Research by Fernandez et al. [64] reported an 85% reduction in PME activity in HPP-treated fruit and vegetable smoothies [64]. Similarly, a 60% reduction in enzymatic activity was observed in mango pulp after treatment at a pressure of 600 MPa at 52°C for 10 minutes [65]. Furthermore, orange juice treated at a pressure of 700 MPa for 1 minute showed stability for more than 50 days [66], and pressure treatment at 800 MPa for 1 minute increased the storage stability for over 60 days [67].

Zhang and coworkers [9] compared the impact of pressure (600 MPa at 4°C), pasteurization, and AA treatment on PPO, sensory properties, and color of fruit pulp (Pawpaw) over a period of approximately 45 days (with intervals of 15 days) incubated at 4°C [9]. A significant decline in PPO activity was detected after HPP treatment. However, no significant changes in the sensory attributes of the product were reported. On the other hand, the untreated samples of fruit pulp exhibited a significant difference in color values after 45 days of refrigerated storage. A recent study [10] explored the impact of HPP treatment (at 350, 450, and 550 MPa) and pasteurization on color attributes, AA, and ACNs activity of pomegranate juice. In comparison to HPP treatment, thermally treated juice showed a significant change in color with a decrease in monomeric ACNs and an increase in polymeric ACNs, browning index (BI), and polymeric color. ACNs are associated with the deep red color in vegetables and fruits and are sensitive to HT, resulting in a brownish color during storage and processing [68]. Previous research [69] reported a decrease in total ACNs, cyanidin-3-rutinoside, and cyanidin-3-glucoside in Mulberry juice through HPP treatment, but in comparison to non-treated juice samples, both ACNs were documented to be higher after 30 days of refrigerated storage. The preservatory effect of color during HPP treatment is mainly contributed by ACNs and the enzymes inactivation such as peroxidases (PODs) and polyphenol oxidases (PPO). Likewise, Yi et al. [11] also revealed parallel findings in terms of the color change of apple juice after HPP however, no substantial change was observed in the color in comparison to untreated samples. Additionally, in contrast to untreated and HPP-treated samples, a significant increase in lightness and decrease in yellowness of thermally treated juice was reported [11]. The change in color of thermally treated apple juice was attributed to the impact on POD and PPO enzymes, leading to non-enzymatic browning reactions [70]. These reports collectively demonstrate the ability of high-pressure treatment to preserve the quality and sensory aspects of juice by sensitizing the microbial cells to acid [71].

2.5.3 SEA FOOD

Sea foods, including fish, shrimp, oysters, and other seafood, are extremely perishable food products with a reduced shelf life in refrigerated environments. The muscle color of seafood plays a significant role in consumer perception of freshness and flavor. Previous consumer studies have shown that consumers associate color with product freshness, indicating better quality, improved flavor, and taste. In a previous study [72], the change in color of salmon was compared after the use of pressure treatment (150 MPa and 300 MPa for 15 minutes) compared to control samples. The results of this analysis demonstrated that an increase in pressure caused the dark muscles of the salmon samples to appear cooked and become opaque. In comparison to the control and samples treated at 150 MPa pressure, the value of lightness/opaque of dark muscles increased with higher pressure [72]. Kaur and coworkers [73] studied the impact of different pressure levels (435, 100, and 270 MPa for 5 minutes) of HPP on black tiger shrimp treated at ambient temperature (25 ± 2°C) and stored at refrigeration temperature for 35 days [73]. They found that the color of the black tiger shrimp changed, with an increase in whiteness index (WI) as the pressure intensity increased, resulting in a brighter and slightly cooked appearance. Moreover, the samples treated at 435 MPa showed an increased shelf life of over 15 days compared to untreated shrimps, which was supported by a reduced microbial population due to the damaging and inhibiting effects of high-pressure treatment on the product.

2.5.4 MEAT AND POULTRY

In the past 20 years, HPP has been progressively utilized in the meat sector as an alternative to thermal methods, mainly for ready-to-eat (RTE) products. The US Agriculture and Food Safety Services (USDA-FSIS) have also published a letter of no-objection (LNO) certificate for the usage of high-pressure treatment as an efficient tool to reduce *Listeria monocytogenes* in RTE poultry and meat products. Similarly, the Canadian health agency has also issued an LNO to control *Listeria monocytogenes* in uncured and cured pork. However, HPP-treated foods are considered novel foods in European countries [74]. HPP in the meat sector is mainly viewed as a suitable alternative to produce safe products, particularly where treatment using heat is not convenient. Most of the food products subjected to HPP are RTE foods.

In food products including meat, the optical traits, and the myoglobin (Mb) substance of the muscle majorly contribute to the color. Whereas, in cured meat, the color is attributed by nitrosyl myoglobin, a reaction product of nitric oxide (NO) and myoglobin. HPP treatments in fresh meat have a greater impact on the color compared to cured meats where the color changes in HPP-treated cured meats are more acceptable based on the water activity (a_w-) value and water content [75]. Discoloration in the HPP-treated meat was mainly attributed to two possible phenomena – (i) Heam replacement or release and/or Myoglobin denaturation induced whitening effect; and (ii) ferric myoglobin – an oxidation product of ferrous myoglobin (above 400 MPa) [76]. Several investigations have been conducted on the application of HPP on fresh meat, although its industrial application remains limited. According to Bajovic, Bolumar, & Heinz [8], the color change in fresh meat is influenced by water-soluble heme proteins and myosin, which are sensitive to pressure [8]. When minced beef was treated with HPP in the range of 200–350 MPa, it resulted in increased L* color values, giving a pink color. On the other hand, a* value decreased at 400–500 MPa, resulting in a gray-brown color [76]. Similarly, research by Bak et al. [77] indicated that treatment above 300 MPa caused pork meat to become significantly more yellow and less red within 24 hours of storage [77]. Marcos et al. [78] observed no effect on color in low acid fermented sausages after 400 MPa color treatments [78]. Studies on the application of HPP in fish products suggest negative impacts on the color of the treated product. Although the impact may vary between species, several studies indicate an increase in yellowness (b*) and lightness (L*) in species with more red muscle [79, 80]. A study by Gómez-Estaca, Gómez-Guillén, & Montero [81] on cold-smoked fish reported an increase in all CIE L*a*b* values [81].

2.5.5 BAKERY

High hydrostatic pressure (HHP) treatment is widely studied and applied in various food matrices such as juices, fresh fruits, meat products, and dairy products. However, very limited studies have been conducted on HPP treatment in bakery applications. Starch gelatinization is known to occur due to thermal treatments. HPP treatments have also been studied to induce gelatinization of starch, but the mechanism is different. The integrity of the HPP-treated starch granule does not change; however, it does induce swelling of the starch granule. As a result, HPP treatment of starches alters their rheological and microstructural properties, which differ from thermal treatments [82]. Additionally, Wang et al. [83] showed that HPP-treated starches have a low gelatinization temperature and enthalpy [83]. They also lose their crystallinity and are likely to aggregate. The impact of HPP on gluten indicates the weakening of gluten at low pressures (200 MPa), whereas high pressures and high temperature have a strengthening effect [84]. Bárcenas, Altamirano-Fortoul, & Rosell [85] studied various HPP treatments ranging from 50 to 250 MPa and concluded that HPP treatment of wheat dough significantly altered the color at a high pressure of 250 MPa [85]. Barcenilla et al. [86] examined the effect of HPP treatments on the cake batter (300 and 600 MPa) in the cake crust and crumb [86]. The authors reported a reduction in L* (darker) and an increased a* (yellowish) in the HPP-treated batter. The authors hypothesized that the change in color was due to decreased gelatinization of starches because of HPP treatments.

2.5.6 DAIRY

The use of HPP has been explored as an alternative to existing heat processes such as pasteurization. The impact of HPP treatment on milk and milk products has been extensively studied [87–89], and a previous study reported increased casein micelle and whey protein denaturation after HPP treatment [90]. HPP treatment also influences the casein micelle structure, resulting in submicelles moving into the serum phase. These effects are characterized by a change in the size of the micelle, a decline in the turbidity of milk, and color lightness [91–93]. A report by Nassar et al. [94] stated a decline in L* values (lightness) as the pressure treatment increased for caprine milk [94]. The authors concluded that the increase in lightness can be ascribed to small fragments of casein micelles resulting from the disruption of casein micelles. A similar trend was noted in a recent study performed by Iturmendi et al. [95]. The authors reported significant differences in L* values of HPP untreated and treated samples of micellar casein concentrates (MCC) solutions at two protein concentrations [95].

The authors reported a decrease in L* values as the pressure treatments increased from 300 MPa to 600 MPa. The disruption of casein micelles negatively impacts their capability to scatter light, which can be identified by a decline in the L* value. Fresh cheeses are particularly of interest for HPP treatment as they are often characterized by a shorter shelf life due to higher moisture, higher pH, and higher water activity. The implementation of HPP in fresh cheeses has been widely explored [96–100]. A previous study examined the impact of HPP treatment (500 MPa for 5 minutes) on starter-free fresh cheese [101]. The authors demonstrated a decline in lightness and a rise in intensity in yellowness in HPP-treated cheeses. The authors attributed these color changes to a more dense and continuous protein network because of HPP treatment. Ribeiro, Leite Júnior, & Cristianini [102] examined the HPP impact (600 MPa for 5 minutes at 25°C) on cheese made from ultrafiltered milk (UF (ultrafiltration)-Cheese) and stated that although HPP-treated cheese has significantly different L*, a*, and b* values, these color changes are not substantial enough to be identified by the human eye [102]. The researchers reported that only ΔE values >1.5 can be perceived by the human eye. Therefore, although the changes in the color of HPP-treated samples can be significant, they may not be considered as noticeable variations to the color of the HPP-treated cheese.

2.6 EFFECT ON MICROBES

The pressure used in HPP is known to destroy pathogenic microorganisms such as species of *Vibrio, Listeria, Escherichia, Salmonella,* and other bacteria, as well as yeasts and molds that cause food spoilage. Microbial spoilage in food and beverages may result in off odor, color, turbidity, flavors, and the production of gas [7]. A previous study [73] compared the microbial profile after the application of two different HPP treatments (150 MPa and 300 MPa for 15 minutes) with a control on Atlantic salmon samples. Treatment 1 (150 MPa for 15 minutes) showed a significant decrease in microbial count to 3 log (day 0 of treatment) and 2 log after 6 days of storage at 4°C, whereas applying a pressure of 300 MPa (Treatment 2) for 15 minutes resulted in total microbial reduction with no growth detected after 6 days of incubation at 4°C. However, for control samples (with no treatment), the microbial counts showed a significant increase in growth after 6 days of incubation at 4°C. The ability of this technology to retain color and other sensorial characteristics of food products, enhanced inactivation of microbes, and spoilage-causing enzymes has encouraged leading food manufacturers and sectors to include high-pressure processed food in the market [4]. Several spoilage-causing vegetative cells and species of yeast and mold can be inactivated by implementing pressure of approximately 300–600 MPa. This pressure can lead to the permeabilization of cell membranes or deformation of cell organelles [4]. In sterilization and pasteurization processes, a combination of high pressure and temperature is often considered the most effective in reducing microbial counts. For instance, a previous study observed the variability in distortion of *Saccharomyces cerevisiae* cells when treated at pressures of approximately 400 and 500 MPa. At 400 MPa, deformation of cytoplasmic organelles followed by leakage of intracellular material was observed, while at 500 MPa, complete distortion of the nucleus and other organelles was observed [103].

Although the application of low pressure induces reversible changes, including alterations in cell morphology, high pressure results in irreversible changes and the death of microbial cells due to pore formation in the cell membrane. Bacterial spores are highly resistant to extreme conditions, including temperature, pressure, and acidic environments. To inactivate bacterial spores, sterilization temperatures above 100°C or pressures exceeding 1,200 MPa may be required [104]. However, meeting these requirements in a product is usually associated with the degradation of nutritional attributes and functionality of the finished products, such as the loss of vitamins, color, texture, taste, and other nutritional components [4, 105]. The impact of HPP on pathogenic and spoilage-causing vegetative organisms in food products varies and depends on various factors, including temperature variations, applied pressure, processing time, water activity, presence of preservatives and/or antimicrobials, and salt levels [8]. It has been reported that pressure levels of 400–600 MPa, with minimal processing time, can reduce the population of common spoilage-associated and pathogenic vegetative microbes in meat and meat products by over four log units [8].

2.7 SUMMARY

Overall, HPP treatment has proven to be effective in extending the shelf life of treated food and food products. Therefore, HPP can be commercially implemented by the food processing industry as a strategy for extending shelf life. However, it is important to closely monitor the appearance of the processed food, such as color and texture changes. Thus, HPP treatment can be combined with other nonthermal treatments to minimize changes in product appearance. Since there is limited research available on the impact of HPP treatment or a combination of nonthermal techniques on food color and the relationship of specific color-contributing enzymes in food, beverages, and juices, further research is needed. Conducting research at the pilot plant level focusing on industrial needs will provide a better understanding of the effects of HPP treatment and combination treatments on colors and enzymes, thereby enhancing market potential.

KEYWORDS

- **bakery and juices**
- **dairy**
- **food colors**
- **high pressure processing**
- **meat and meat products**
- **microbes**
- **nonthermal treatments**
- **sea food**

REFERENCES

1. Andrés, V., Villanueva, M. J., & Tenorio, M. D., (2016). The effect of high-pressure processing on color, bioactive compounds, and antioxidant activity in smoothies during refrigerated storage. *Food Chemistry*, *192*, 328–335. https://doi.org/10.1016/j.foodchem.2015.07.031.
2. Thakur, B. R., & Nelson, P. E., (1998). High-pressure processing and preservation of food. *Food Reviews International*, *14*(4), 427–447. https://doi.org/10.1080/87559129809541171.
3. Hugas, M., Garriga, M., & Monfort, J. M., (2002). New mild technologies in meat processing: High pressure as a model technology. *Meat Science*, *62*(3), 359–371. https://doi.org/10.1016/S0309-1740(02)00122-5.

4. Rastogi, N. K., Raghavarao, K. S. M. S., Balasubramaniam, V. M., Niranjan, K., & Knorr, D., (2007). Opportunities and challenges in high pressure processing of foods. *Critical Reviews in Food Science and Nutrition*, *47*(1), 69–112. https://doi.org/10.1080/10408390600626420.
5. Clydesdale, F. M., (1991). Color perception and food quality. *Journal of Food Quality*, *14*(1), 61–74. https://doi.org/10.1111/j.1745-4557.1991.tb00047.x.
6. Sadilova, E., Stintzing, F. C., Kammerer, D. R., & Carle, R., (2009). Matrix dependent impact of sugar and ascorbic acid addition on color and anthocyanin stability of black carrot, elderberry, and strawberry single strength and from concentrate juices upon thermal treatment. *Food Research International*, *42*(8), 1023–1033. https://doi.org/10.1016/j.foodres.2009.04.008.
7. Bull, M. K., Zerdin, K., Howe, E., Goicoechea, D., Paramanandhan, P., Stockman, R., Sellahewa, J., et al., (2004). The effect of high-pressure processing on the microbial, physical, and chemical properties of Valencia and navel orange juice. *Innovative Food Science & Emerging Technologies*, *5*(2), 135–149. https://doi.org/10.1016/j.ifset.2003.11.005.
8. Bajovic, B., Bolumar, T., & Heinz, V., (2012). Quality considerations with high pressure processing of fresh and value-added meat products. *Meat Science*, *92*(3), 280–289. https://doi.org/10.1016/j.meatsci.2012.04.024.
9. Zhang, L., Dai, S., & Brannan, R. G., (2017). Effect of high-pressure processing, browning treatments, and refrigerated storage on sensory analysis, color, and polyphenol oxidase activity in pawpaw (*Asimina triloba* L.) pulp. *LWT*, *86*, 49–54. https://doi.org/10.1016/j.lwt.2017.07.023.
10. Yuan, L., Cheng, F., Yi, J., Cai, S., Liao, X., Lao, F., & Zhou, L., (2022). Effect of high-pressure processing and thermal treatments on color and in vitro bio-accessibility of anthocyanin and antioxidants in cloudy pomegranate juice. *Food Chemistry*, *373*, 131397. https://doi.org/10.1016/j.foodchem.2021.131397.
11. Yi, J., Kebede, B. T., Doan, N. H. D., Buve, C., Grauwet, T., Van, L. A., Hu, X., & Hendrickx, M., (2017). Quality change during high pressure processing and thermal processing of cloudy apple juice. *LWT-Food Science and Technology*, *75*, 85. https://doi.org/10.1016/j.lwt.2017.03.040.
12. Elamin, W. M., Endan, J. B., Yosuf, Y. A., Shamsudin, R., & Ahmedov, A., (2015). High pressure processing technology and equipment evolution: A review. *Journal of Engineering Science & Technology Review*, *8*(5), 75–83.
13. Yoruk, R., & Marshall, M. R., (2003). Physicochemical properties and function of plant polyphenol oxidase: A review. *Journal of Food Biochemistry*, *27*(5), 361–422. https://doi.org/10.1111/j.1745-4514.2003.tb00289.x.
14. Zhao, G., Zhang, R., & Zhang, M., (2017). Effects of high hydrostatic pressure processing and subsequent storage on phenolic contents and antioxidant activity in fruit and vegetable products. *International Journal of Food Science & Technology*, *52*(1), 3–12. https://doi.org/10.1111/ijfs.13203.
15. Spence, C., Levitan, C. A., Shankar, M. U., & Zampini, M., (2010). Does food color influence taste and flavor perception in humans? *Chemosensory Perception*, *3*(1), 68–84. https://doi.org/10.1007/s12078-010-9067-z.

16. Bayarri, S., Calvo, C., Costell, E., & Durán, L., (2001). Influence of color on perception of sweetness and fruit flavor of fruit drinks. *Food Science and Technology International*, *7*(5), 399–404. https://doi.org/10.1106/JJWN-FFRQ-JBMC-LQ5R.
17. Bhanduriya, K., Mealy, L., Anand, S., & Metzger, L., (2022). Effect of midday pasteurizer washing on thermoduric organisms and their progression through Cheddar cheese manufacturing and ripening. *Journal of Dairy Science*, *105*(1), 109–122.
18. U.S. Food and Drug Administration. Overview of Food Ingredients, Additives, and Colors. http://www.fda.gov/Food/IngredientsPackagingLabeling/FoodAdditivesIngredients/ucm094211.htm#qa.
19. Hutchings, J. B., (1994). Chemistry of food color. In: Hutchings, J. B., (ed.), *Food Color and Appearance* (pp. 367–469). Springer US. https://doi.org/10.1007/978-1-4615-2123-5_10.
20. Schwartz, S. J., & Lorenzo, T. V., (1990). Chlorophylls in foods. *Critical Reviews in Food Science and Nutrition*, *29*(1), 1–17. https://doi.org/10.1080/10408399009527511.
21. Jadhav, R. V., & Bhujbal, S. S., (2020). *A Review on Natural Food Color*, *2*(2), 12–20.
22. Elbe, J. H. V., Huang, A. S., Attoe, E. L., & Nank, W. K., (2002). *Pigment Composition and Color of Conventional and Veri-Green Canned Beans* (Vol. 34, No. 1, pp. 52–54). ACS Publications. https://doi.org/10.1021/jf00067a014.
23. Andersen, Ø. M., & Francis, G. W., (2004). Techniques of pigment identification. In: *Plant Pigments and Their Manipulation* (Vol. 14, pp. 293–341).
24. Kayesh, E., Shangguan, L., Korir, N. K., Sun, X., Bilkish, N., Zhang, Y., Han, J., et al., (2013). Fruit skin color and the role of anthocyanin. *Acta Physiologiae Plantarum*, *35*(10), 2879–2890. https://doi.org/10.1007/s11738-013-1332-8.
25. Samanta, A., Das, G., & Das, S. K., (2011). Roles of flavonoids in plants. *Carbon*, *100*(6), 12–35.
26. Yoshida, K., Oyama, K., & Kondo, T., (2012). Chemistry of flavonoids in color development. In: *Recent Advances in Polyphenol Research* (pp. 99–129). John Wiley & Sons, Ltd. https://doi.org/10.1002/9781118299753.ch4.
27. Wrolstad, R. E., Durst, R. W., & Lee, J., (2005). Tracking color and pigment changes in anthocyanin products. *Trends in Food Science & Technology*, *16*(9), 423–428. https://doi.org/10.1016/j.tifs.2005.03.019.
28. Lee, H. S., & Coates, G. A., (1999). Thermal pasteurization effects on color of red grapefruit juices. *Journal of Food Science*, *64*(4), 663–666. https://doi.org/10.1111/j.1365-2621.1999.tb15106.x.
29. Alper, N., Bahçeci, K. S., & Acar, J., (2005). Influence of processing and pasteurization on color values and total phenolic compounds of pomegranate juice. *Journal of Food Processing and Preservation*, *29*(5, 6), 357–368. https://doi.org/10.1111/j.1745-4549.2005.00033.x.
30. Pandalaneni, K., Bhanduriya, K., Amamcharla, J. K., Marella, C., & Metzger, L. E., (2019). Influence of milk protein concentrates with modified calcium content on enteral dairy beverage formulations: Storage stability. *Journal of Dairy Science*, *102*(1), 155–163. https://doi.org/10.3168/jds.2018-15239.
31. Crinó, M. A., Heenan, C. N., Nguyen, M. H., & Stathopoulos, C. E., (2013). The stability of natural red/pink food colors in ultrahigh-temperature (UHT) products.

Journal of the Science of Food and Agriculture, *93*(8), 2022–2027. https://doi.org/10.1002/jsfa.6009.

32. Jittanit, W., Wiriyaputtipong, S., Charoenpornworanam, H., & Songsermpong, S., (2011). Effects of varieties, heat pretreatment and UHT conditions on the sugarcane juice quality. *Chiang Mai Journal of Science*, *38*(1), 116–125.
33. Liu, T., Lv, B., Zhao, W., Wang, Y., Piao, C., Dai, W., Hu, Y., et al., (2020). Effects of ultrahigh temperature pasteurization on the liquid components and functional properties of stored liquid whole eggs. *BioMed Research International*, *2020*, e3465465. https://doi.org/10.1155/2020/3465465.
34. Hayakawa, K. I., & Timbers, G. E., (1977). Influence of heat treatment on the quality of vegetables: Changes in visual green color. *Journal of Food Science*, *42*(3), 778–781. https://doi.org/10.1111/j.1365-2621.1977.tb12601.x.
35. Ngo, T., Wrolstad, R. E., & Zhao, Y., (2007). Color quality of Oregon strawberries—Impact of genotype, composition, and processing. *Journal of Food Science*, *72*(1), C025–C032. https://doi.org/10.1111/j.1750-3841.2006.00200.x.
36. Ştefănoiu, G. A., Tănase, E. E., Miteluţ, A. C., & Popa, M. E., (2016). Unconventional treatments of food: Microwave vs. radiofrequency. *Agriculture and Agricultural Science Procedia*, *10*, 503–510. https://doi.org/10.1016/j.aaspro.2016.09.024.
37. Pellegrini, N., Chiavaro, E., Gardana, C., Mazzeo, T., Contino, D., Gallo, M., Riso, P., et al., (2010). Effect of different cooking methods on color, phytochemical concentration, and antioxidant capacity of raw and frozen brassica vegetables. *Journal of Agricultural and Food Chemistry*, *58*(7), 4310–4321. https://doi.org/10.1021/jf904306r.
38. Önal, B., Adiletta, G., Di Matteo, M., Russo, P., Ramos, I. N., & Silva, C. L. M., (2021). Microwave and ultrasound pre-treatments for drying of the "Rocha" pear: Impact on phytochemical parameters, color changes and drying kinetics. *Foods*, *10*(4), 853. https://doi.org/10.3390/foods10040853.
39. Nahimana, H., & Zhang, M., (2011). Shrinkage and color change during microwave vacuum drying of carrot. *Drying Technology*, *29*(7), 836–847. https://doi.org/10.1080/07373937.2011.573753.
40. Ghanem, N., Mihoubi, D., Kechaou, N., & Mihoubi, N. B., (2012). Microwave dehydration of three citrus peel cultivars: Effect on water and oil retention capacities, color, shrinkage, and total phenols content. *Industrial Crops and Products*, *40*, 167–177. https://doi.org/10.1016/j.indcrop.2012.03.009.
41. Yost, M., Abu-Ali, J. M., & Barringer, S. A., (2006). Kinetics of potato color and texture development during baking, frying, and microwaving with the addition of liquid smoke. *Journal of Food Science*, *71*(9), E364–E369. https://doi.org/10.1111/j.1750-3841.2006.00171.x.
42. Pereira, R. N., Rodrigues, R. M., Genisheva, Z., Oliveira, H., De Freitas, V., Teixeira, J. A., & Vicente, A. A., (2016). Effects of ohmic heating on extraction of food-grade phytochemicals from colored potato. *LWT*, *74*, 493–503. https://doi.org/10.1016/j.lwt.2016.07.074.
43. Parmar, P., Singh, A. K., Meena, G. S., Borad, S., & Raju, P. N., (2018). Application of ohmic heating for concentration of milk. *Journal of Food Science and Technology*, *55*(12), 4956–4963. https://doi.org/10.1007/s13197-018-3431-4.

44. Mercali, G. D., Schwartz, S., Marczak, L. D. F., Tessaro, I. C., & Sastry, S., (2014). Ascorbic acid degradation and color changes in acerola pulp during ohmic heating: Effect of electric field frequency. *Journal of Food Engineering*, *123*, 1–7. https://doi.org/10.1016/j.jfoodeng.2013.09.011.
45. Bhargava, N., Mor, R. S., Kumar, K., & Sharanagat, V. S., (2021). Advances in application of ultrasound in food processing: A review. *Ultrasonics Sonochemistry*, *70*, 105293. https://doi.org/10.1016/j.ultsonch.2020.105293.
46. Cao, X., Cai, C., Wang, Y., & Zheng, X., (2019). Effects of ultrasound processing on physicochemical parameters, antioxidants, and color quality of bayberry juice. *Journal of Food Quality*, *2019*, e7917419. https://doi.org/10.1155/2019/7917419.
47. Cortés, C., Esteve, M. J., & Frígola, A., (2008). Effect of refrigerated storage on ascorbic acid content of orange juice treated by pulsed electric fields and thermal pasteurization. *European Food Research and Technology*, *227*(2), 629–635. https://doi.org/10.1007/s00217-007-0766-x.
48. Chiozzi, V., Agriopoulou, S., & Varzakas, T., (2022). Advances, applications, and comparison of thermal (pasteurization, sterilization, and aseptic packaging) against non-thermal (ultrasounds, UV radiation, ozonation, high hydrostatic pressure) technologies in food processing. *Applied Sciences*, *12*(4), 2202. https://doi.org/10.3390/app12042202.
49. Wallner-pendleton, E. A., Sumner, S. S., Froning, G. W., & Stetson, L. E., (1994). The use of ultraviolet radiation to reduce *salmonella* and psychrotrophic bacterial contamination on poultry carcasses. *Poultry Science*, *73*(8), 1327–1333. https://doi.org/10.3382/ps.0731327.
50. Birmpa, A., Sfika, V., & Vantarakis, A., (2013). Ultraviolet light and ultrasound as non-thermal treatments for the inactivation of microorganisms in fresh ready-to-eat foods. *International Journal of Food Microbiology*, *167*(1), 96–102. https://doi.org/10.1016/j.ijfoodmicro.2013.06.005.
51. Krebbers, B., Matser, A. M., Koets, M., & Van, D. B. R. W., (2002). Quality and storage-stability of high-pressure preserved green beans. *Journal of Food Engineering*, *54*(1), 27–33. https://doi.org/10.1016/S0260-8774(01)00182-0.
52. Van, L. A., Ooms, V., Weemaes, C., Van, D. B. I., Ludikhuyze, L., Indrawati, Denys, S., & Hendrickx, M., (1998). Thermal and pressure−temperature degradation of chlorophyll in broccoli (*Brassica oleracea* L. italica) juice: A kinetic study. *Journal of Agricultural and Food Chemistry*, *46*(12), 5289–5294. https://doi.org/10.1021/jf980505x.
53. Matser, A. M., Krebbers, B., Van, D. B. R. W., & Bartels, P. V., (2004). Advantages of high-pressure sterilization on quality of food products. *Trends in Food Science & Technology*, *15*(2), 79–85. https://doi.org/10.1016/j.tifs.2003.08.005.
54. Oey, I., Lille, M., Van, L. A., & Hendrickx, M., (2008). Effect of high-pressure processing on colour, texture and flavor of fruit- and vegetable-based food products: A review. *Trends in Food Science & Technology*, *19*(6), 320–328. https://doi.org/10.1016/j.tifs.2008.04.001.
55. McInerney, J. K., Seccafien, C. A., Stewart, C. M., & Bird, A. R., (2007). Effects of high-pressure processing on antioxidant activity, and total carotenoid content and availability, in vegetables. *Innovative Food Science & Emerging Technologies*, *8*(4), 543–548. https://doi.org/10.1016/j.ifset.2007.04.005.

56. Rodrigo, D., Van, L. A., & Hendrickx, M., (2007). Combined thermal and high pressure color degradation of tomato puree and strawberry juice. *Journal of Food Engineering*, *79*(2), 553–560. https://doi.org/10.1016/j.jfoodeng.2006.02.015.
57. De Ancos, B., Rodrigo, M. J., Sánchez-Moreno, C., Pilar, C. M., & Zacarías, L., (2020). Effect of high-pressure processing applied as pretreatment on carotenoids, flavonoids, and vitamin C in juice of the sweet oranges "Navel" and the red-fleshed "Cara Cara." *Food Research International*, *132*, 109105. https://doi.org/10.1016/j.foodres.2020.109105.
58. Saikaew, K., Lertrat, K., Meenune, M., & Tangwongchai, R., (2018). Effect of high-pressure processing on color, phytochemical contents, and antioxidant activities of purple waxy corn (*Zea mays* L. var. Ceratina) kernels. *Food Chemistry*, *243*, 328–337. https://doi.org/10.1016/j.foodchem.2017.09.136.
59. Kouniaki, S., Kajda, P., & Zabetakis, I., (2004). The effect of high hydrostatic pressure on anthocyanins and ascorbic acid in blackcurrants (*Ribes nigrum*). *Flavor and Fragrance Journal*, *19*(4), 281–286. https://doi.org/10.1002/ffj.1344.
60. Gomes, M. R. A., & Ledward, D. A., (1996). Effect of high-pressure treatment on the activity of some polyphenol oxidases. *Food Chemistry*, *56*(1), 1–5. https://doi.org/10.1016/0308-8146(95)00141-7.
61. Heinz, V., & Buckow, R., (2010). Food preservation by high pressure. *Journal für Verbraucherschutz und Lebensmittelsicherheit*, *5*(1), 73–81. https://doi.org/10.1007/s00003-009-0311-x.
62. Awasti, N., Chaudhary, P., & Anand, S., (2019). Manufacturing low-spore-count skim milk powders by combining optimized raw milk holding conditions and hydrodynamic cavitation. *Dairy Science Publication Database*, *102*(Suppl.1). https://openprairie.sdstate.edu/dairy_pubdb/2097
63. Van, D. B. I., Ludikhuyze, L. R., Van, L. A. M., & Hendrickx, M. E., (2000). Inactivation of orange pectin esterase by combined high-pressure and -temperature treatments: A kinetic study. *Journal of Agricultural and Food Chemistry*, *48*(5), 1960–1970. https://doi.org/10.1021/jf990659s.
64. Fernandez, M. V., Denoya, G. I., Agüero, M. V., Jagus, R. J., & Vaudagna, S. R., (2018). Optimization of high-pressure processing parameters to preserve quality attributes of a mixed fruit and vegetable smoothie. *Innovative Food Science & Emerging Technologies*, *47*, 170–179. https://doi.org/10.1016/j.ifset.2018.02.011.
65. Kaushik, N., Rao, P. S., & Mishra, H. N., (2016). Process optimization for thermal-assisted high-pressure processing of mango (*Mangifera indica* L.) pulp using response surface methodology. *LWT – Food Science and Technology*, *69*, 372–381. https://doi.org/10.1016/j.lwt.2016.01.062.
66. Goodner, J. K., Braddock, R. J., & Parish, M. E., (1998). Inactivation of *Pectin esterase* in orange and grapefruit juices by high pressure. *Journal of Agricultural and Food Chemistry*, *46*(5), 1997–2000. https://doi.org/10.1021/jf9709111.
67. Nienaber, U., & Shellhammer, T. H., (2001). High-pressure processing of orange juice: Combination treatments and a shelf-life study. *Journal of Food Science*, *66*(2), 332–336. https://doi.org/10.1111/j.1365-2621.2001.tb11342.x.
68. Vegara, S., Martí, N., Mena, P., Saura, D., & Valero, M., (2013). Effect of pasteurization process and storage on color and shelf-life of pomegranate juices.

LWT – Food Science and Technology, *54*(2), 592–596. https://doi.org/10.1016/j.lwt.2013.06.022.
69. You, Y., Li, N., Han, X., Guo, J., Zhao, Y., Liu, G., Huang, W., & Zhan, J., (2018). Influence of different sterilization treatments on the color and anthocyanin contents of mulberry juice during refrigerated storage. *Innovative Food Science & Emerging Technologies*, *48*, 1–10. https://doi.org/10.1016/j.ifset.2018.05.007.
70. Terefe, N. S., Buckow, R., & Versteeg, C., (2014). Quality-related enzymes in fruit and vegetable products: Effects of novel food processing technologies, part 1: High-pressure processing. *Critical Reviews in Food Science and Nutrition*, *54*(1), 24–63. https://doi.org/10.1080/10408398.2011.566946.
71. Pagán, R., Jordan, S., Benito, A., & Mackey, B., (2001). Enhanced acid sensitivity of pressure-damaged *Escherichia coli* O157 cells. *Applied and Environmental Microbiology*, *67*(4), 1983–1985. https://doi.org/10.1128/AEM.67.4.1983-1985.2001.
72. Yagiz, Y., Kristinsson, H. G., Balaban, M. O., Welt, B. A., Ralat, M., & Marshall, M. R., (2009). Effect of high-pressure processing and cooking treatment on the quality of Atlantic salmon. *Food Chemistry*, *116*(4), 828–835. https://doi.org/10.1016/j.foodchem.2009.03.029.
73. Kaur, B. P., & Srinivasa, R. P., (2018). Effect of storage temperature and packaging on quality and shelf-life of high pressure processed black tiger shrimp (*Penaeus monodon*). *Journal of Food Processing and Preservation*, *42*(1), e13366. https://doi.org/10.1111/jfpp.13366.
74. Campus, M., (2010). High pressure processing of meat, meat products and seafood. *Food Engineering Reviews*, *2*(4), 256–273. https://doi.org/10.1007/s12393–010–9028-y.
75. Ferrini, G., Comaposada, J., Arnau, J., & Gou, P., (2012). Color modification in a cured meat model dried by quick-dry-slice process® and high pressure processed as a function of NaCl, KCl, K-lactate and water contents. *Innovative Food Science & Emerging Technologies*, *13*, 69–74. https://doi.org/10.1016/j.ifset.2011.09.005.
76. Carlez, A., Veciana-Nogues, T., & Cheftel, J. C., (1995). Changes in color and myoglobin of minced beef meat due to high pressure processing. *LWT – Food Science and Technology*, *28*(5), 528–538. https://doi.org/10.1006/fstl.1995.0088.
77. Bak, K. H., Lindahl, G., Karlsson, A. H., Lloret, E., Ferrini, G., Arnau, J., & Orlien, V., (2012). High pressure effect on the color of minced cured restructured ham at different levels of drying, pH, and NaCl. *Meat Science*, *90*(3), 690–696. https://doi.org/10.1016/j.meatsci.2011.10.015.
78. Marcos, B., Aymerich, T., Dolors, G. M., & Garriga, M., (2007). Assessment of high hydrostatic pressure and starter culture on the quality properties of low-acid fermented sausages. *Meat Science*, *76*(1), 46–53. https://doi.org/10.1016/j.meatsci.2006.09.020.
79. Chevalier, D., Le Bail, A., & Ghoul, M., (2001). Effects of high-pressure treatment (100–200 MPa) at low temperature on turbot (*Scophthalmus maximus*) muscle. *Food Research International*, *34*(5), 425–429. https://doi.org/10.1016/S0963-9969(00)00187-3.
80. Hurtado, J. L., Montero, P., & Borderias, A. J., (2000). Extension of shelf life of chilled hake (*Merluccius capensis*) by high pressure/prolongación de la vidaútil de merluza (*Merluccius capensis*) sometidaaaltaspresionesconservadaenrefrigeración.

Food Science and Technology International, *6*(3), 243–249. https://doi.org/10.1177/108201320000600307.
81. Gómez-Estaca, J., Gómez-Guillén, M. C., & Montero, P., (2007). High pressure effects on the quality and preservation of cold-smoked dolphin fish (*Coryphaenahip purus*) fillets. *Food Chemistry*, *102*(4), 1250–1259. https://doi.org/10.1016/j.foodchem.2006.07.014.
82. Gomes, M. R. A., Clark, R., & Ledward, D. A., (1998). Effects of high pressure on amylases and starch in wheat and barley flours. *Food Chemistry*, *63*(3), 363–372. https://doi.org/10.1016/S0308-8146(98)00014-4.
83. Wang, B., Li, D., Wang, L., Chiu, Y. L., Chen, X. D., & Mao, Z., (2008). Effect of high-pressure homogenization on the structure and thermal properties of maize starch. *Journal of Food Engineering*, *87*(3), 436–444. https://doi.org/10.1016/j.jfoodeng.2007.12.027.
84. Kieffer, R., Schurer, F., Köhler, P., & Wieser, H., (2007). Effect of hydrostatic pressure and temperature on the chemical and functional properties of wheat gluten: Studies on gluten, gliadin and glutenin. *Journal of Cereal Science*, *45*(3), 285–292. https://doi.org/10.1016/j.jcs.2006.09.008.
85. Bárcenas, M. E., Altamirano-Fortoul, R., & Rosell, C. M., (2010). Effect of high-pressure processing on wheat dough and bread characteristics. *LWT – Food Science and Technology*, *43*(1), 12–19. https://doi.org/10.1016/j.lwt.2009.06.019.
86. Barcenilla, B., Román, L., Martínez, C., Martínez, M. M., & Gómez, M., (2016). Effect of high-pressure processing on batters and cakes properties. *Innovative Food Science & Emerging Technologies*, *33*, 94–99. https://doi.org/10.1016/j.ifset.2015.11.011.
87. Styles, M. F., Hoover, D. G., & Farkas, D. F., (1991). Response of *Listeria monocytogenes* and vibrio parahaemolyticus to high hydrostatic pressure. *Journal of Food Science*, *56*(5), 1404–1407. https://doi.org/10.1111/j.1365-2621.1991.tb04784.x.
88. Erkmen, O., & Karataş, Ş., (1997). Effect of high hydrostatic pressure on *Staphylococcus aureus* in milk. *Journal of Food Engineering*, *33*(3), 257–262. https://doi.org/10.1016/S0260-8774(97)00021-6.
89. Gervilla, R., Capellas, M., Ferragut, V., & Guamis, B., (1997). Effect of high hydrostatic pressure on *Listeria innocua* 910 CECT inoculated into ewe's milk. *Journal of Food Protection*, *60*(1), 33–37. https://doi.org/10.4315/0362-028X-60.1.33.
90. Huppertz, T., Kelly, A. L., & Fox, P. F., (2002). Effects of high pressure on constituents and properties of milk. *International Dairy Journal*, *12*(7), 561–572. https://doi.org/10.1016/S0958-6946(02)00045-6.
91. Devi, A. F., Buckow, R., Singh, T., Hemar, Y., & Kasapis, S., (2015). Color change and proteolysis of skim milk during high pressure thermal–processing. *Journal of Food Engineering*, *147*, 102–110. https://doi.org/10.1016/j.jfoodeng.2014.09.017.
92. Kim, H. Y., Kim, S. H., Choi, M. J., Min, S. G., & Kwak, H. S., (2008). The effect of high pressure–low temperature treatment on physicochemical properties in milk. *Journal of Dairy Science*, *91*(11), 4176–4182. https://doi.org/10.3168/jds.2007-0883.
93. Tran, M., Roberts, R., Felix, T. L., & Harte, F. M., (2018). Effect of high-pressure-jet processing on the viscosity and foaming properties of pasteurized whole milk. *Journal of Dairy Science*, *101*(5), 3887–3899. https://doi.org/10.3168/jds.2017-14103.

94. Nassar, K. S., Zhang, S., Lu, J., Pang, X., Ragab, E. S., Yue, Y., & Lv, J., (2019). Combined effects of high-pressure treatment and storage temperature on the physicochemical properties of caprine milk. *International Dairy Journal*, *96*, 66–72. https://doi.org/10.1016/j.idairyj.2019.03.003.
95. Iturmendi, N., García, A., Galarza, U., Barba, C., Fernández, T., & Maté, J. I., (2020). Influence of high hydrostatic pressure treatments on the physicochemical, microbiological, and rheological properties of reconstituted micellar casein concentrates. *Food Hydrocolloids*, *106*, 105880. https://doi.org/10.1016/j.foodhyd.2020.105880.
96. Capellas, M., Mor-Mur, M., Sendra, E., & Guamis, B., (2001). Effect of high-pressure processing on physico-chemical characteristics of fresh goats' milk cheese (Mató). *International Dairy Journal*, *11*(3), 165–173. https://doi.org/10.1016/S0958-6946(01)00088-7.
97. Daryaei, H., Coventry, M. J., Versteeg, C., & Sherkat, F., (2008). Effect of high-pressure treatment on starter bacteria and spoilage yeasts in fresh lactic curd cheese of bovine milk. *Innovative Food Science & Emerging Technologies*, *9*(2), 201–205. https://doi.org/10.1016/j.ifset.2007.06.011.
98. Sandra, S., Stanford, M. A., & Goddik, L. M., (2004). The use of high-pressure processing in the production of queso fresco cheese. *Journal of Food Science*, *69*(4), FEP153–FEP158. https://doi.org/10.1111/j.1365-2621.2004.tb06340.x.
99. Okpala, C. O. R., Piggott, J. R., & Schaschke, C. J., (2010). Influence of high-pressure processing (HPP) on physico-chemical properties of fresh cheese. *Innovative Food Science & Emerging Technologies*, *11*(1), 61–67. https://doi.org/10.1016/j.ifset.2009.10.003.
100. Evert-Arriagada, K., Hernández-Herrero, M. M., Juan, B., Guamis, B., & Trujillo, A. J., (2012). Effect of high pressure on fresh cheese shelf-life. *Journal of Food Engineering*, *110*(2), 248–253. https://doi.org/10.1016/j.jfoodeng.2011.05.011.
101. Evert-Arriagada, K., Hernández-Herrero, M. M., Guamis, B., & Trujillo, A. J., (2014). Commercial application of high-pressure processing for increasing starter-free fresh cheese shelf-life. *LWT – Food Science and Technology*, *55*(2), 498–505. https://doi.org/10.1016/j.lwt.2013.10.030.
102. Ribeiro, L. R., Leite, Jr. B. R. De. C., & Cristianini, M., (2018). Effect of high-pressure processing on the characteristics of cheese made from ultrafiltered milk: Influence of the kind of rennet. *Innovative Food Science & Emerging Technologies*, *50*, 57–65. https://doi.org/10.1016/j.ifset.2018.10.012.
103. Farr, D., (1990). High pressure technology in the food industry. *Trends in Food Science & Technology*, *1*, 14–16. https://doi.org/10.1016/0924-2244(90)90004-I.
104. Knorr, D., (1995). Hydrostatic pressure treatment of food: Microbiology. In: Gould, G. W., (ed.), *New Methods of Food Preservation* (pp. 159–175). Springer US. https://doi.org/10.1007/978-1-4615–2105-1_8.
105. Alsaleem, K. A., Hammam, A. R. A., & Awasti, N., (2021). Safety, regulatory aspects, and environmental impacts of using non-thermal processing techniques for dairy industries. In: *Non-Thermal Processing Technologies for the Dairy Industry*. CRC Press.

CHAPTER 3

APPLICATION OF ULTRASOUND AND ITS IMPACT ON COLOR

DAKSHAYANI RAJENDRAN[1] and JAGANMOHAN RANGARAJAN[2]

[1]*Food Science and Technology, Department of Food Product Development, National Institute of Food Technology, Entrepreneurship, and Management, Thanjavur (NIFTEM-T), Tamil Nadu, India*

[2]*School of Sensory Science, Department of Food Product Development, National Institute of Food Technology, Entrepreneurship, and Management, Thanjavur (NIFTEM-T), Tamil Nadu, India*

ABSTRACT

Thermal processing of food undergoes degradation of the food product quality due to heat, and its thermal effects may be desirable or undesirable. Thus, nonthermal methods come into existence, which overcome the demerits of thermal methods. Ultrasound, being a nonthermal technique, has a wide scope of application in the food, pharmaceutical, biotechnology, medical, and clinical fields. On the other hand, color, being a major attribute for the preservation of food and a consumer trait for choosing food, has to be preserved during its processing. Therefore, this chapter will discuss ultrasound and its effect on food color during processing. Numerous studies have noted that ultrasound preserves the color of the product when treatment conditions are kept minimal and based on the sensitivity of the food product being treated. This is mainly due to the

Nonthermal Food Processing Technologies: Impact on Color Profile.
Kaavya Rathnakumar and R. Pandiselvam (Eds.)

cavitation effect created during ultrasound processing of food. Ultrasound produces a synergistic effect on food and preserves the product quality, ensuring that the food meets consumer acceptance levels.

3.1 INTRODUCTION

The traditional methods of processing foods, such as drying, frying, salting, smoking, soaking, pickling, etc., are still widely used in the food industry on a large scale. However, their application has become limited in recent times due to the advancements in modern food processing techniques. The primary objective of traditional methods was to extend the shelf life of products by reducing microbial growth through thermal treatments, specifically by applying heat. In recent years, traditional methods have been gradually replaced or reduced due to their drawbacks, including thermal degradation of the end product, high energy consumption, low production efficiency, and longer processing time [4].

To overcome these disadvantages, alternative approaches have been developed, such as high pressure processing (HPP), supercritical fluid extraction (SFE), cold plasma (CP), pulsed electric field (PEF), ultraviolet (UV), and ultrasound. These methods will be discussed in detail in other chapters. Ultrasound (US) refers to waves generated beyond the threshold limit of human hearing (>16 kHz) and can be categorized into high-energy and low-energy waves. The range of low-energy ultrasound falls between 20 kHz and 500 MHz. The application of ultrasound can be done in three ways: coupling with a device, direct application to the product, or submerging in a sonic water bath [22]. Low-energy ultrasound helps in understanding the interaction between matter, while high-energy ultrasound enables a deeper examination of the structure, composition, and dimensions of the product [4].

This chapter will cover the basics of ultrasound systems, the generation mechanism of ultrasound, and its application in various products. It will also discuss the color measurement system and the effect of ultrasound on food products, including its impact on color.

3.2 GENERATION OF ULTRASOUND

The device capable of generating and/or detecting ultrasound waves is known as an ultrasound transducer. The transducer is a key component of ultrasound technology that converts electrical pulses into acoustic energy of the required intensity. The major types of electrical transducers are piezoelectric, electrostatic, magnetostrictive, and electromagnetic (EM). Piezoelectric transducers are based on the piezoelectric effect, magnetostrictive transducers utilize the magnetostriction effect in ferromagnetic materials, electrostatic or conductive transducers are plane condensers with two electrodes placed close to each other, one being a very thin membrane and the other rigid. EM transducers make use of the interaction between the magnetic field of a permanent magnet and the alternating electric current in a moving coil to convert electric oscillations into sonic vibrations. Mechanical transducers, used as transmitters, are known for their ability to produce high amplitude vibrations [23].

3.3 TYPES OF SONICATION

The US can be divided into two types: (i) based on its operation as batch and continuous types; and (ii) based on the mode of operation into probe and bath types. Sonotrode, which is a probe type, is used where the acoustic energy produced from the transducer is directly applied to the product undergoing treatment or analyzed for its morphology or composition. The probe type is mainly used for cutting and slicing actions, and it contains a generator to produce alternating current and the required frequency for US. The transducer is used to convert electrical oscillation to mechanical displacement, and an amplifier transmits the mechanical vibration to the sonotrode as sound waves. US water bath is a bath type system mainly used for sanitation and cleaning in food. US cleaners use piezoelectric transducers, which are either positioned at the bottom of processing vessels or immersed in the cleaning solution, to convert low-frequency AC to high-frequency sound waves. In order to generate cavitation, the transducer sends out high-intensity waves to the solvent. These waves compress the liquid and form millions of microscopic cavities. When these bubbles suddenly collapse in the cleaning solvent, they generate a tremendous amount of energy at high pressure and temperature, which quickly removes the dirt from goods submerged in the cleaning solution [4]. Apart from this, the US can be combined with other preservation techniques as follows: Mano sonication (MS), which is the combination of US and pressure (MS); Mano-thermo-sonication, which combines ultrasound, pressure, and heat (MTS); and thermos-sonication, which combines US and heat (TS) [17].

3.4 INACTIVATION OF MICROORGANISMS

Microbes and enzymes are two primary factors contributing to food spoilage, quality degradation, and safety concerns. Microbes rely on food as a source of nutrients for their growth, while enzymes are naturally present in food to break down these nutrients. To suppress or eliminate them, high-power ultrasound is used, which disrupts the biological cells. When combined with heat treatment (HT), ultrasound can potentially accelerate the sterilization process, reducing the time and intensity of heat required to damage the cells (as illustrated in Figure 3.1). This is achieved by thinning the cell membranes, localizing the heat effect, and generating free radicals. Ultrasound cavitation occurs in two states: transient and stable. Transient cavitation occurs when gas or vapor-filled bubbles undergo irregular oscillations and implode. This results in increased local temperature and pressure, which denatures enzymes and disintegrates biological cells. The implosion of bubbles also produces liquid jets and high shear forces in the surrounding solvent, which can damage the cell wall or cell membrane. Stable cavitation refers to the oscillation of bubbles in a regular manner, for multiple cycles. These bubbles create microstreaming in the liquid and its surroundings, resulting in stress on microbial species. Enzymatic inactivation is achieved through depolymerization, where the use of ultrasound generates intracellular cavitation. These mechanical shocks have the potential to disrupt cell structure and functional properties at the site of cell breakdown [6].

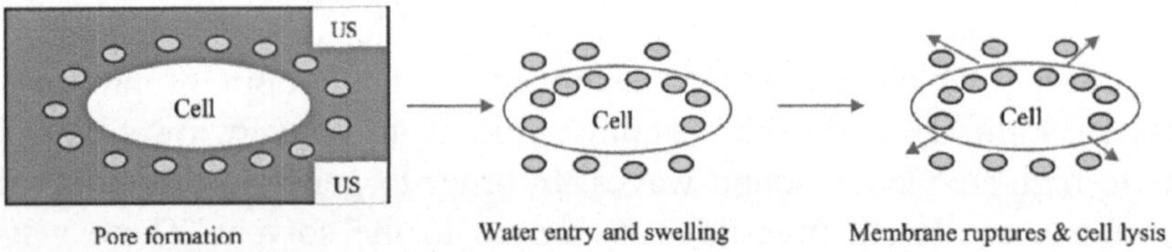

FIGURE 3.1 US impact on cell damage.
Source: Adapted from Ref. [34]

3.5 MECHANISM OF ACTION

Ultrasound in a liquid medium produces acoustic cavitation, which involves the generation, growth, and collapse of bubbles (Figure 3.2). As the ultrasound waves propagate, the bubbles begin to oscillate and eventually collapse, resulting in thermal effects (pressure of 1,000 atm, temperature of 5,000 K), chemical effects (generation of free radicals), and mechanical effects (turbulence, collapse pressure, and shear stresses). The variation in ultrasound frequency leads to the development of alternating negative and positive pressure, ultimately causing expansion or compression of materials and cell disruption. Ultrasound causes water to hydrolyze, leading to the creation of free radicals, namely OH^- and H^+ ions. During sonication, the generated bubbles can be classified into two categories based on their morphology: stable cavitation bubbles, which are non-linear and form large bubbles that reach an equilibrium size during pressure cycles, and internal (transient) cavitation bubbles, which rapidly collapse and disintegrate into smaller bubbles. The small bubbles formed disintegrate quickly [22].

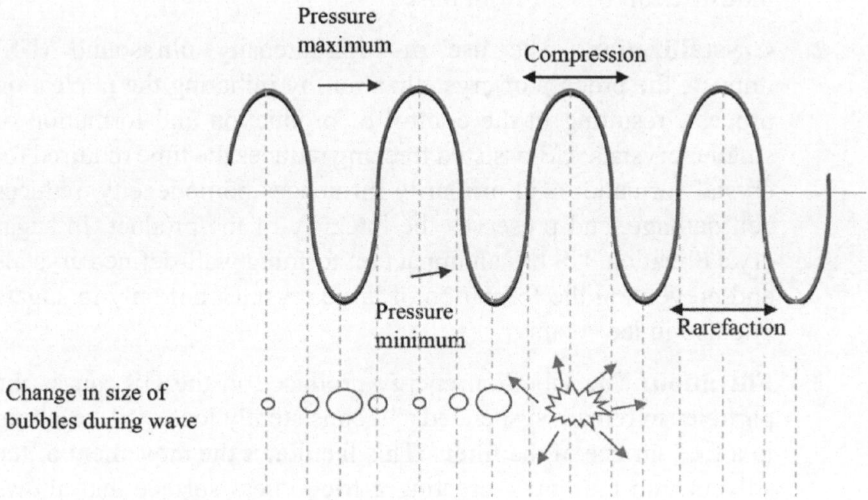

FIGURE 3.2 Cavitation caused by ultrasound.

3.6 ULTRASOUND IN FOOD APPLICATIONS

Ultrasound has a huge impact on food applications such as mixing, foaming, filtration, extraction, dehydration, homogenization, structural and functional modification. Degassing produces antifoaming action, particle size reduction, temporary or permanent shifting of viscosity, cell destruction, aggregate dispersion, microorganism inactivation, equipment, and glassware sterilization, and fermentation. In food applications, ultrasound helps maintain the quality of food products and ensures safe consumption by reducing the negative impact on sensory characteristics of food [12].

The effect and principle of ultrasound in various unit operation of food technology are as follows:

1. **Emulsion:** Surfactants increase stability in emulsions and dispersions by preventing the entrapment of dispersed material in the mobile phase. In the US, the liquid is degassed, reducing the complexity of stabilized air pockets, decreasing the amount of air bubbles, and aiding in the emulsification process. A highly stable emulsion is formed at low energy due to the bursting of cavitation bubbles at the interface between two immiscible liquids. The petrochemical, chemical, pharmaceutical, cosmetic, and food industries all benefit from this.
2. **Crystallization:** The use of high-intensity ultrasound (US) impacts the process of crystallization by initiating the nucleation process, resulting in the control of production and formation of smaller crystals. US-assisted freezing reduces the time required for crystal formation and primarily influences homogeneity, reduces cell damage, and preserves the integrity of the product. In sugar crystallization, US has an impact on forming well-defined crystals and preventing the formation of large crystals, not only in sugars but also in ice cream.
3. **Filtration:** The vibration energy produced in the US causes the particles to remain suspended and consistently localized, resulting in a free surface of the filters. This facilitates the movement of the solvent into the pores, creating a frictionless surface that allows smaller particles or liquid to pass through easily. Ultimately, this enhances the flow and reduces the processing period. It also helps prevent the deposition of filter cake.
4. **Separation:** The application of high-energy US waves to a low-frequency (30 kHz) emulsion aids in the separation of the emulsion into water and oil constituents. This technique is mainly used in the dairy industry for cleaning the membrane used for protein concentration and separation. Therefore, US waves have proven to be effective in cleaning the membrane, especially when applied at low frequencies.

5. **Viscosity:** The US influences viscosity in both increasing and decreasing trends, based on its intensity. Its application with respect to viscosity is quite notable in the field of dairy products such as milk, yogurt, curd, casein, and calcium concentration.
6. **Antifoaming Action:** The use of degassing and antifoaming agents in the US promotes the release of trapped gas from liquid samples, reducing the dispersion of gas below the liquid level. This process also helps prevent foam formation in carbonated drinks, ultimately enhancing product quality.
7. **Fermentation:** The US monitors both high and low frequency to observe the progress of fermentation and influence its progression. It also eliminates microorganisms that hinder the process.

It also helps in meat tenderization and checking the quality of perishable products, namely fruits, vegetables, poultry, dairy, and fish products, through a non-destructive mechanism.

It also has other applications, and during these various applications, the treated samples undergo numerous physical, microbial, and functional variations in the product. The major physical variation is color, which will be further discussed in this chapter.

3.7 COLOR MEASUREMENT

The color of food is the primary visual perception for consumers, influencing their preferences and satisfaction. Color is a fundamental characteristic of food that indicates ripeness or rawness of fruits, helps determine the correct time for harvesting, and indicates the safety of consuming roasted meat. It also motivates us to eat. In the early 17th century, numerous scientists studied the mechanization of vision and color perception in humans, leading to the development of equipment for measuring color. According to Maxwell, Young, and Helmholtz in the 19th century, individuals with average color perception possess three ocular pigments in their eyes that can sense the short, mid, and long wave areas of the visible light spectrum. These advancements in the early 1990s paved the way for the Commission Internationale de l'Eclairage (CIE) color grading system. The CIE method of color measurement utilizes the spectral energy distribution of the light source and the color matching capabilities of standard observers to represent or reproduce the object's spectrum in a three-dimensional color space.

The system is based on the trichromatic concept but uses fictitious positive primaries X, Y, and Z instead of real red, green, and blue primaries. The primary Y, also known as luminous reflectance or transmittance, contains the full brightness stimulus. However, it was determined that the CIE system of measuring color was not visually instructive. The next color space system introduced was L, a, and b, where a and b represent the red/green and yellow/blue opponent coordinates, and L is the square root of the tristimulus value Y [21]. These color standards are still used today to evaluate the color value of food.

3.8 ULTRASOUND IMPACT ON COLOR MEASUREMENT

3.8.1 BLANCHING

Several applications of ultrasound (US) have been discussed, although its impact on one of the main attributes, namely color, will be further examined. A study was conducted to investigate the influence of US, blanching, and the combined effect of both on vacuum-packed potato strips. It was found that during storage, the lightness (L^*) of the blanched samples decreased by 12% ($p < 0.05$), resulting in potato strips with a higher hue value and less yellow color. On the other hand, US had a minimal impact on the color of the potato strips, as the hue content remained unaffected. This can be attributed to the fact that US has a lesser enzymatic effect on the samples. The blanched potato strips appeared darker than the ones treated with US during the 10th day of storage at $p < 0.05$ [9]. Similarly, Lagnika et al. [20] investigated the impact of US and blanching on the sensory assessment of dried cashew apple. They found that US treatment influenced the preservation of whiteness in the dried cashew apple. The color difference between the samples treated with US + blanching and the untreated samples was significantly lower. This can be attributed to the limited access of air to the plant tissues when they are immersed in a liquid medium.

3.8.2 DRYING

Chen et al. [7] investigated the novel technique of drying, along with ultrasound (US) and vacuum dehydration, in order to reduce the drying time and improve the quality of carrot slices (CS). The study concluded that the CS remained darker, with less red and yellow hues, compared to the ultrasound vacuum sample. This may be attributed to the longer exposure time to heat during drying. The treatments were greatly influenced and represented by the ΔE value, where the vacuum dried carrot slices had a higher ΔE compared to the US vacuum slices, possibly due to the loss of beta-carotene and prolonged drying time. Another study on seaweed drying demonstrated that ultrasound-treated seaweed for drying resulted in improved color quality and reduced time and power costs [19]. Tao et al. [29] conducted a study on white cabbage, using US-assisted air drying with water blanching, and noted its impact on color. Blanching, drying, and their combination significantly reduced the L* values. The blanched and dried samples had a darker surface compared to the directly dried sample, which may be attributed to factors such as enzymatic reactions, oxidation of vitamin C, caramelization, and the Maillard reaction. Similarly, blanching as well as drying had various effects on the a* and b* values of cabbage. However, the treated sample with ultrasound (US) showed similar properties in terms of L*, a*, and b* values compared to the untreated sample. Ultrasound is recommended not only for preserving the color of the product but also for reducing processing time. In a study by Wiktor et al. [31], the influence of microwave (MW) assisted drying along with pulsed electric field (PEF) and ultrasound on carrots was examined. It was noted that the total color changes ranged from 2.8 to 5.8. The L* value indicated a lighter color in both treated and untreated dried carrots, with ultrasound showing a significant difference regardless of the sequence. Carotenoid retention was higher in the treated samples, while a* and b* values decreased due to the effect of PEF and ultrasound, which partially inactivated the enzymes responsible for browning of the dried carrot tissue. Sunil et al. [28] investigated the impact of ultrasound pre-treatment on microwave drying of okra and found that ultrasound was effective in maintaining the color of the okra, while microwave drying resulted in a darker color due to the degradation of green chlorophyll to olive green or brown-green pheophytin and pyro pheophytin. Sledz et al. [27] studied the impact of ultrasound on the color, carotenoid content, and structural properties of fresh and dried carrot tissue. They found that a 30-minute exposure to ultrasound influenced the L*, a*, and b* values, with longer exposure times leading to a darkening of the carrot color. Osae et al. [25] studied the effect of US, osmotic dehydration (OD), and osmosonication (OS) pre-treatments on dried ginger slices and found that US pre-treatment retained better color quality in the dried sample compared to OS and OD treatment. Therefore, US pre-treatment was recommended for the study. There was no significant variance ($p < 0.05$) in the a* and b* values between the control and treated samples. However, the L* value showed negligible change when treated with US and OD, while OS had a significant effect on the L* value, resulting in a very dark color compared to the fresh ginger sample.

Jin, Zhang, & Shi [18] studied the effect of US and drying methods (hot air drying (HAD) and radio frequency assisted hot air drying (RFHAD)) on bitter melon (*Momordica charantia* L.). The ΔE value of bitter melon with and without US pre-treatment showed insignificant difference ($p < 0.05$) within the same drying method. However, with respect to the drying method, the L* value of all the dried samples was higher than the fresh one, indicating that the dried samples became brighter compared to the fresh sample. The a* value decreased for RFHAD and USRFHAD compared to HAD, USHAD, which can be attributed to high chlorophyll decomposition during prolonged drying duration of HAD. The b* value of RFHAD and USRFHD was lesser than HAD and UHAD, indicating that the hot air-dried samples exhibited a more yellow color compared to the RF samples dried.

3.8.3 EXTRACTION

Albuquerque et al. [2] optimized the extraction of natural colorant from jaboticaba epicarp using high temperature (HT) and ultrasound (US) to extract the colorant (anthocyanin (ACN)) into macrons. The color value was observed for a 6-day shelf life and it was found that the ACN treated with US had a distinct color compared to the control ($p<0.05$). These variations were attributed to the different ACN content. Wizi et al. [32] studied the influence of ultrasound microwave-assisted extraction (UMAE) for extracting natural pigments from the outer layer of sorghum husk using different solvents. It was found that the colorant extracted by UMAE had a high content of apigeninidin and luteolin din, as well as higher thermal stabilities compared to sorghum husk extract with water.

3.8.4 MICROBES

The effects of ultrasound (US) treatment on the physical and chemical properties, microbial load, and microstructures of *bovine longissimus dorsi* (BLD) were examined. The results showed that US did not have a significant influence on radiance (P = 0.42), redness (a*, P = 0.45), or yellowness (b*, P = 0.94). This may be attributed to the short exposure time of US, which was insufficient to initiate denaturation and oxidize the color pigment, myoglobin. However, the hue angle of the meat color showed a significant change with US treatment during storage. Specifically, the meat treated with 16 Wcm^{-2} exhibited a major deprivation in color. On the other hand, the intensity of 28 Wcm^{-2} and 90 Wcm^{-2} suggested a similar color to the non-sonicated sample during the 7-day storage at 4°C.

In a separate study, Ferrario & Guerrero [11] investigated the impact of a constant flow over a pulsed light system (PLS) combined with US on *Escherichia coli, Salmonella enteritidis,* and *Saccharomyces cerevisiae* in commercial and freshly made apple juice. The combined treatment effectively prevented enzymatic browning during storage, possibly due to the inactivation of enzymes responsible for browning by US and PLS. Similarly, Hashemi [16] examined the impact of continuous and pulsed US treatment on microbial and other quality parameters of Mirabelle plum fruit. The study found that US treatment decreased ΔE content during storage, which may be attributed to the mode of sonication and treatment time (TT).

3.8.5 FRYING

Devi et al. [10] investigated the impact of US and MW-assisted vacuum frying on the quality parameters of mushroom chips. The color, being an important attribute of the chips, was found to be highest in US-assisted vacuum frying compared to untreated samples. This can be attributed to the effect of US, which reduces non-enzymatic browning due to a shorter heating period, resulting in a slower start of the non-enzymatic reaction in the samples. However, the L* value decreased in all subsequent frying processes. The a* value increased in all frying temperatures and showed significant differences for all treatments ($p \leq 0.05$). In comparison, there was a reduced redness of the samples in the US treatment due to the non-enzymatic reaction. The b* value increased in all three processes, with the US-treated samples showing significantly higher values ($p \leq 0.05$).

3.8.6 MEAT – MARINATION, FREEZING, AND THAWING

Gomez-Salazar et al. [15] conducted a study on the acid marination of meat combined with US treatment to assess the quality parameters of the meat. The results showed that the L* and a* values decreased when NaCl was used as the marination solution along with US treatment, possibly due to the effects of cavitation, which generate free radicals and result in lipid or protein oxidation. The change in the redness index (a*) was attributed to the instability of heme pigment and the secondary activity of lipid oxidation, which tends to decrease the stability of oxymyoglobin. De Lima Alves et al. [8] conducted research on Italian Salami and its effect on physicochemical and microbial load when treated with US. The findings suggest that US treatment can contribute to the red color of meat, which is important for consumer acceptance. Aksoy et al. [1] investigated the effect of ultrasound-assisted vacuum drying (USAVD), vacuum drying (VD), and freeze drying (FD) on the drying of minced meat. They found that the drying techniques influenced the total color difference (ΔE), with higher drying temperatures resulting in higher ΔE values. USAVD showed minimal color change compared to VD due to the shorter drying time at all temperatures, but overall, FD was suggested as the preferred method for drying minced meat.

3.8.7 BEVERAGE

Tomadoni et al. [30] examined the changes in non-clarified strawberry juice (NCSJ) during storage at 5°C after undergoing US treatment and thermal treatment. It was found that US treatment maintained the color and showed no significant difference compared to the untreated sample. On the other hand, thermal treatment resulted in a lower L* value and a higher hue angle than the control, indicating a lower quality sensory product compared to ultrasound treatment. This effect of ultrasound may be attributed to cavitation, which can accelerate diffusion rate, chemical reactions, dispersion, formation of aggregates, and particle breakdown. Gomes [14] investigated the impact of High-pressure processing (HPP) and US on prebiotic cranberry juice quality. The study found that both treated and untreated juice showed negligible color change. The L value exhibited minimal variation with respect to treatment, indicating good preservation of luminosity. Similarly, hue and chroma variations were also negligible. HPP + US was recommended for preserving the quality of prebiotic cranberry juice. Zhang et al. [33] studied the effect of US irradiation on the color of wine during storage. The study observed that US-treated and untreated wines had very similar color, suggesting that this technology could be adopted to increase the production of red wines with the same quality as traditionally aged wine in a shorter span of time. Thus, US treatment promotes the color of red wine by generating high pressure and temperature during cavitation, which mediates free radicals and chain reactions, resulting in more colored pigments. Campoli et al. [5] investigated the effect of US processing on guava juice, focusing on its structural, physical, and lycopene content. The study found that the physical stability of the juice increased after US treatment without changing its color during storage.

3.8.8 FRESH PRODUCE – STORAGE

Gani et al. [13] investigated the impact of ultrasound (US) treatment on the chemical, physical, nutraceutical, and microbial quality of fresh strawberries. The results suggested that a 30-minute US treatment retained better color of the strawberries during storage. However, both the L* and a* values of the treated and untreated samples were significantly affected during the 15-day storage period, possibly due to the increase in TT color degradation caused by the cavitation mechanism of US. In a study by Meurer et al. [24], the technological properties of chickpea cooking water were examined with and without US treatment. It was found that US improved the color of the product. The foam of French meringue prepared using US-treated chickpea cooking water showed no significant difference ($p > 0.05$) compared to the untreated sample. Quintero-Lira et al. [26] studied the impact of US on liquefying crystallized honey and analyzed its quality characteristics. The results showed a significant change ($p < 0.05$) in honey after US treatment. The L* value of the US-treated honey showed a difference in behavior and increased due to the reduction in crystal size, while the a* and b* values tended towards red and yellow color. The total color change (ΔE) in all honey samples was attributed to the cavitation effect of US, which resulted in physicochemical reactions.

The sonication conditions and treatment effects of the above-mentioned research are summarized in Table 3.1.

TABLE 3.1 Various Food Product and its Color Measurement System with Sonication Treatments

Sl. No.	Product	Equipment Used	Ultrasound Parameter	References
1.	Cashew apple	Chroma-meter (CR-410, Konica Minolta Sensing, Tokyo, Japan).	Probe diameter – 10 mm, power – 500 W, Frequency – 20 k Hz, 10 minutes.	[20]
2.	Carrot slices	Spectrophotometer (PC-1. Shanghai Jia Bao Testing instrument Co. Ltd.), Hunter Lab Scale.	Bath – ±1°C, 10 sec on 10 sec off. Vacuum drying 65 to 75°C.	[7]
3.	*Ascophyllum nodosum*	Hunter-lab DP-9000 colorimeter (Hunter Associates Laboratory, Virginia, USA).	Frequency – 20 kHz, probe depth – 25 mm, probe diameter – 13 mm.	[19]
4.	Cabbage leaves	Konica Minolta chroma meter (CR-400, Konica Minolta, Japan), CIE L*a*b* system.	Transducer – 20 kHz, diameter – 10 mm, pulse model – 4s on, 2 s off.	[29]
5.	Carrot slices	Chromameter (CM-5, Konica-Minolta, Japan).	Immersive sonication – Treatment time – 20 minutes, frequency – 21 kHz, 180 W. Contact sonication – 24 kHz, approximately 250 W.	[31]
6.	Okra	–	30 kHz, 600 W, voltage 230 V, time – 30 and 60 minutes, PZT type ceramics transducers.	[28]
7.	Parsley leaves	CIE L*a*b* system (Konica-Minolta, CR-5).	300 W, bath frequency-21 kHz, 1:40 (w/w) solid to liquid ratio (leaves:water).	[27]
8.	Dried ginger slices	Konica Minolta colorimeter (CR-400, Osaka, Japan. L*a*b* values.	US – 1:10 ginger slice: water ratio, bath – 33 kHz, 600 W, 30 minutes, 30°C, duration 10s on, 5s off. Osmosonication – 1:10 sample: sucrose solution, 33 kHz, 600 W, 30 minutes, 30°C, duration 10 s on, 5 s off.	[25]
9.	Bitter melon	Digital camera DigiEye system (VeriVide Ltd, Leicester, UK).	1,200 W, 15 minutes in bath (25 ± 0.5°C).	[18]
10.	Jabuticaba epicarp – anthocyanin	CR-400, Konica Minolta Sensing Inc, Tokyo, Japan CIE L*a*b* color space recorded by spectra magic Nx software (Version CM-S100W2.03.0006).	100–500 W, 20 kHz, 50 g/L was S/L ratio, 30–35°C.	[2]

TABLE 3.1 *(Continued)*

Sl. No.	Product	Equipment Used	Ultrasound Parameter	References
11.	Sorghum husk – natural colorant	AATCC test method 8–2007.	Probe diameter – 15 mm, 360 W, 10, 20, and 30 minutes, 45, 55, and 65°C, microwave power – 14 to 33 W.	[32]
12.	Apple juice	Handheld tristimulus reflectance Spectro colorimeter Minolta Co. Model CM-508-d, Japan.	600 W, 20 kHz, 13 mm diameter, 25 ± 1°C.	[11]
13.	Mirabelle plum	Chromometer (CR-410, Minolta Co. Ltd, Osaka, Japan).	100 W, 15, 30, 45, and 60 minutes, 10s on, 5s off.	[16]
14.	Mushroom chips	Chromameter model no. CR-400 Konica Minolta Sensing, Inc (Osaka, Japan).	120 W, 28 kHz	[10]
15.	Rabbit meat	Color grand scheme colorimeter – color flex, hunter lab, Reston VA, hunter parameter L*a*b*.	4°C, 120 minutes, S/L 1:5 sample: solution.	[15]
16.	Italian salami	–	Bath – 25 kHz, 500 W, 3, 6, 9 minutes, optimized condition – 3 minutes, 128 W.	[8]
17.	Minced meat	Chromameter – CR-13, Konica Minolta, Tokyo, Japan.	Ultrasonic vacuum water bath – 25, 35, and 45°C.	[1]
18.	Kiwi fruit juice	Colorimeter – Lovibond, RT series, England, CIE – L*a*b*color space.	40 kHz, 180 W, 20 ± 1°C, 30 minutes.	[30]
19.	Prebiotic cranberry juice	Minolta colorimeter (color quest XE hunter lab Northants, UK) – read on CIELAB scale L*a*b*.	Probe type – power density 600 and 1,200 W/L, 5 minutes at room temperature.	[14]
20.	Wine	TU-1810 UV-visible spectrophotometer (Beijing Persee General Instrument Co. Ltd., Beijing, China – 420, 520, and 620 nm.	Time – 14 and 28 min, 16–22°C	[33]
21.	Guava juice	Colorimeter – Konica Minolta, Chroma Meter, CR-400, Mahwah NJ, USA, CIELab technique, measurement area 64 mm^2, angle 10?.	1,000 W, 20 kHz, 1.26 cm^2 titanium tip, 3, 6, and 9 minutes.	[5]
22.	Strawberry	Color Flex Spectro colorimeter (Hunter Lab Colorimeter D-25, Hunter Associates Laboratory, Ruston, USA).	Bath type – 33 kHz, 60 W, 0, 10, 20, 30, 40, and 60 minutes, 25°C.	[13]

TABLE 3.1 *(Continued)*

Sl. No.	Product	Equipment Used	Ultrasound Parameter	References
23.	Chickpea cooking water, foam, and meringues.	Hunter Lab ColorQuest XE colorimeter (Hunter Associates Laboratory, Reston, VA), CIE Lab coordinates.	Probe type – 13 mm diameter titanium alloy probe, 20 kHz, 50 and 100% power capacity, intensities – 34 and 67 W/cm^2.	[24]
24.	Honey	Colorimeter – Minolta CM-%08d, Osaka, Japan.	Bath type – 300, 600, and 900s, 20°C, 42 kHz.	[26]
25.	Potato strips	Colorimeter – Konica Minolta CR-400, Japan, CIE Lab scale D_{65}light source, 10?.	Bath type – 40 kHz, 200 W, 3 minutes.	[3]

3.9 CONCLUSION

Thus, the use of ultrasound (US) has been found to be very effective in preserving the color of food products. However, there are various factors that can affect the preservation effect, such as time, temperature, sensitivity of the product and pigment, liquid medium, solid to liquid ratio, working condition, power level, frequency, sample size, and probe diameter. US acts as a hurdle technology in various applications. For example, it can be used for juice homogenization and microbial reduction simultaneously. The above studies indicate that US at low frequency and with less exposure time can effectively preserve the color of food compounds. Further research should focus on the application of US treatment to various food commodities and its impact on enzymes responsible for color, particularly on enzymatic and non-enzymatic changes. US treatment has proven to be an effective technique for preserving the color of food compounds with a hurdle effect.

KEYWORDS

- **cavitation**
- **CIE color space**
- **color measurement**
- **drying**
- **LAB color space**
- **microorganisms**
- **sonication**
- **ultrasound**

REFERENCES

1. Millicent G. Managa, Yasmina Sultanbawa, & Dharini Sivakumar (2020). Effects of Different Drying Methods on Untargeted Phenolic Metabolites, and Antioxidant Activity in Chinese Cabbage (*Brassica rapa* L. subsp. chinensis) and Nightshade (*Solanum retroflexum* Dun.), pp. 1–23.
2. Albuquerque, B. R., Pinela, J., Barros, L., Oliveira, M. B. P. P., & Ferreira, I. C. F. R., (2020). Anthocyanin-rich extract of jabuticaba epicarp as a natural colorant: Optimization of heat- and ultrasound-assisted extractions and application in a bakery product. *Food Chemistry*, *316*, 126364. https://doi.org/10.1016/j.foodchem.2020.126364.
3. Amaral, R. D. A., Benedetti, B. C., Pujolà, M., Achaerandio, I., & Bachelli, M. L. B., (2016). A first approach of using ultrasound as an alternative for blanching in vacuum-packaged potato strips. *Food and Bioprocess Technology*, *9*(10), 1794–1801. https://doi.org/10.1007/s11947-016-1758-2.
4. Bhargava, N., Mor, R. S., Kumar, K., & Sharanagat, V. S., (2021). Advances in application of ultrasound in food processing: A review. *Ultrasonics Sonochemistry*, *70*(2020), 105293. https://doi.org/10.1016/j.ultsonch.2020.105293.
5. Campoli, S. S., Rojas, M. L., Do Amaral, J. E. P. G., Canniatti-Brazaca, S. G., & Augusto, P. E. D., (2018). Ultrasound processing of guava juice: Effect on structure, physical properties, and lycopene in vitro accessibility. *Food Chemistry*, *268*, 594–601. https://doi.org/10.1016/j.foodchem.2018.06.127.
6. Chemat, F., Zill-E-Huma, & Khan, M. K., (2011). Applications of ultrasound in food technology: Processing, preservation, and extraction. *Ultrasonics Sonochemistry*, *18*(4), 813–835. https://doi.org/10.1016/j.ultsonch.2010.11.023.
7. Chen, Z. G., Guo, X. Y., & Wu, T., (2016). A novel dehydration technique for carrot slices implementing ultrasound and vacuum drying methods. *Ultrasonics Sonochemistry*, *30*, 28–34. https://doi.org/10.1016/j.ultsonch.2015.11.026.
8. De Lima, A. L., Stefanello, Da. S. M., Martins, F. D. R., Rodrigues, A. D., Roggia, R. A., Da Silva, B. D., Fagundes, B. V. S., et al., (2018). Effect of ultrasound on the physicochemical and microbiological characteristics of Italian salami. *Food Research International*, *106*, 363–373. https://doi.org/10.1016/j.foodres.2017.12.074.
9. Dehghannya, J., Naghavi, E. A., & Ghanbarzadeh, B., (2016). Frying of potato strips pretreated by ultrasound-assisted air-drying. *Journal of Food Processing and Preservation*, *40*(4), 583–592. https://doi.org/10.1111/jfpp.12636.
10. Devi, S., Zhang, M., & Law, C. L., (2018). Effect of ultrasound and microwave assisted vacuum frying on mushroom (*Agaricus bisporus*) chips quality. *Food Bioscience*, *25*, 111–117. https://doi.org/10.1016/j.fbio.2018.08.004.
11. Ferrario, M., & Guerrero, S., (2016). Effect of a continuous flow-through pulsed light system combined with ultrasound on microbial survivability, color, and sensory shelf life of apple juice. *Innovative Food Science and Emerging Technologies*, *34*, 214–224. https://doi.org/10.1016/j.ifset.2016.02.002.
12. Gallo, M., Ferrara, L., & Naviglio, D., (2018). Application of ultrasound in food science and technology: A perspective. *Foods*, *7*(10), 1–18. https://doi.org/10.3390/foods7100164.

13. Gani, A., Baba, W. N., Ahmad, M., Shah, U., Khan, A. A., Wani, I. A., Masoodi, F. A., & Gani, A., (2016). Effect of ultrasound treatment on physico-chemical, nutraceutical, and microbial quality of strawberry. *LWT – Food Science and Technology, 66*, 496–502. https://doi.org/10.1016/j.lwt.2015.10.067.
14. Gomes, W. F., Tiwari, B. K., Rodriguez, Ó., De Brito, E. S., Fernandes, F. A. N., &Rodrigues, S., (2017). Effect of ultrasound followed by high pressure processing on prebiotic cranberry juice. *Food Chemistry*, *218*, 261–268. https://doi.org/10.1016/j.foodchem.2016.08.132.
15. Gómez-Salazar, J. A., Ochoa-Montes, D. A., Cerón-García, A., Ozuna, C., & Sosa-Morales, M. E., (2018). Effect of acid marination assisted by power ultrasound on the quality of rabbit meat. *Journal of Food Quality*, *2018*. https://doi.org/10.1155/2018/5754930.
16. Hashemi, S. M. B., (2018). Effect of pulsed ultrasound treatment compared to continuous mode on microbiological and quality of Mirabelle plum during postharvest storage. *International Journal of Food Science and Technology*, *53*(3), 564–570. https://doi.org/10.1111/ijfs.13629.
17. Jesus, M. S., Ballesteros, L. F., Pereira, R. N., Genisheva, Z., Carvalho, A. C., Pereira-Wilson, C., Teixeira, J. A., & Domingues, L., (2020). Ohmic heating polyphenolic extracts from vine pruning residue with enhanced biological activity. *Food Chemistry*, *316*(2019), 126298. https://doi.org/10.1016/j.foodchem.2020.126298.
18. Jin, W., Zhang, M., & Shi, W., (2019). Evaluation of ultrasound pretreatment and drying methods on selected quality attributes of bitter melon (*Momordica charantia* L.). *Drying Technology*, *37*(3), 387–396. https://doi.org/10.1080/07373937.2018.1458735.
19. Kadam, S. U., Tiwari, B. K., & O'Donnell, C. P., (2015). Effect of ultrasound pre-treatment on the drying kinetics of brown seaweed *Ascophyllum nodosum*. *Ultrasonics Sonochemistry*, *23*, 302–307.
20. Lagnika, C., Madjid, A. A. O., Sanni, A., & Lagnika, L., (2019). Effect of blanching and ultrasound on drying time, physicochemical and bioactive compounds of dried cashew apple. *American Journal of Food Science and Technology*, *7*(6), 227–233. https://doi.org/10.12691/ajfst-7-6-10.
21. MacDougall, D., (2002). *Color in Food: Improving Quality*. Woodhead Publishing.
22. Majid, I., Nayik, G. A., & Nanda, V., (2015). Ultrasonication and food technology: A review. *Cogent Food and Agriculture*, *1*(1), 1–11. https://doi.org/10.1080/23311932.2015.1071022.
23. Villamiel, M., García-Pérez, J. V., Montilla, A., Carcel, J. A., & Benedito, J., (2017). *Ultrasound in Food Processing* (Vol. 59). In Institute of Food Science and Technology (first). John Wiley and Sons Ltd.
24. Meurer, M. C., De Souza, D., & Ferreira, M. L. D., (2020). Effects of ultrasound on technological properties of chickpea cooking water (aquafaba). *Journal of Food Engineering*, *265*(2019), 109688. https://doi.org/10.1016/j.jfoodeng.2019.109688.
25. Osae, R., Zhou, C., Xu, B., Tchabo, W., Tahir, H. E., Mustapha, A. T., & Ma, H., (2019). Effects of ultrasound, osmotic dehydration, and osmosonication pretreatments on bioactive compounds, chemical characterization, enzyme inactivation, color, and antioxidant activity of dried ginger slices. *Journal of Food Biochemistry*, *43*(5), 1–14. https://doi.org/10.1111/jfbc.12832.

26. Quintero-Lira, A., Ángeles, S. A., Aguirre-Álvarez, G., Reyes-Munguía, A., Almaraz-Buendía, I., & Campos-Montiel, R. G., (2017). Effects of liquefying crystallized honey by ultrasound on crystal size, 5-hydroxymethylfurfural, color, phenolic compounds and antioxidant activity. *European Food Research and Technology*, *243*(4), 619–626. https://doi.org/10.1007/s00217-016-2775-0.
27. Sledz, M., Wiktor, A., Rybak, K., Nowacka, M., & Witrowa-Rajchert, D., (2016). The impact of ultrasound and steam blanching pre-treatments on the drying kinetics, energy consumption and selected properties of parsley leaves. *Applied Acoustics*, *103*, 148–156. https://doi.org/10.1016/j.apacoust.2015.05.006.
28. Sunil, C. K., Kamalapreetha, B., Sharathchandra, J., Aravind, K. S., & Rawson, A., (2017). Effect of ultrasound pre-treatment on microwave drying of okra. *Journal of Applied Horticulture*, *19*(1), 58–62. https://doi.org/10.37855/jah.2017.v19i01.09.
29. Tao, Y., Han, M., Gao, X., Han, Y., Show, P. L., Liu, C., Ye, X., & Xie, G., (2019). Applications of water blanching, surface contacting ultrasound-assisted air drying, and their combination for dehydration of white cabbage: Drying mechanism, bioactive profile, color, and rehydration property. *Ultrasonics Sonochemistry*, *53*(2018), 192–201. https://doi.org/10.1016/j.ultsonch.2019.01.003.
30. Tomadoni, B., Moreira, M. D. R., Espinosa, J. P., & Ponce, A., (2017). Individual and combined effects of pomegranate extract and ultrasonic treatments on kiwifruit juice quality parameters. *Journal of Food Process Engineering*, *40*(1), 1–10. https://doi.org/10.1111/jfpe.12339.
31. Wiktor, A., & Witrowa-Rajchert, D., (2020). Drying kinetics and quality of carrots subjected to microwave-assisted drying preceded by combined pulsed electric field and ultrasound treatment. *Drying Technology*, *38*(1, 2), 176–188. https://doi.org/10.1080/07373937.2019.1642347.
32. Wizi, J., Wang, L., Hou, X., Tao, Y., Ma, B., & Yang, Y., (2018). Ultrasound-microwave assisted extraction of natural colorants from sorghum husk with different solvents. *Industrial Crops and Products*, *120*, 203–213. https://doi.org/10.1016/j.indcrop.2018.04.068.
33. Zhang, Q. A., & Wang, T. T., (2017). Effect of ultrasound irradiation on the evolution of color properties and major phenolic compounds in wine during storage. *Food Chemistry*, *234*, 372–380. https://doi.org/10.1016/j.foodchem.2017.05.022.
34. Nikitenko, Sergey & Chemat, Farid. (2015). Ultrasound in Process Engineering. 10.1201/b18533-9.

CHAPTER 4

IMPACT OF MICROWAVE PROCESSING ON FOOD COLOR

U. GNANA MOORTHY ESWARAN,[1] SANGEETHA KARUNANITHI,[1] AKSHAY R. PATIL,[2] and PREM PRAKASH SRIVASTAV[1]

[1]*Department of Agricultural and Food Engineering, Indian Institute of Technology, Kharagpur, West Bengal, India*

[2]*National Institute of Food Technology, Entrepreneurship, and Management, Thanjavur, Tamil Nadu, India*

ABSTRACT

A cutting-edge heating technique called microwave (MW) processing has been utilized in the industrial, agricultural, and food industries. MW cooking offers numerous advantages, including rapid heating, high temperatures, fast cooking methods, clean energy, and low operating costs. Furthermore, this method can enhance the standards and safety of food products. As new electro-technology is implemented in the agricultural and food processing industries, the utilization of MW measurements and material dielectric characteristics is expanding. Despite the fact that direct heating by MWs can provide advantages over traditional heat transfer, the unique energy transfer mechanism in MW heating has presented new challenges in processing. Food appearance assessments are employed to uphold food quality both during and after processing. One of the most critical visual characteristics of food products is color, as it impacts consumer acceptance. Abnormal colors, particularly those associated with

Nonthermal Food Processing Technologies: Impact on Color Profile.
Kaavya Rathnakumar and R. Pandiselvam (Eds.)

deterioration in eating quality or spoilage, result in consumer rejection of the product. Changes in the color of food components can be attributed to the concept that cell membrane rupture occurs during MW heating, followed by the release of various chemical substances.

4.1 INTRODUCTION

The food processing sector has traditionally utilized thermal technologies for food preservation and obtaining the proper texture, flavor, and color. The challenge for the food processing industry is to effectively utilize thermal technology to achieve these goals without compromising important sensory and nutritional qualities. Most conventional heating techniques rely on convection or conduction mechanisms, but they have drawbacks such as longer processing times, reduced energy efficiency, and thermal damage from overheating, which mainly affects the surface properties of the materials. These issues can be addressed through the use of control and monitoring systems, as well as innovative equipment design. In response to growing environmental concerns and consumer demand for affordable, high-quality food, efforts are being made to make these processes more energy-efficient, practical, and environmentally friendly [10]. Electromagnetic (EM) technologies, which have the potential to replace conventional practices, have also gained interest in the food processing industry. One revolutionary heating technique that has been widely used in industry and agriculture, particularly in the food industry, is microwave (MW) processing. Microwave cooking offers numerous benefits, including rapid heating, high cooking temperatures, short cooking times, clean energy, and low operating costs. The processing conditions in most activities dictate the product quality, process economics, and environmental impact. Additionally, this technology has the ability to improve the safety aspects and quality of food products [18]. A microwave (MW) is a type of electromagnetic wave that is non-ionizing and has a frequency range of 300 MHz to 300 GHz (gigahertz). Because MW energy is swiftly transformed to heat and no radioactive material is created, MW-processed foods are safe to consume. Although it was originally believed that MW processing was a special form of thermal processing, it is now commonly used in both homes and the food industry. The attractiveness of MW heating is largely a result of its rapidity, which aligns well with people's contemporary fast-paced lifestyles. Instantaneous heating, low maintenance, and ease of use are other benefits of MW heating [33]. Combining MW technology with conventional drying techniques (such as hot air) can speed up the drying process, produce foods of superior quality, and allow a wider variety of foods to be dried, making it commonly utilized in the food business [36]. Dielectric processing of food is a technique that heats polar molecules by displacing them from their equilibrium sites in an electric field using internal molecular friction and charged ions. It is typically characterized as an efficient and superior heating technique based on the principle of volumetric heating mechanism [30]. Microwave (MW) and radio frequency (RF) are commonly employed processing parameters in industrial-scale applications. Some of the sectors where MW processing is gaining traction are metallurgy, chemical engineering, food processing, and paper manufacturing. Although dielectric treatment in the food processing industry offers several advantages, the full potential of dielectric energy in industrial applications has not yet been realized. The efficiency of an MW heating system is determined by two factors: processing time and uniformity in processing, which includes hot spot development [47]. In MW sample processing, thermal runaway is a major concern. It typically results from larger absorption of MW power or generation of heat in hot environments, which increases dielectric loss factors. Degradation in quality attributes of processed foods is due to non-uniform heating caused by thermal runaway during MW treatments. However, a challenge in dielectric heating is the occurrence of hot spots in various zones due to product specifications such as shape, size, dielectric characteristics, moisture content, and design of equipment such as cavity space, power, EM field, and electrode configuration. Microwave (MW) power absorption is influenced by the physical-chemical properties of a material, including its water content, chemical composition, and temperature. These processing factors have an impact on the material's dielectric properties, which in turn modify the penetration depth and wavelength characteristics of energy absorption in the sample. Furthermore, the specific effects of dielectric-assisted processes cannot be predicted, and dielectric techniques applied to agricultural goods consistently achieve their objectives due to the heat effect. However, it is undeniable that dielectric treatment applied to food causes some modifications. Foods are complex systems due to their diverse ingredients. In addition to physical changes such as expansion, charring, or dehydration that occur after MW treatment, researchers have shown particular interest in the effects on food components and the relationship between their physical qualities and chemical alterations [42].

4.2 BASIC THEORY AND PRINCIPLES

The electromagnetic (EM) wavelength range between 10 and 103 m is commonly used to define the dielectric range. In agricultural crops, these EM waves can lead to dipolar rotation. The fundamental units of matter are atoms, each of which is made up of an enclosing cloud of electrons with a net positive charge. Both the dipole moment and electric field display deflection if the charge intensity of the electric field gets distorted, making them uncoincided. EM waves are produced in this scenario, which causes the dielectric molecule to vibrate before returning to its initial state. Such a molecule is referred to as a dipolar molecule and the relaxation period is measured in seconds [27]. There are two techniques to produce thermal energy in an EM field. Firstly, it starts with dipolar polarization, and it is essential for heating agricultural produce in an EM field under MW circumstances. Polar water molecules can become dipolar polarized in an EM field. Other molecules may experience pressures from the electric field and develop "induced" dipoles. Dipolar polarization is a type of polarization that may be produced in any molecule and allows for the distortion of asymmetric nuclei. It naturally happens in polar molecules (orientation polarization and distortion polarization). Dipolar molecules respond to an external electric field by rotating towards the direction of polarization, but the produced electric field strives to align the charges within each permanent dipole; in other words, each permanent dipole remains constant between charges that are connected to chemical bonding. The dipoles come back to their respective random orientation when the electric field relaxes to zero, only to be brought back into alignment when the electric field escalates to its opposite polarity. The electrical field changes energy as a result of this. The material stores frictional energy as potential energy, which is later converted into irrational kinetic or thermal energy [51]. Ionic conduction, or polarization brought about by the relative displacements of positive and negative ions, is a distinct method. The result is that the non-ionized water molecules collide with the polarity-changing ions as they accelerate in the opposite direction. This can lead to multiple collisions and energy transfers due to friction, occurring millions of times per second. The heat released is a result of the interaction between the rate of polarization change, which represents the current displacement in the dielectric, and the applied field. When averaged over a cycle, the amount of power "lost" in the material is determined by the phase angle between polarization and applied field. Therefore, the majority of dielectric lag is controlled by the flexibility of molecules, which harbor dipoles and experience randomization effects due to temperature [30]. The system's configuration and the interfaces between processed materials and the surrounding space define the correct boundary conditions, which in turn govern the distribution of EM energy in MW heating systems. This significantly affects how well EM energy is coupled with materials, how EM fields are distributed, and how effectively EM energy is converted to heat energy within those materials [35].

The composition of food, including its moisture and salt content, as well as the frequency of ingestion, affects how temperature impacts the dielectric properties of food. Moist foods with a small amount of salt exhibit dielectric characteristics that are primarily influenced by water at microwave (MW) frequencies. The loss factors for highly moist alkaline foods increase as MW frequencies decrease and temperatures rise, resulting in a phenomenon known as "thermal runaway" [44]. Selective heating of food products in an electromagnetic (EM) field can lead to thermal runaway, causing uneven heating. As the moisture content of food drops below a certain point, the dielectric properties rapidly decrease. High temperatures facilitate the migration of bound water by reducing the required moisture content. Dehydrated foods have a reduced ability to convert EM energy into heat energy, as the loss factor decreases with increasing moisture content. However, during the MW process, the moist portion of the product may convert more MW energy into thermal energy compared to the dry portion. Hot air drying (HAD) often results in an uneven distribution of moisture, as the interior is dampened more than the exterior. This accelerates the drying process.

4.3 MICROWAVE (MW) APPLICATORS

Conventionally, MW ovens operate at a frequency range of 2,450 MHz, while industrial applications operate at a frequency range of 915 MHz. The components of a MW system include the magnetron, oven cavity, step-up transformer, applicator, waveguide, and food sample [38]. The step-up transformer in the MW system converts the frequency of the alternating current from a domestic power line, which runs at 50 to 60 Hz, to 2,450 MHz. The magnetron converts electrical energy into MW energy using 4,000 V. The magnetron generates MWs, which are then transmitted to the MW cavity through the waveguide. The food to be heated is placed in the MW chamber, which has walls that reflect MW energy. In a conventional home MW oven, the applicator is a rectangular cavity that resembles a resonator.

The three types of applicators that are most frequently employed in MW systems are single mode applicators (SMA), multimode applicators (MMA), and traveling wave applicators (TWA). System dependability, effective power transfer, and low cost are the three main characteristics for an efficient MW system.

4.3.1 SINGLE MODE APPLICATOR (SMA)

Only one way of propagation has been allowed inside the single-mode cavity in a specific region with a specified volume (Figure 4.1). This property enables efficient microwave energy dispersion via the single-mode resonant cavity applicator. A single-mode cavity, either rectangular or cylindrical, has the material positioned at set locations with strong electric fields. The components of a single-mode applicator include a feeding waveguide, a small microwave resonator, and a tuning plunger with required dimensions near the microwave wavelength [43]. In a single-mode applicator, the target sample is placed where there is a strong field. The material's diameter should be less than half of the incoming microwave wavelength, making the applicator's small size a significant drawback. The tuning plunger must be reciprocally rotated to make the necessary adjustments for the time-average field to be distributed evenly throughout the substance.

4.3.2 MULTIMODE APPLICATOR (MMA)

Microwave-assisted heating (MMA) is commonly utilized in approximately 60% of industrial systems. Despite its similarity to domestic systems, MMA is designed to handle larger volumes. The substance is placed inside a rectangular conductive metal box, resembling a container. The applicator receives microwave (MW) radiation from the waveguide after it

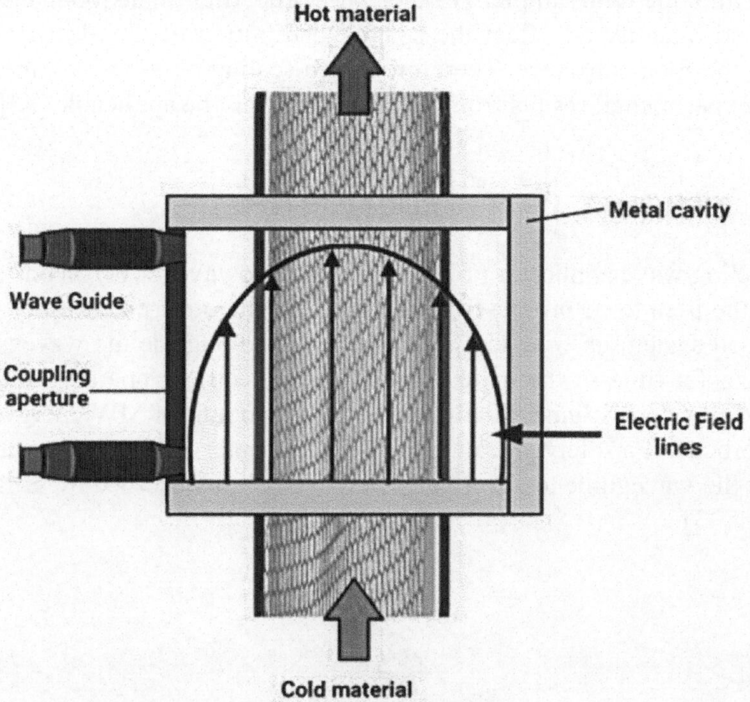

FIGURE 4.1 Representation of single mode applicator.

has undergone multiple reflections by the hollow walls. The food material begins to interact with the reflecting waves, inducing an electrical field that permeates through the multimode cavity and the food system. MMA generates steady electric field propagation modes at various locations. The field distribution of the system is influenced by the dielectric properties of the food, its quantity, and its placement within the MMA. It performs optimally with food systems that have a high dielectric loss factor and broader applicator regions (50%). However, for food items with a reduced loss factor and occupying a limited area of the MMA (20%), the electrical field is not adequately diffused, resulting in an uneven temperature distribution. Consequently, the food sample may exhibit hot and cold spots. To maintain an even temperature distribution, it is necessary to incorporate a mode stirrer (rotating reflector) or a continually spinning turntable into the applicator. This ensures uniformity and efficient dispersion of the electric field within the food sample. The dielectric properties of the food, its location, and quantity all affect the spatial distribution of the electric field within the food substance. Therefore, when scaling up to the commercial level, experimental results from one study may not be applicable [53].

4.3.3 TRAVELING WAVE APPLICATOR (TWA)

A traveling wave applicator (TWA) has the microwave (MW) on one side, while the item to be processed passes through the other end (Figure 4.2). The food system absorbs the MW as it moves alongside the waveguide. The transfer time of the material used, its dielectric properties, and the sample size and volume all contribute to the amount of MW energy that is absorbed. As a safety precaution, a standard water load is present at one end of the waveguide to absorb any MW that is not absorbed by the food system [33].

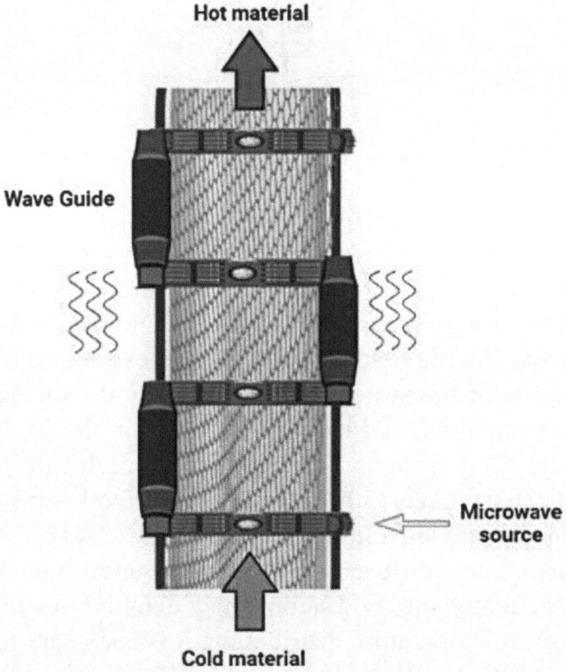

FIGURE 4.2 Representation of traveling wave applicator.

4.4 FOOD COLORS

Colors are significant quality indicators that affect consumer acceptability of food. The market for synthetic colorants has recently decreased as natural colorants have grown in popularity. Fruits and vegetables contain natural colorants. However, natural colors have disadvantages such as higher cost and less stability [48]. Thermal procedures are frequently used to prepare plant-based food, extending shelf life, enhancing flavor, and improving palatability. Processing and storing food results in significant changes in the color of food products. Good quality food is usually distinguished by its color compared to poor quality food. Although it may seem obvious, there are significant repercussions for the food industry. The ability to perceive color has been shown to be a useful tool for assessing food quality [31]. In countries with low labor costs, manual sorting based on hand and sight is still common in the food industry. In industries where labor costs have increased, automated approaches have been adopted. All optical sorting equipment must now be able to detect and remove any cross contaminants due to growing consumer awareness of the cleanliness of food. By eliminating imperfect, discolored, and product shape-deformed products, optical sorting also enhances the product's appearance. Today's consumers demand higher quality, which has led to an increase in lawsuits. Two of the most crucial elements in judging a vegetable's overall quality are color and color consistency. Freshness, flavor, and nutritional value are typically assessed using color. Color influences the pleasantness and acceptability of food, as well as judgments of sweetness, bitterness, saltiness, and flavor intensity [22].

4.4.1 COLOR MEASUREMENT

The CIELAB scale is used to quantify color changes that occur in processed food samples by the variables L*, a*, and b*. L* represents brightness or luminosity, a* represents the green (–) to red (+) scale, and b* indicates the blue (–) to yellow (+) scale, as clearly depicted in Figure 4.1. The variation of the color value before treatment (L*) and after treatment (Lo*) is necessary for calculating the overall color change (ΔE), whiteness index (WI), Chroma (C*), as well as hue (H*), which are determined using Equations (1)–(4), respectively [52].

$$\Delta E = \sqrt{\left(L^* - L_O^*\right)^2 + \left(a^* - a_O^*\right)^2 + \left(b^* - b_O^*\right)^2} \tag{1}$$

$$WI = 100 - \sqrt{\left(100 - L^*\right)^2 + \left(a^*\right)^2 + \left(b^*\right)^2} \quad (2)$$

$$C^* = 100 - \sqrt{\left(a^*\right)^2 + \left(b^*\right)^2} \quad (3)$$

$$\mathrm{H} = \tan^{-1} \frac{b^*}{a^*} \quad (4)$$

4.5 IMPACT OF MICROWAVE (MW) ON COLOR OF FRUITS AND VEGETABLES

The color of fruits and vegetables is mainly determined by four primary groups of natural pigments: carotenoids, chlorophylls, flavonoids (such as flavones, anthocyanins (ACNs), and flavonols), and betalains (such as betacyanins and betaxanthins) [58]. The post-harvest period refers to the time between harvest and consumption or further preservation of a vegetable [12]. On the other hand, minimal processing refers to any short-time procedure that enhances the shelf life and quality of a product. Plant tissue is alive, breathing, and transpiring. The physiology of microbial growth, plant tissue, and physical damage incurred during handling and processing post-harvest result in a variety of chemical reactions that interact during the course of the post-harvest period. Color may be affected by various industrial procedures involving heating. Blanched, boiled, cooked, tinned, dehydrated, pasteurized, or dried vegetables are all options [39]. Maintaining the color of green vegetables during heat processing remains an industrial difficulty. Customers are highly interested in heated green vegetables that retain a clear, bright green color. Vegetables must be heated to stabilize them and deactivate enzymes that reduce quality. At temperatures greater than 60°C, the living cells are destroyed, pectic substances disintegrate, and the cell structure becomes permanently altered [11]. Depending on the vegetable species, pigment concentration, heating technique, and processing conditions, thermal heating has different effects on color degradation and isomerization (yellow, green, orange, red, etc.) of the vegetables. The effects of temperature, duration of treatment, and the presence of oxygen on pigment stability are well recognized. During heating, ACNs undergo several changes, resulting in yellowish or brownish colors and the formation of a carbinol base called chalcone, while heat degradation of betanin follows a first-order reaction [48].

Microwave (MW) drying drives liquid onto the surface of solid food, where it is often transformed into vapor. This method allows for drying without creating surface overheating, preserving the product's surface color. However, MW drying can easily cause localized overheating due to rapid energy absorption and high drying rates. This overheating can result in significant thermal damage to the final product's quality. The technology for drying is particularly challenging to control due to the risk of overheating.

Studies on kale subjected to boiling, steaming, and MW heating have shown that after 20 minutes and 30 minutes of MW treatment, kale retained the highest color value, with L values ranging from 40.26 to 34.55 and 40.26 to 33.26, respectively [4]. Dos Reis et al. [13] evaluated the extraction efficiency of carotenoids, phenolic compounds, flavonoids, and chlorophylls from broccoli and cauliflower under different cooking conditions. MW cooking resulted in a decrease in the L value of fresh cauliflower in the range of 56.7 to 49.4, and fresh broccoli in the range of 46.1 to 39.9. Zhong et al. [61] investigated the effect of steamable bag MW technique on frozen broccoli florets and compared it with conventional microwaving and steaming methods. While steamable bag MW technology sustained the extreme darkness, preserved the b* value, and enhanced the a* value of broccoli florets, conventional MW processing raised the L* value, b* value, and a* values. The cause of the color shift in the texture and color of broccoli may be attributed to the different matrix structure between broccoli cooked using conventional microwave (MW) technique and steamable bag MW technique. A greater ΔE value was observed in broccoli florets cooked under conventional MW, while broccoli florets cooked using steamable bag MW showed similar color to that of uncooked florets. The primary components of Knol-Khol and radish are anthoxanthins, which are colorless pigments. They had a slight green tint that disappeared after cooking. The color levels did not show significant differences when cooked using various techniques. According to Aamir, Muhammad, & Waraporn [1], the impact of MW drying on the color attribute of okra was assessed. The values for MW power demonstrated a substantial impact on the color and sensory characteristics of dried okra. As the power (440, 618, and 800 W) and duration increased, the values for L*, a*, and b* decreased, while the overall color change ΔE increased from 6.20 to 24.19. Unlike L* and a*, which followed zero-order kinetic parameters, b* and ΔE followed first-order kinetic parameters. This indicates that an increase in MW output power leads to an increase in the rate of deterioration. This may be due to the fact that the temperature of the food rises as a result of the transmission of high energy within the product. The color measurement findings reveal a significant increase in L*, a*, and b* during the cooling process in drying beetroot cubes. Almost all of the color values determined for the dehydrated samples differ significantly from those obtained for samples with higher moisture content under the same drying conditions. The maximum values of L* and a* were obtained for MW freeze-dried beetroot cubes, and the lowest for cubes barely pre-dried by convective technique when examining the most dehydrated materials under various drying settings. It is important to note that the lowest moisture level was found in freeze-dried cubes, which helped attain the higher color parameter values. However, high L* and a* values may be the result of a Maillard reaction that took place at a high temperature or was brought on by prolonged exposure to hot air in samples that have been efficiently pre-dried or dried by convection. Jha et al. [26] investigated the effect of MW drying at different power levels of banana blossoms and the color parameters of the respective dried powder. MW power has a significant impact on color values such as L*, a*, and b*. For all powdered banana flowers, the mean L* value ranged from 43.29 to 54.43. All of the banana blossom powder samples that were dried in the MW had substantially different brightness (L*) levels, with the powder dried at 640 W having a much greater L* value than other samples. The material dried at 160 with 800 W showed no discernible difference in the L* value. MW drying at 800 W and 160 W both reduced the brightness value of the powder sample compared to the sample dried using 640 W.

Özcan et al. [45] studied the comparative effect of MW and oven drying properties on Pepino and Kiwi fruit slices. Fruit slices were dried using two different ovens, conventional technique for 20 hours at 70°C and MW for 3 minutes at 720 W. Change in color has been used as an indicator in controlling the overheating and undesirable characteristics of the drying process. Öztürk et al. [46] examined the combined effect of MW and MW-infrared drying of banana and kiwifruit. Varied MW powers (180 W, 270 W, and 360 W) for banana and (350 W, 420 W, and 490 W) for kiwifruit were used to dry samples with different initial moisture concentrations. The application parameters for banana combination drying were set to 360 W and 600 W for upper and lower halogen lights, while those for kiwifruit were set to 420 W and 600 W, respectively. It was discovered that there is no effect on L* values of banana samples due to their initial moisture content, whereas it has a considerable impact on both a* and b* values. Furthermore, the effect of MW power on all color values was determined to be minimal.

The low processing times of MW applications can explain this. While having no impact on a* values, combination drying also enhanced L* and b* values. Combination drying clearly gave findings that were more similar to those of the fresh sample when the L* and b* values were contrasted to those of other samples. This is because of the rapid drying time, which preserves color better and causes less harm to color pigments. Miraei et al. [40] studied the application of hybrid MW drying of nectarine slices. The nectarine slices were treated at different air-MW combination methods (80, 160, 240, and 320 W). According to a color analysis of dried samples, MW dried samples preserve a surface color that is more similar to fresh samples than hot air treated materials. In hybrid MW drying of nectarine slices, an increase in MW power level results in a reduction of the L* value from 71.25 to 57.05, and a significant difference of 4.67 in the a* value compared with fresh samples, whereas the b* value decreases from 45.08 to 54.50. If the MW heating system's power is increased, ACN and carotenoids may be extracted from the material's texture more effectively. This could lead to increases in redness due to the disintegration of cellular structures. It's important to understand that variations in a*, b*, and L* can be attributed to a variety of chemical processes that take place during MW processing, such as chlorophyll decomposition, pigment degradation, ascorbic acid (AA) browning, Maillard reaction-based browning, and caramelization. The influence of intermittent drying kinetics and color quality of red beetroot was investigated utilizing a combination of hot air with high power ultrasound and MW. From the observed result, convective MW drying shows the lowest discoloration in red beetroot with a lower value of color change (ΔE) of 19.65 ± 3.51. This may be due to the generation of heat within the sample by MW radiation. Beetroots dried using intermittent-hybrid MW processes lost less of their natural color. Furthermore, similar values of ΔE were observed for these samples, averaging 15.57 ± 1.00. Additionally, the findings on betanin content demonstrated that dried red beetroot had greater betanin retention with shorter MW radiation exposure times [52].

Imaizumi et al. [24] summarized the application of MW drying and freeze drying (FD) on the color properties of holy basil. To evaluate the color parameters of holy basil, the MW power levels of 300, 450, 600, and 800 W were used to dry 50 g of the sample, respectively. The results show that the L* value shows a higher significant difference compared to the fresh ones. The a* value increased, and a negative a* value represents green color, with rising values towards zero representing a loss of green coloration, i.e., decomposition of green color due to drying. A decrease in the b* value has also been reported, indicating yellowness, which has also been correlated to chlorophyll degradation by MW treatment. The highest degradation of chlorophyll was noted during FD and was linked to enzymatic reactions on chlorophyll reduction, whereas MW drying inactivates enzymes, thus retaining the greenness of holy basil leaves.

Xuet al. [59] demonstrated the effect of MW (0–500 W) drying on carrots in combination with HAD (60 to 80°C). The color properties of carrots have been studied as one of the major parameters for assessing the quality of dried products. These results indicate that carrot cubes processed through low temperature MW drying retain their color better than those processed at high temperature and high voltage. The reduction in color change values suggests a decrease in undesirable enzymatic changes and significant damage to cell structures in the dried products. Heat and oxygen can cause damage to carotenoids, which are primarily degraded through oxidation and temperature. This is due to their increased susceptibility to oxidative reactivity and the isomerization of *trans*-carotenoids into less brightly colored *cis* forms.

4.6 IMPACT OF MICROWAVE (MW) PROCESSING ON MEAT PRODUCTS

Since the 1970s, several batch and continuous industrial MW systems have been developed for thawing, tempering, and drying. The meat processing business extensively utilizes MW cooking and drying. In the thermal process of high-frequency heating, water molecules oscillate, resulting in friction and the production of heat. Meat can be heated using radiation by harnessing EM energy. Energy is transferred as waves when electrons in an atom transition from a higher to a lower energy state. These waves have different frequencies and energies. The lower-energy EM radiation emits extraordinarily long waves with frequencies ranging from 300 MHz to 300 GHz. When exposed to a magnetic field, polar molecules align with it. The polar molecules alternate at the MW frequency to maintain this alignment as the field changes. By breaking the H-bonds of neighboring water molecules as they rotate, heat is generated. The food system needs to consider dielectric properties significantly to develop efficient processes, but various methods exist to enhance uniform MW heating [32]. Frozen food can be tempered to prevent drip loss and facilitate slicing. Thawing frozen meat is done at a temperature of –2 to –4°C before using MW energy to temper the meat at refrigeration temperature. At temperatures slightly below 0°C, the outer layer of the meat can absorb a significant amount of energy, warming the surface area. Items that are frozen and thawed in the MW may experience "runaway heating" due to partial heating of the liquid phase. In comparison to 2,450 MHz, 915 MHz has a greater power penetration, making it more effective for heating thick masses of materials. Microwave (MW) thawing utilizes electromagnetic (EM) waves directed at the product through waveguides, without the need for electrodes. However, the use of MW thawing is limited due to thermal instability. Frozen food can be partially cooked using MW while the remaining portions remain frozen. When the temperature is below –5°C, the absorption of EM radiation by frozen food can increase the temperature. The absorption of energy is higher when a part of the material is much hotter than its surroundings during irradiation. Heating inequality worsens as enthalpy and absorption increase. To prevent overheating, power density reduction is used to enable thermal conduction and stabilize the distribution of enthalpy throughout the food. Cooking fresh meat in the MW can be challenging due to the lack of browning caused by the cool air surrounding the meat. As the meat heats up, fluid content from the interior transfers to the exterior, resulting in a tough and flavorless product. The use of MW cooking in the meat sector has only recently been adopted. Roasts cooked in the microwave required less time to reach endpoint temperatures than roasts prepared traditionally. Due to the quick heating period, meat cooked in the microwave does not have the customary brown surface associated with other cooking methods. Instead, it provides greater color and a softer texture to the meat. While evaporative and total losses were decreased, drip loss did not differ between microwave-cooked and broiler-cooked chops. Chops that were broiled received the highest flavor ratings, which did not change with respect to microwave power level. Hu et al. [21] investigated the effect of microwave heating on the quality of frozen meat. The cut meat sample, measuring 3*3*3 cm^3, was subjected to microwave dehydration at different power levels of 50, 100, and 150 W. The sample's L* value was reduced by the 50 W microwave pre-dehydration treatment, as the surface was heated abruptly during the heating process. After freezing and thawing, the amount of light reflected by the water decreased, resulting in no change in color but a lower L* value. At the same time, after thawing, some of the moisture was lost, and the surface now reflects normal light. Under other power conditions, there was no significant color change observed in the brightness value of the pork tenderloin. Furthermore, the MW treatment does not influence the color difference in a* and b* values of pork tenderloin, ensuring that the color of the sample is better protected. MW treatment at 120 W pre-dehydration and MW 150 W pre-dehydration, on the other hand, had superior retention than the control group in terms of influence on ΔE value. The following factors contribute to the change in a* value: Free radicals induce oxidation, which causes heme pigment instability. Thermal impacts can change the structure of myoglobin and hemoglobin dyes [16]. Furthermore, lipid oxidation is the primary cause of an increase in b* value of meat. When lipid oxidation happens, a pre-oxidant is formed, which combines with oxymyoglobin, resulting in the creation of hemimyoglobin.

4.7 IMPACT OF MICROWAVE (MW) PROCESSING ON SEA-FOOD PRODUCTS

Alternatives to animal protein that are considered healthier include fish and fish products. Due to their readily digested protein, including a predominant amount of ω-3-Polyunsaturated fatty acids, notably eicosapentaenoic acid and docosahexaenoic acid, high macro- and micro-mineral composition, and low calorific density, seafood is seen as a superior nutritional product [19]. Fish has become a popular and significant food product worldwide, thanks to technological developments in fish processing and growing consumer knowledge of the health advantages of fish. Around 156 million tons, or 20.5 kg (kilogram) per person, of the 179 million tons of fish obtained in 2018 were consumed for human utilization (Food and Agricultural Organization (FAO)). Fish and fish-based products were the third most popular protein source consumed globally, accounting for 17.1% of the world's protein supply, after cereals and milk. On the other hand, due to its cancer-causing properties, processed red meat is becoming less popular [49, 55]. Fish processing and consumption are therefore projected to increase in the coming years. However, the food industry effectively utilizes MW energy, either as a replacement for or in addition to conventional heat treatment (HT) methods. Scientific investigation into MW heating of food has been ongoing since the 1940s. The food industry has extensively employed MW heating for various processes such as baking, roasting, tempering, pasteurizing, rehydrating, thawing, and drying [14]. The development of MW energy in the food processing industry has made the widespread usage of fish processing technology possible. Numerous studies conducted over the past decade have demonstrated that MW heating is a viable method to replace or supplement existing procedures in the production of high-quality fish products. MW technology is well-known for its applications in cooking [17], drying [50, 54, 57], and thawing [5, 6, 8, 15]. Undoubtedly, the rapid penetration and heating capabilities of MW can save time and energy when cooking fish. Due to their higher moisture content compared to other foods, fish and shellfish are particularly suitable for MW cooking. However, the duration of treatment, product form, and MW frequency all affect the development of hot spots. The composition and structure of fish muscle have an impact on its dielectric properties, including the loss factor and dielectric constant [56]. Xue et al. [60] conducted an investigation on the color quality of MW pasteurization of Walbaum filets. The color change that occurs in heated products is linked to non-enzymatic browning reaction, protein denaturation, fat exudation, as well as water exudation. The treated groups had higher lightness (L*) values than the control groups. Connective tissue being damaged and many of the proteins were broken down during the heating process. The increase in L* values after pasteurization operations may be related to the breakdown and gelation of collagen.

Kipcak, Azmi, and Osman [29] examined the impact of heating fish items in a microwave on their preparation. Except for the values of brightness (L*) and hue, the fish ratio impact caused a significant difference in the color parameters. Specifically, R2 burgers had a significantly higher b* index than R1 burgers. In contrast, R1 samples exhibited higher a* values than R2 burgers for the redness index. The color characteristics were significantly affected by storage time. After 30 days of storage, the yellowness scores increased considerably and then remained stable, while the lightness and redness values decreased. The color degradation during storage time may be attributed to enzymatic and non-enzymatic processes that cause degradation of myofibrillar proteins and disorganization of myofibrils. The brightness and color parameters did not differ significantly between the two recipes and the raw samples. However, the redness score of R1 burgers was much higher than that of R2 burgers, while the yellowness score of R2 samples was significantly higher than that of R1 samples. The storage impact was also noticeable on the color of the cooked samples, with the exception of the redness rating. At 60 and 90 days, there was a significant decrease in the L* value, while a considerable increase in the b* value was observed during the second half of the storage period. The color appearance in cooked frozen food is influenced by various factors, including the Maillard reaction, protein degradation, and fat and water exudation during heating. In a study by Lee et al. [34], the application of novel MW-assisted induction heating (MAIH) on cooking parameters of white shrimp was evaluated. The MAIH system operated at temperatures of 90°C and 130°C with power outputs of 1,300 W and 2,450 W, respectively, for a duration of 60 to 120 seconds. Using a colorimeter, the CIE color features of peeled shrimp heated using the MAIH apparatus were measured at the two different temperatures. Regardless of the temperature, the levels of L* and a* values increased with the lengthening of the heating process. This pattern indicates that prolonged heating leads to the decomposition of pigments and fading of the color. However, it was found that increasing the heating time significantly decreased the b* (yellowness) readings for the heated samples. The b* values ranged from 1.45 to 2.55 and from 1.01 to 1.86 after heating at 130°C and 90°C, respectively. These findings suggest that as the heating time increases, the appearance of peeled shrimp gradually turns white. This is attributed to the denaturation of heme proteins, such as hemoglobin and myoglobin, and the formation of cooked seafood flesh hemoprotein, which occurs simultaneously with the whitening phase.

4.8 IMPACT OF MICROWAVE (MW) ON OTHER FOOD SYSTEM

İzli, Gökçen, Gulcin, & Senel [25] conducted a study on the quality retention attributes during the drying of pumpkin powder using MW convective drying technique. Approximately 11 pumpkin powders were tested to observe the impact of MW heating and MW convection drying on their color values. The lightness ratio of raw pumpkins was found to be 70.43, while the dried pumpkin powders showed an increase in the L* value. Compared to certain pumpkin powders, the dried pumpkin powder particles exhibited significantly higher L* values. The MW convective dried pumpkin powders displayed a higher lightness, resulting in a lighter appearance compared to other dried samples. This could be attributed to the reduction of non-enzymatic browning. It is believed that the color changes observed after drying pumpkin samples are caused by the heat-induced breakdown of pigments, as well as enzymatic or non-enzymatic (Maillard) browning processes. The a* readings of the dehydrated pumpkin granules were considerably higher compared to the fresh sample, indicating the development of browning. In this study, it was demonstrated that the pumpkin samples dried using MW at 100 W and convective drying at 60°C for the longest durations of 196 minutes and 162 minutes, respectively, exhibited the most significant color changes. Overall, the MW dried pumpkin powders had a* values that were most similar to those of the raw pumpkin, with the nearest b* value to the raw pumpkin. In comparison to freshly cut pumpkin slices, the C parameter, which represents the color intensity of pumpkin powders, showed much greater values. In short, the values for all color criteria of MW-convective dehydrated pumpkin granules that were similar to the raw pumpkin samples were identified (7.85). Color is a crucial quality indicator for assessing the superiority of dried food products. Serious browning, which can have either an enzymatic or non-enzymatic cause, can be differentiated from color loss in the majority of fruits. Browning becomes worse at higher drying temperatures. The hues that most closely resemble the natural shade of raw pumpkin slices are the most popular. These examples may be categorized as more favored and desirable foods by customers in terms of color quality. From that perspective, MW-convective drying is a gratifying approach for producing dried pumpkin powders because it produces samples with color characteristics that are the closest to those of fresh pumpkin fruits when compared to other drying techniques.

Keseret et al. [28] characterized the color parameters of carrot powder obtained through intermittent MW drying technique. Three times each of 7 distinct MW power levels, such as 150, 200, 250, 300, 350, and 400 W, were used to dry fresh carrot samples ranging in size from 10 to 15 mm. The dried samples had greater L* readings than the fresh samples, indicating that the dried samples are lighter. Darker product colors resulted from a considerable decline in L* at higher electrical MW power. The a* value of the dried samples was significantly lower than that of the fresh samples, and it sharply dropped with increasing MW power levels. Except for the samples processed at higher MW powers, the b* values were nearly identical to those of the raw samples (400 and 450 W). When carrots were dried at a lower MW power (150 W), the greatest value of E was discovered. Color is a crucial quality indicator for determining the quality of dry food materials. Serious browning, which may have enzymatic or non-enzymatic causes in the majority of vegetables, is a defining characteristic of color deterioration. MW drying uses less energy than traditional drying while producing a food system with improved sensory qualities (Table 4.1).

TABLE 4.1 An Overview of the Impact of Different Microwave Processing on the Color of Various Food Products

Food Material	MW Processing Parameter	Color Value	Observation	References
Microwave cooking of Kale (*Brassica oleracea* L. var. *acephala*).	900 W – 20 min	L* – 34.55 ± 0.22 a* – –8.12 ± 0.30 b* – 18.59 ± 1.28 ΔE – 13.82 ± 0.95	Microwave cooking does not cause much change in lightness values. Decrease in greenness value of Kale was due to formation of pheophytin through degradation of chlorophyll. But more significant change in color has been observed with prolonged time of treatment which changes the reflectance property of the Kale's surface by formation of degraded products as a result of microwave cooking.	[3]
	900 W – 30 min	L* – 33.26 ± 0.77 a* – –5.41 ± 0.16 b* – 19.43 ± 0.30 ΔE – 15.97 ± 0.27		
Microwave drying of Okra	440 W – 7 min	L* – 21.11 a* – –0.82 b* – 2.52 ΔE – 6.20	The difference in color values is a result of increase in output MW power level resulting in decomposition of carotenoid and chlorophyll pigments. Consequently, non-enzymatic Maillard Browing results in the synthesis of brown color pigments. This may be related to the fact that food products heat up as high energy waves pass through them.	[1]
	618 W – 5 min	L* – 24.57 a* – –0.52 b* – 5.05 ΔE – 22.02		
	800 W – 4 min	L* – 22.41 a* – –0.12 b* – 7.15 ΔE – 24.19		
Microwave assisted dehydration of apple cubes.	700 W – 10%	L* – 67.72 a* – 6.16 b* – 34.94 ΔE – 27.87	Highest reduction in L* value has been reported with increase in concentration of solution. Also, brightness of the microwave dehydrated has been significantly reduced due to shrinking of cytoplasm thus causes leaching out of water from apple cubes.	[37]
	700 W – 20%	L* – 86.16 a* – 6.48 b* – 35.93 ΔE – 15.85		
	700 W – 30%	L* – 45.61 a* – 6.70 b* – 34.96 ΔE – 38.93		

TABLE 4.1 *(Continued)*

Food Material	MW Processing Parameter	Color Value	Observation	References
Microwave drying of powdered banana blossoms.	800 W	L* – 43.83 ± 0.07 a* – 4.93 ± 0.09 b* – 6.94 ± 0.07	Increase in the treatment power level in turn increases the greenness, lightness, and yellowness of the powdered banana. At higher MW power level maximum a* value has been obtained through enzymatic and non-enzymatic browning reaction.	[26]
–	640 W	L* – 54.43 ± 0.46 a* – 4.19 ± 0.03 b* – 10.71 ± 0.29		
–	480 W	L* – 50.88 ± 1.15 a* – 4.56 ± 0.38 b* – 10.15 ± 0.99		
Microwave drying of Kiwi and banana fruits.	180 W	L* – 42.11 ± 2.46 a* – 7.48 ± 0.33 b* – 17.10 ± 0.84	Result shows that shorter the drying time higher the perseverance of color values. The investigation on the b* values of the kiwifruit samples revealed a beneficial result of MW and combination drying. This increase in b* value is caused by the samples' yellow color, which was a result of chlorophyll breakdown.	[46]
	270 W	L* – 43.74 ± 3.38 a* – 6.41 ± 0.37 b* – 16.35 ± 2.29		
	360 W	L* – 43.59 ± 2.77 a* – 5.96 ± 1.59 b* – 16.06 ± 0.57		
	350 W	L* – 55.39 ± 1.82 a* – 2.86 ± 1.13 b* – 21.63 ± 1.73		
	420 W	L* – 49.45 ± 1.09 a* – 4.63 ± 1.56 b* – 21.92 ± 1.08		
	490 W	L* – 45.69 ± 2.98 a* – 4.60 ± 1.53 b* – 19.34 ± 1.39		

TABLE 4.1 *(Continued)*

Food Material	MW Processing Parameter	Color Value	Observation	References
Microwave drying of nectarine slices.	80 W	L* – 69.95 ± 2.81 a* – 39.35 ± 1.25 b* – 60.18 ± 2.77 ΔE – 36.68 ± 2.90	According to a color analysis of dried samples, the surface colors of MW-dried products are more similar to fresh produce. The increased MW power of the hybrid drying technology results in higher E values. This study implies that increased power speeds up degradation of color values by raise in temperature of product because of higher energy levels transferred. The surface color of the MW-dried samples, however, was better since there were fewer color fluctuations. The hybrid method's quick energy absorption of samples reduces drying time greatly, preventing the formation of too many brown pigments on the sample surfaces. It's important to note that fluctuations in a*, b*, and L* are connected to variety of internal chemical reactions that occur during drying, such as the breakdown of pigments, the breakdown of chlorophyll, the browning caused by ascorbic acid, Maillard reaction and caramelization.	[40]
	160 W	L* – 68.20 ± 3.12 a* – 40.65 ± 3.44 b* – 57.05 ± 4.53 ΔE – 37.51 ± 4.79		
	240 W	L* – 65.20 ± 3.09 a* – 41.15 ± 2.91 b* – 56.30 ± 4.77 ΔE – 39.63 ± 4.48		
	320 W	L* – 57.05 ± 2.84 a* – 45.80 ± 4.10 b* – 54.50 ± 3.48 ΔE – 44.27 ± 3.86		
Microwave dehydration of pork.	50 W	L* – 57.40 ± 1.74 a* – 4.65 ± 0.29 b* – 6.06 ± 0.27 ΔE – 37.84 ± 0.76	The 50 W MW pre-dehydration treatment lowers the sample's L* value. After freezing and thawing, the water loss reflects less light, resulting in a lower L* value, but the color does not noticeably alter. In addition, some of the moisture was lost as it thawed and now reflects ordinary light. After pre-dehydration, the brightness value of pork tenderloin did not considerably alter. The pork loin's color difference a* and b* values are unaffected by the microwave treatment. However, microwave pre-dehydration of 120 W and 150 W has a greater retention than the control group in terms of its impact on the ΔE value. The following factors account for the majority of the change in a* value: Free radicals encourage oxidation, which makes heme pigment unstable; heat impacts may change the structures of myoglobin and hemoglobin dyes. In addition, oxidation of lipids is the primary cause of the rise in meat's b* value; oxidation of lipids results in the development of a pre-oxidant that combines with oxymyoglobin to produce hemimyoglobin.	[20]
	100 W	L* – 57.96 ± 0.74 a* – 4.62 ± 0.97 b* – 5.90 ± 1.07 ΔE – 38.43 ± 0.50		
	150 W	L* – 55.40 ± 0.91 a* – 4.97 ± 0.68 b* – 6.23 ± 0.58 ΔE – 35.98 ± 0.76		

TABLE 4.1 *(Continued)*

Food Material	MW Processing Parameter	Color Value	Observation	References
Microwave pasteurization of Walbaum filets.	7 kW	L* – 69.01 ± 0.59 a* – 12.95 ± 0.80 b* – 20.03 ± 1.40 ΔE – 12.22 ± 0.66	The color change in heated products is a result of the protein denaturation, water exudation, fat exudation, and Maillard reaction. During the heating process, some of the protein was broken down and connective tissue was damaged. After pasteurization procedures, L* values increased, which may be related to the breakdown and gelation of collagen. The largest ΔE values phenomenon might result from microwave pasteurization's quicker heating process.	[60]
Microwave assisted induction heating of white shrimp cooking.	130°C – 80 s 90°C – 100 s	L* – 49.49 ± 0.58 a* – 10.28 ± 0.44 b* – –2.38 ± 0.23 ΔE – 48.40 ± 0.55 L* – 51.49 ± 0.85 a* – 15.46 ± 0.56 b* – –1.35 ± 0.08 ΔE – 49.07 ± 0.89	Astaxanthin, a red carotenoid, is found in crustaceans like crabs and shrimp. In the bodies of living crustaceans, this chemical links to proteins to create crustacyanin, which has a brown or dark brown appearance. The protein is denatured during the heating process and releases astaxanthin, which gives the shell and meat their red color. These findings suggest that as heating time is increased; the peeled shrimp's appearance turns increasingly white. This results from the denaturation of heme proteins (such as hemoglobin and myoglobin), and the creation of the cooked seafood flesh hemoprotein occurs concurrently with the whitening phase.	[34]
Microwave convective drying of pumpkin powder.	100 W 200 W	L* – 66.16 ± 2.96 a* – 26.78 ± 0.85 b* – 58.46 ± 2.07 ΔE – 17.5 ± 1.17 L* – 67.85 ± 3.52 a* – 20.75 ± 0.68 b* – 53.22 ± 1.94 ΔE – 9.87 ± 1.23	It is assumed that heat-induced pigment degradation and browning reactions are what causes the color changes that appear after drying pumpkin samples. As a result of increase in power levels employed during MW drying, the browning process gets enhanced. Furthermore, research show that higher temperatures and prolonged drying durations lead to increased pigment degradation and color changes in food.	[22]

TABLE 4.1 *(Continued)*

Food Material	MW Processing Parameter	Color Value	Observation	References
Microwave drying of carrot powder.	150 W	L* – 74.33 ± 1.06 a* – 16.79 ± 4.61 b* – 37.87 ± 1.73 ΔE – 25.04 ± 1.29	Darker product colors resulted from a considerable drop in L* at higher applied MW powers. The a* values of the dried carrot powder were lower, and they sharply decreased as the MW power levels increased. With the exception of the samples dried at higher MW powers, the b* values were nearly identical to those of the fresh samples.	[28]
	200 W	L* – 72.95 ± 0.69 a* – 16.12 ± 4.60 b* – 39.54 ± 1.53 ΔE – 23.84 ± 0.46		
	250 W	L* – 70.50 ± 2.65 a* – 12.87 ± 2.08 b* – 39.66 ± 1.84 ΔE – 22.53 ± 2.31		
	300 W	L* – 68.48 ± 2.97 a* – 12.74 ± 1.09 b* – 39.07 ± 1.19 ΔE – 20.71 ± 2.01		
	350 W	L* – 67.38 ± 0.52 a* – 11.87 ± 2.55 b* – 39.04 ± 0.55 ΔE – 20.00 ± 0.83		
	400 W	L* – 67.56 ± 1.73 a* – 13.44 ± 1.50 b* – 39.17 ± 0.92 ΔE – 19.61 ± 1.77		
	450 W	L* – 63.35 ± 4.26 a* – 10.33 ± 2.64 b* – 36.68 ± 0.81 ΔE – 17.90 ± 2.27		

4.9 SUMMARY

Despite the widespread usage of MWs in the food processing industry, several areas require further future research. A long-standing technological obstacle to consistent food processing is the non-uniformity of the MW field, which can generally result in unfavorable and uneven aesthetic appearance of foods. To reduce the detrimental effects on food quality and safety, research in this field needs to be continued. Although there is little information on utilizing MWs to create low temperature cooking and preserve food colors, more work can be done to comprehend and utilize this approach. Because MW processing has the distinct properties of high efficiency, future research should focus on developing novel assisted techniques to increase efficiency while maintaining product quality and safety. Combining MW processing with other conventional techniques can effectively overcome the disadvantages of conventional techniques in terms of preserving color changes. Furthermore, the question of whether nonthermal impacts occur during MW sterilization is still up for debate.

KEYWORDS

- **food system**
- **microwave applicators**
- **microwave processing**
- **multimode applicator**
- **sea-food products**
- **single mode applicator**
- **traveling wave applicator**

REFERENCES

1. Aamir, M., & Waraporn, B., (2017). Effect of microwave drying on quality kinetics of okra. *Journal of Food Science and Technology, 54*(5), 1239–1247.
2. Alvi, T., Zarnab, A., & Muhammad, K. I. K., (2022). Clean label extraction of bioactive compounds from food waste through microwave-assisted extraction technique-A review. *Food Bioscience*, 101580.
3. Armesto, J., Gómez-Limia, L., Javier, C., & Sidonia, M., (2019). Effects of different cooking methods on the antioxidant capacity and flavonoid, organic acid, and mineral contents of Galega Kale (*Brassica oleracea* var. acephala cv. Galega). *International Journal of Food Sciences and Nutrition, 70*(2), 136–149.
4. Benucci, I., Claudio, L., Caterina, M., & Marco, E., (2022). Natural colorants from vegetable food waste: Recovery, regulatory aspects, and stability—A review. *Comprehensive Reviews in Food Science and Food Safety*.

5. Cai, L., Minjie, C., Ailing, C., Joe, R., Jianrong, L., & Rongfa, G., (2018). Ultrasound or microwave vacuum thawing of red sea bream (*Pagrus major*) fillets. *Ultrasonics Sonochemistry, 47*, 122–132.
6. Cai, L., Minjie, C., Joe, R., & Ailing, C., (2019). Recent advances in food thawing technologies. *Comprehensive Reviews in Food Science and Food Safety, 18*(4), 953–970.
7. Cai, L., Wendi, Z., Ailing, C., Minjie, C., & Jianrong, L., (2019). Effects of ultrasonics combined with far infrared or microwave thawing on protein denaturation and moisture migration of *Sciaenops ocellatus* (red drum). *Ultrasonics Sonochemistry, 55*, 96–104.
8. Chen, X., Jun, Y., Mingyue, S., Yi, C., Qiang, Y., & Jianhua, X., (2022). Structure, function, and advance application of microwave-treated polysaccharide: A review. *Trends in Food Science & Technology*.
9. Chen, Z., Yulin, L., Likun, W., Shuyi, L., Keke, W., Jun, S., & Bin, X., (2017). Evaluation of the possible non-thermal effect of microwave radiation on the inactivation of wheat germ lipase. *Journal of Food Process Engineering, 40*(4), e12506.
10. Contreras, C., Benlloch-Tinoco, M., Rodrigo, D., & Martínez-Navarrete, N., (2017). Impact of microwave processing on nutritional, sensory, and other quality attributes. In: *The Microwave Processing of Foods* (pp. 65–99). Woodhead Publishing.
11. Danowska-Oziewicz, M., Agnieszka, N., Anna, D., & Natalia, M., (2020). The effects of cooking method on selected quality traits of broccoli and green asparagus. *International Journal of Food Science & Technology, 55*(1), 127–135.
12. Dhingra, D., & Sangeeta, C., (2022). Fresh-cut fruits and vegetables-A review. *Agricultural Engineering Today, 46*(1), 17–24.
13. dos, R., Luzia, C. R., Viviani, R. D. O., Martine, E. K. H., André, J., Simone, H. F., & Alessandro, De. O. R., (2015). Carotenoids, flavonoids, chlorophylls, phenolic compounds, and antioxidant activity in fresh and cooked broccoli (*Brassica oleracea* var. Avenger) and cauliflower (*Brassica oleracea* var. AlphinaF1). *LWT-Food Science and Technology, 63*(1), 177–183.
14. Doymaz, I., Azmi, K., & Sabriye, P., (2015). Microwave drying of green bean slices: Drying kinetics and physical quality. *Czech Journal of Food Sciences, 33*(4).
15. Flora-Glad, C. E., Da-Wen, S., Zhang, H., & Jun-Hu, C., (2017). Microwave-assisted food processing technologies for enhancing product quality and process efficiency: A review of recent developments. *Trends in Food Science & Technology, 67*, 58–69.
16. Fan, H., Daming, F., Jianlian, H., Jianxin, Z., Bowen, Y., Shenyan, M., Wenguo, Z., & Hao, Z., (2020). Cooking evaluation of crayfish (*Procambarus clarkia*) subjected to microwave and conduction heating: A visualized strategy to understand the heat-induced quality changes of food. *Innovative Food Science & Emerging Technologies, 62*, 102368.
17. Guo, Q., Da-Wen, S., Jun-Hu, C., & Zhong, H., (2017). Microwave processing techniques and their recent applications in the food industry. *Trends in Food Science & Technology, 67*, 236–247.
18. Guzik, P., Piotr, K., Marzena, Z., & Władysław, M., (2021). Microwave applications in the food industry: An overview of recent developments. *Critical Reviews in Food Science and Nutrition*, 1–20.

19. Chung-Su, H., Tae-Hwan, K., Jeong-Hyeon, L., Jin-Ho, W., Byeong-Hyo, C., & Sung-Chan, C., (2016). Drying characteristics of sea tangle using combination of microwave and far-infrared dryer. *Journal of Biosystems Engineering, 41*(1), 43–50.
20. Horuz, E., Hüseyin, B., Haluk, K., & Medeni, M., (2017). Drying kinetics of apricot halves in a microwave-hot air hybrid oven. *Heat and Mass Transfer, 53*(6), 2117–2127.
21. Hu, R., Min, Z., & Arun, S. M., (2022). Application of infrared and microwave heating prior to freezing of pork: Effect on frozen meat quality. *Meat Science, 189*, 108811.
22. Husarova, O., & Natalija, D., (2022). *Influence of Microwave Processing of Apples on the Process of Convective Drying* (pp. 51, 52). Collection of scientific papers «SCIENTIA» Chicago, USA.
23. Husein, Y., Giuliana, P., & Giacomo, P., (2019). *Application of the Mechanical Separation Process in Different Fish Species for the Development of a New Product Based on Fish.*
24. Imaizumi, T., Pongphen, J., Natta, L., & Nattapon, K., (2021). Effect of microwave drying on drying characteristics, volatile compounds, and color of holy basil (*Ocimumtenuiflorum* L.). *Agriculture and Natural Resources, 55*(1), 1–6.
25. İzli, G., Gulcin, Y., & Senel, E. B., (2022). Quality retention in pumpkin powder dried by combined microwave-convective drying. *Journal of Food Science and Technology, 59*(4), 1558–1569.
26. Jha, P., Murlidhar, M., & Pramod, K. P., (2021). Microwave drying of banana blossoms (*Musa acuminata*): Mathematical modeling and drying energetics. *Journal of Food Processing and Preservation, 45*(9), e15717.
27. Jiang, H., Zhigang, L., & Shaojin, W., (2018). Microwave processing: Effects and impacts on food components. *Critical Reviews in Food Science and Nutrition, 58*(14), 2476–2489.
28. Keser, D., Gamze, G., Hasim, K., Muharrem, K., Yurtsever, S., Yunus, E. S., Aysel, A., & Serkan, S., (2020). Characterization of aroma and phenolic composition of carrot (*Daucus carota* 'Nantes') powders obtained from intermittent microwave drying using GC–MS and LC–MS/MS. *Food and Bioproducts Processing, 119*, 350–359.
29. Kipcak, A. S., & Osman, İ., (2021). Microwave drying of fish, chicken, and beef samples. *Journal of Food Science and Technology, 58*(1), 281–291.
30. Knoerzer, K., Regier, M., & Schubert, H., (2017). Measuring temperature distributions during microwave processing. In: *The Microwave Processing of Foods* (pp. 327–349). Woodhead Publishing.
31. König, L. M., & Britta, R., (2018). Colorful= healthy? Exploring meal color variety and its relation to food consumption. *Food Quality and Preference, 64*, 66–71.
32. Kumari, S., & Sujoy, K. S., (2022). The efficient thermal processing of cylindrical multiphase meat: A study on the selection of microwave heating strategy. *International Journal of Food Engineering*.
33. Jean-Claude, L., & Hamoud-Agha, M. M., (2022). Microwave food processing: Principles and applications. *Thermal Food Engineering Operations*, 301–347.

34. Yi-Chen, L., Chih-Ying, L., Cheng-I, W., Hung-Nan, T., Kuohsun, C., & Yung-Hsiang, T., (2021). Preliminary evaluation of a novel microwave-assisted induction heating (MAIH) system on white shrimp cooking. *Foods, 10*(3), 545.
35. Lentz, R. R., Juming, T., & Fermin, P. R. Jr., (2020). Electromagnetic basis of microwave heating. In: *Development of Packaging and Products for Use in Microwave Ovens* (pp. 3–71). Woodhead Publishing.
36. Li, M., Bo, W., Yuchuan, W., Jiguang, L., & Min, Z., (2022). Evaluation of the uniformity, quality, and energy cost of four types of vegetables and fruits after pilot-scale pulse-spouted bed microwave (915 MHz) freeze-drying. *Drying Technology*, 1–18.
37. Manzoor, A., Mohammad, A. K., Muhammad, A. M., & Rayees, A. S., (2021). Comparative study of microwave assisted and conventional osmotic dehydration of apple cubes at a constant temperature. *Journal of Agriculture and Food Research, 5*, 100176.
38. Meda, V., Orsat, V., & Raghavan, V., (2017). Microwave heating and the dielectric properties of foods. In: *The Microwave Processing of Foods* (pp. 23–43). Woodhead Publishing.
39. Mehmood, A., & Alam, Z., (2020). Effects of different cooking techniques on bioactive contents of leafy vegetables. *International Journal of Gastronomy and Food Science, 22*, 100246.
40. Miraei, A., Seyed-Hassan, B. S., & Abozar, N., (2018). Effects of hot-air and hybrid hot air-microwave drying on drying kinetics and textural quality of nectarine slices. *Heat and Mass Transfer, 54*(4), 915–927.
41. Mishra, R. R., & Apurbba, K. S., (2016). Microwave–material interaction phenomena: Heating mechanisms, challenges, and opportunities in material processing. *Composites Part A: Applied Science and Manufacturing, 81*, 78–97.
42. Nguyen, T. P., & Sirichai, S., (2022). Microwave processing technology for food safety and quality: A review. *Agriculture and Natural Resources, 56*(1), 57–72.
43. Okress, E. C., (2017). *Microwave Power Engineering: Generation, Transmission, Rectification* (Vol. 1). Academic Press.
44. Orsat, V., Raghavan, G. S. V., & Krishnaswamy, K., (2017). Microwave technology for food processing: An overview of current and future applications. *The Microwave Processing of Foods*, 100–116.
45. Özcan, M. M., Fahad, Al. J., Isam, A. M. A., Nurhan, U., Elfadil, E. B., & Kashif, G., (2020). Effect of microwave and oven drying processes on antioxidant activity, total phenol, and phenolic compounds of kiwi and pepino fruits. *Journal of Food Science and Technology, 57*(1), 233–242.
46. Öztürk, S., Özge, Ş., & Yaşar, Ö. A., (2017). Dielectric properties and microwave and infrared-microwave combination drying characteristics of banana and kiwifruit. *Journal of Food Process Engineering, 40*(3), e12502.
47. Perino-Issartier, S., Jean-François, M., & Farid, C., (2010). Microwave food processing. In: *Alternatives to Conventional Food Processing* (pp. 415–458).
48. Ranganath, K. G., (2022). Pigments that color our fruits: An overview. *Erwerbs-Obstbau*, 1–13.

49. Rashid, M. A., Saifullah, M., Habib, M. A., Md Iqbal, H., & Sattar, M. A., (2016). The effect of different cooking methods on nutrient content of selected Bangladeshi fish. *Int. J. Adv. Res. Chem. Sci., 3*, 9–14.
50. Regier, M., Knoerzer, K., & Schubert, H., (2017). Introducing microwave-assisted processing of food: Fundamentals of the technology. In: *The Microwave Processing of Foods* (pp. 1–22). Woodhead Publishing.
51. Resurreccion, Jr. F. P., Luan, D., Tang, J., Liu, F., Tang, Z., Pedrow, P. D., & Cavalieri, R., (2015). Effect of changes in microwave frequency on heating patterns of foods in a microwave assisted thermal sterilization system. *Journal of Food Engineering, 150*, 99–105.
52. Silva, M. M., Fernando, H. R., & Fernando, C. L., (2022). Food color additives: A synoptical overview on their chemical properties, applications in food products, and health side effects. *Foods, 11*(3), 379.
53. Simunovic, J., & Sandeep, K. P., (2022). Key technological advances and industrialization of continuous flow microwave processing for foods and beverages. *Food Engineering Innovations Across the Food Supply Chain*, 149–162.
54. Szadzińska, J., Dominik, M., Andrzej, P., Grzegorz, M., Reihaneh, P., & Abdolreza, K., (2019). Ultrasound-and microwave-assisted intermittent drying of red beetroot. *Drying Technology*.
55. Tacon, A. G. J., Daniel, L., & Marc, M., (2020). Fish for health: Improved nutritional quality of cultured fish for human consumption. *Reviews in Fisheries Science & Aquaculture, 28*(4), 449–458.
56. Viji, P., Jesmi, D., & Ravishankar, C. N., (2022). Research developments in the applications of microwave energy in fish processing: A review. *Trends in Food Science & Technology*.
57. Wei, L., Weiqiang, Q., Yingshan, J., Ruilin, Z., KyungHo, R., Yudong, C., & Yinzhe, J., (2020). Effects of different drying methods on quality changes and energy characteristics of tilapia fillets. *Journal of Microwave Power and Electromagnetic Energy, 54*(3), 186–209.
58. Wu, X., Yang, Z., David, B. H., Pei, C., & Pamela, R. P., (2019). Effects of domestic cooking on flavonoids in broccoli and calculation of retention factors. *Heliyon, 5*(3), e01310.
59. Xu, W., Xiaohuang, C., Guanyu, Z., Ziming, X., & Dongyun, W., (2020). Effect of temperature difference on the aroma and quality of carrots processed through microwave drying combined with hot air drying. *Food and Bioproducts Processing, 120*, 58–68.
60. Xue, Q., Changhu, X., Donglei, L., Yunqi, W., Shijie, B., Zihao, W., & Haijin, M., (2021). Comprehensive investigation into quality of pasteurized *Oncorhynchus keta* Walbaum fillets and non-thermal effects of microwave. *LWT, 146*, 111466.
61. Zhong, X., Muhammad, S., Dalbir, S. S., Bruce, H., Kirk, D. D., & Eva, A., (2017). Effect of microwave steamable bag design on the preservation of ascorbic acid and antioxidant capacity and on the physical properties of cooked frozen vegetables: A case study on broccoli (*Brassica oleracea*). *LWT-Food Science and Technology, 83*, 165–171.

CHAPTER 5

EFFECT OF OZONE PROCESSING PARAMETERS ON THE COLOR PROFILE OF FOOD

PRATISHTHA VERMA,[1] ARSHDEEP SINGH,[2] and GURU YESHWANTH SUBBIAH PRABHAKARAN[3]

[1]*Department of Dairy and Food Science, South Dakota State University, Brookings, South Dakota, USA*

[2]*Division of Food, Nutrition and Exercise Sciences, University of Missouri, Columbia, Missouri*

[3]*Department of Food Science, University of Wisconsin, Madison, Wisconsin, USA*

ABSTRACT

According to FAO estimates, globally, about one-third of the total food produced is wasted from the pre-harvest to the post-harvest stage due to improper farm practices, processing, storage, or transport. A significant portion of the post-harvest and processing losses may be avoided using efficient techniques. Ozonation is one such technology that holds great promise in reducing such losses. It is a cost-efficient and environmentally-friendly technology whose application is gaining momentum in the food industry. Ozone gas is beneficial because it has one and a half times the oxidation potential of chlorine and helps decompose ethylene, which is the primary culprit in the quick ripening of ethylene-sensitive horticultural produce. Its half-life in water is extremely short and does not leave any residue in the treated product. It has a short contact time and can be

Nonthermal Food Processing Technologies: Impact on Color Profile.
Kaavya Rathnakumar and R. Pandiselvam (Eds.)

applied in both aqueous and gaseous phases. It can decompose some of the significant pesticides, thus making foods all the safer. However, it has some drawbacks that can be overcome if properly used. The adverse effect of ozone on natural color pigments of foods is also one of its shortcomings. Several review articles have been previously published on different aspects of ozone applications. A special effort has been made in the present review to bring together comprehensive information on the effects of ozone on the natural coloring pigments of foods.

5.1 INTRODUCTION

Consumer perception is an essential aspect of food marketing as sensory properties such as appearance, color, taste, aroma, and texture of raw ingredients and food products determine the final quality and influence the acceptance of food products by consumers. The saying "Consumers buy with their eyes" implies that color is one of the physical characteristics of food products that attracts customers' attention. Therefore, color is a crucial factor that can determine the popularity of a food product. In today's food processing industry, the main objective is to ensure food safety and quality while preserving the natural color and pigment retention in the final product. Both thermal and nonthermal processing methods used in the production of food products can affect their color. Thermal processing involves subjecting food to heat treatments to enhance its shelf life, but these treatments may have an impact on the nutritional value and sensory qualities of the food. Prolonged exposure to heat during thermal processing can cause noticeable changes to the food and reduce product quality [7, 8]. Various nonthermal technologies, such as ozone processing, pulsed electric field (PEF), cold plasma (CP), ultrasonication, etc., are available to process food in a very short time while affecting the sensory and textural properties favorably or unfavorably [9, 10]. This chapter aims to describe the effect of ozone processing, one of the most commonly used nonthermal food processing technologies, on natural pigments responsible for food color. The four natural pigments responsible for color in food products: carotenoids, anthocyanins (ACNs), chlorophyll, and betalains, are discussed, along with the properties of ozone, and mechanisms and components of ozone processing are presented. Additionally, the effects of ozone processing on the four natural pigments are summarized.

5.2 PRINCIPLES OF OZONE PROCESSING

Ozone (O_3) is a triatomic molecule consisting of three oxygen atoms. Schoenbein discovered and named ozone in the year 1840 [11]. Ozone is a homonuclear molecule and a powerful oxidizing agent used for water and air treatments, especially in disinfection processes in the food industry. According to regulations set forth by the Food and Drug Administration (FDA) [12], ozone was classified as generally recognized as safe (GRAS) and was permitted for use in direct contact with food and food products [13, 14].

Ozone can be generated using electrolysis, thermal, radiochemical, chemical, UV radiation, or electric (corona) discharge methods [12–14]. Among these, UV radiation, usually at a wavelength of 188 nm, and the corona discharge technique can be employed to produce ozone and free radical oxygen.

In the UV method, ozone is generated by exposing oxygen to UV radiation at a wavelength of 140–190 nm. Some oxygen molecules undergo photodissociation, creating radical oxygen atoms that bind with other oxygen molecules to generate ozone. The Department of Occupational Safety and Health Administration (OSHA) regulates acceptable exposure levels in work environments, and these levels must be controlled and monitored in commercial manufacturing facilities. The use of this method is limited due to the associated high cost and poor ozone yield [12, 13].

Corona discharge is the most commonly used method. Despite its high electricity consumption, it is capable of producing the necessary quantities of ozone for commercial use [13]. In this method, oxygen is passed through two parallel or concentric electrodes – one high-voltage and the other low-voltage electrode, separated by a ceramic dielectric material. During the ozone generation process, a diatomic oxygen molecule splits and generates free radical oxygen, which then reacts with molecular oxygen to produce a triatomic ozone molecule [12, 15]. The generation of free radical oxygen requires a significant amount of energy. In the food processing industry, an ozone manufacturing plant typically consists of an oxygen source, ozone generators, an electric power source, a reactor, a cooling system, an excess gas removal unit, and an ozone analyzer [16–18].

5.3 APPLICATION OF OZONE TECHNOLOGY/PROCESSING

Ozone gas is applied in both the aqueous and gaseous phases in the food industry. It possesses a strong antimicrobial property and is primarily utilized to prolong the shelf life of food products [19–22]. Ozone treatments are widely employed in post-harvest handling and food processing facilities [22–26]. When deactivating microbial organisms, ozone eradicates their genetic material through two methods. Firstly, ozone oxidizes certain amino acids, the sulfhydryl group of enzymes, and proteins into smaller peptides. Secondly, ozone induces oxidation of polyunsaturated fatty acids into acids and peroxides. The oxidation of microbial genetic material ultimately leads to cell leakage and subsequent cell death. Ozone demonstrates effectiveness against various pathogenic bacteria, including *Salmonella, Escherichia coli, Listeria monocytogenes, Staphylococcus aureus, Bacillus cereus, Pseudomonas aeruginosa, Klebsiella* spp., etc. [27, 28]. Reports indicate that gram-negative bacteria are more susceptible to ozone compared to gram-positive bacteria [13].

5.3.1 AQUEOUS OZONE

In its aqueous form, ozone is utilized for water disinfection and purification. The stability of ozone is influenced by the pH of water. When the pH is below 7, ozone remains as a molecule and does not react with water. However, as the pH increases, the decay of ozone becomes much faster. At a pH greater than 7.5, ozone produces highly reactive free radicals such as hydroxyl, oxygen, and hydroxide ions after decomposition [16]. It is important to note that liquid ozone can be explosive at low temperatures [29]. The efficacy of ozone depends on its temperature in the aqueous medium. As the temperature increases, ozone stability decreases, resulting in a decrease in its biocidal potency in water [30]. Typically, ozone has a half-life of 30–40 minutes in water [23]. Additionally, ozone can be used for surface disinfection of aseptic food packaging materials [12].

5.3.2 GASEOUS OZONE

Gaseous ozone is considered relatively safer than liquid ozone. However, a high concentration (>10%) can cause an explosive chain reaction, resulting in the complete conversion of ozone into oxygen in the container [31]. Some research studies have shown that the efficacy of ozone treatments at different levels decreases as the pH of the ozone increases during post-harvest ozone treatments of fresh produce [32, 33].

Various studies have reported that gaseous ozone is less effective as an antimicrobial compound compared to liquid ozone [22, 34, 35]. The half-life of ozone in the air is reported to be 2–3 hours [23]. The antimicrobial activity of ozone gas is also affected by different levels of relative humidity (RH). Even at lower concentrations, as the RH increases, the efficacy of ozone gas against *E. coli* increases [36]. Increasing the RH can eventually enhance the effectiveness of ozone gas by shortening the treatment time (TT) [34]. Gaseous ozone can be used as an alternative to chemical treatment to prevent the growth of fungus, mycotoxins, and control insects in stored grains and other foods [12].

5.4 OVERVIEW OF FOOD COLORS – ANTHOCYANINS (ACNS), BETALAINS, CAROTENOIDS, AND CHLOROPHYLLS

The color of a food product affects food palatability and acceptance. All kinds of foods, especially fruits and vegetables, are naturally colored by four types of pigments: ACNs, betalains, carotenoids, and chlorophylls are responsible for green, yellow-orange-red, red-blue-purple, and red colors, respectively [37]. A summary of these natural pigments with their properties and factors influencing them is presented in Table 5.1. The chemical structures of ACNs, betalains, carotenoids, and chlorophylls are presented in Figures 5.1–5.7. These natural pigments end up in the food chain by directly incorporating food derived from plants or using plants as animal feed. The degradation of these natural color pigments is affected by extrinsic factors like temperature, light, pH, the presence of oxygen, enzymes, metal ions, and phenolic compounds [4].

TABLE 5.1 Properties of Natural Pigments (Anthocyanins, Betalains, Carotenoids, and Chlorophylls) as Affected by Different Conditions

Basic Properties	Natural Pigments			
	Anthocyanins	**Betalains**	**Carotenoids**	**Chlorophylls**
Chemical formula	Flavylium ($C_{15}H_{11}O^{+}$)	Betanin ($C_{14}H_{17}N_3O_7$)	β-Carotene ($C_{40}H_{56}$)	Chlorophyll-a ($C_{55}H_{72}O_5N_4Mg$)
		Vulgaxanthin-I ($C_{14}H_{17}N_3O_7$)	Capsaicin ($C_{18}H_{27}NO_3$)	Chlorophyll-b ($C_{55}H_{70}O_6N_4Mg$)

TABLE 5.1 *(Continued)*

Basic Properties	Natural Pigments			
	Anthocyanins	**Betalains**	**Carotenoids**	**Chlorophylls**
Solubility	Soluble in water	Soluble in water	Soluble in oils with some exceptions.	Soluble in oils and organic substances.
Response to light	Degradation in the presence of light during storage.	Light present at lower temperature degrades betalains, whereas fluorescent light had a negligible effect on betalains degradation.	Sensitive to light.	Sunlight is required for the photosynthesis process.
Effect of temperature	Depending on the temperature range, the stability of anthocyanins varies. Acylated-, glycosylated-, and methylated-anthocyanins are more stable than hydroxylated-anthocyanins at high temperatures.	Betalains are the most stable at refrigeration temperatures. The degradation levels are directly proportional to the rise in temperature.	Higher temperature results in the degradation of carotenoids.	At high temperatures, Mg^{+} (magnesium) is replaced by hydrogen ions.
Effect of pH	Below pH 2, red is a dominant color. From pH 3–6, color varies from colorless to yellow. At pH >7, the blue color is generated.	Below pH 3, the color is violet. Yellow color is better stable at pH ranging from 5–6. Above pH 7, the color shifts toward blue.	Neutral and basic pH conditions decrease the stability of carotenoids. It is comparably more stable in acidic conditions.	Under acidic conditions, chlorophyll changes into pheophytin, which gives food brown color.
Effect of the presence of oxygen.	Very sensitive to the presence of oxygen.	Betalains degrade in the presence of oxygen.	Highly sensitive to oxygen.	No effect.

Source: Adapted and modified from various sources [4, 37–41].

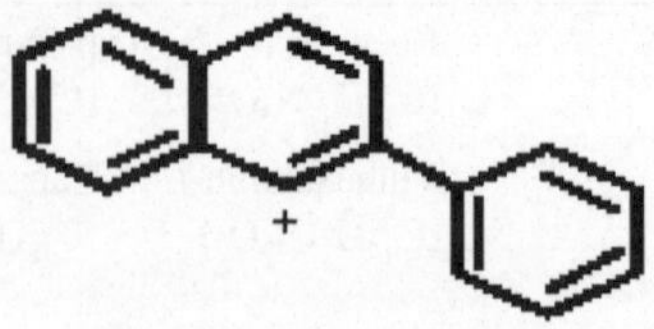

FIGURE 5.1 Flavylium ion (2-phenyl chromenylium).

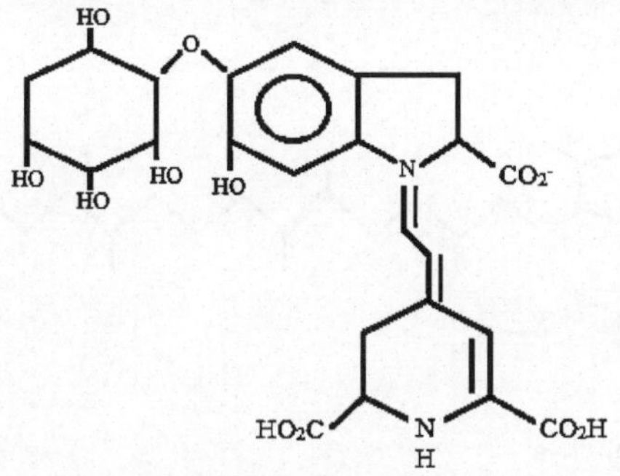

FIGURE 5.2 Betanin.

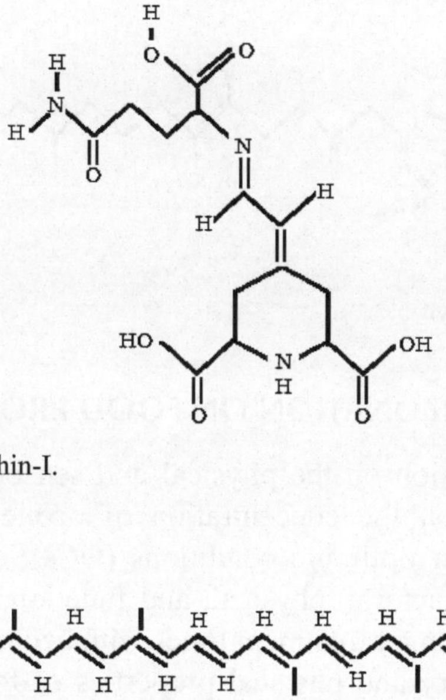

FIGURE 5.3 Vulgaxanthin-I.

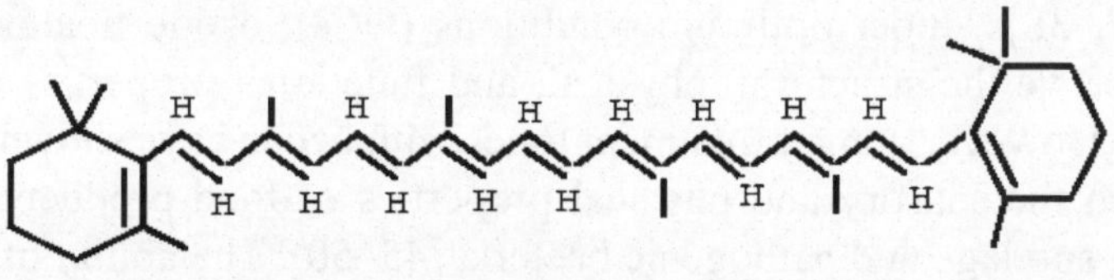

FIGURE 5.4 β-Carotene.

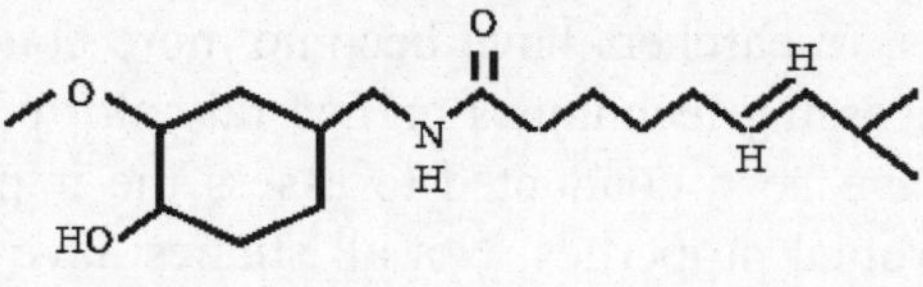

FIGURE 5.5 Capsaicin.

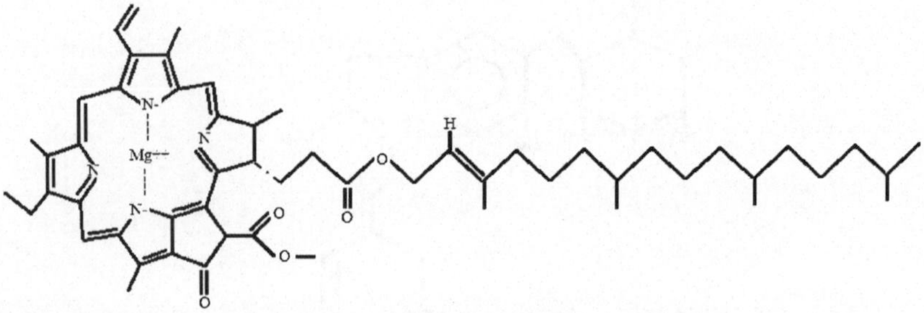

FIGURE 5.6 Chlorophyll-a.

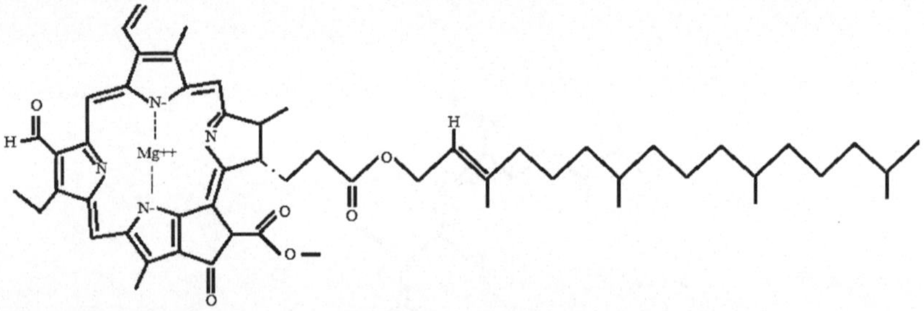

FIGURE 5.7 Chlorophyll-b.

5.5 EFFECT OF OZONATION ON FOOD PROPERTIES

The effect of ozonation on the physical and sensory properties of food depends primarily on the concentration of ozone and treatment time (TT) [37, 38]. Under optimum conditions (OCs), ozone treatments can help improve the structural, physical, and functional properties of foods [39–44]. However, longer treatment times with ozone have shown adverse effects on the sensory and physical properties of food products, despite reducing spoilage and pathogenic bacteria [45–50]. The ability of humans to categorize food based on touch, perception of color by the eyes, and organoleptic properties in the mouth and olfactory region make food texture and sensory qualities crucial features of foods.

In recent years, researchers have become more concerned about the effect of food processing techniques on natural color pigments in foods. Various studies have been conducted to assess the impact of ozonation on color and microbial properties. Not all studies have reported a negative impact of ozonation on the color profile of the food matrix. Table 5.2

TABLE 5.2 Effect of Ozone Treatments on the Color of Fruits, Vegetables, Meat, Eggs, and Milk

Commodity	Type	Ozone Treatment Parameters	Storage and Holding Parameters	Effect on Color	References
Fruits	Apple juice	An ozone generator was used to generate ozone gas. The concentration of ozone gas varied from 1 to 4.8% w/w of oxygen.	A 250 mL (milli liter) bubble diffuser was used to deliver ozone for treatments with a holding time of up to 10 min at 20 ± 0.5°C.	The color of ozone-treated samples was lighter as compared to untreated samples.	[45]
	Strawberry juice	An ozone generator with a varying concentration of ozone gas from 0–7.8% w/w of oxygen.	A 250 mL bubble diffuser was used to deliver ozone for treatments. Samples were treated with ozone solution for up to 10 min.	Mean redness (a*) and yellowness (b*) values in strawberry juice decreased in comparison to the control samples after treatment with ozone.	[47]
	Raspberry	An ozone generator with a glass-metal corona discharge lamp coated with an alloy of platinum-iridium dust. The ozone concentration varied from 0.3–0.9 mg/L.	The raspberry and control samples were treated for 60 and 120 min and then stored at 5°C for 8 days.	No statistical difference was reported in the fruit surface color of ozone-treated and control samples.	[51]
	Apple	An ozone generator producing 0.3–0.8 ppm of ozone solution.	The apple and control samples were treated in ozone solution and then stored for six months at 5°C.	The ozone-treated apple samples were significantly lighter in color than the control samples.	[52]
	Grape juice	An ozone generator prepared ozone solution with extrinsic parameters of ozone concentration of 0–7.8% w/w oxygen.	A 100 mL bubble diffuser was used to generate ozone. Samples were treated for up to 10 min.	The color of ozone-treated samples was degraded as compared to control samples because of the oxidation of anthocyanins.	[53]

TABLE 5.2 *(Continued)*

Commodity	Type	Ozone Treatment Parameters	Storage and Holding Parameters	Effect on Color	References
	Blackberry juice	Ozone solutions (0–7.8% w/w oxygen) were prepared using an ozonizer.	A 250 mL bubble diffuser was used to prepare ozone solutions, and treatment time was up to 10 min.	Anthocyanin content in ozone-treated samples was less as compared to control samples. An increase in lightness values was noted with ozone concentration and treatment time.	[46]
Vegetables	Broccoli	A prototype ozone generator was used to generate 0.04 μL/L ozone.	Ozone-treated and control samples were stored at 3 ± 1°C for 21 days and 10 ± 1°C for 7 days.	The color change was less conspicuous in ozone-treated samples as compared to the control. Increased base browning was also observed in ozone-treated samples.	[54]
–	Carrots	Two individual experiments were conducted with ozone gas and ozone dissolved in water. In the ozone gas experiment, an acrylic chamber of 0.075 m^3 with perforated shelves enabling a free flow of ozone gas was used. For ozone dissolved in water, a circular chamber of PVC containing 10 L of deionized water was used for treating carrots. The ozone concentration in gas was 0–5.0 mg/L and in water solution 0–10.0 mg/L.	Treatment time for gas and aqueous form varied from 9.5 to 110.5 min, and holding temperature ranged from 3.9 to 24.1°C. After treatment, control and treated samples were kept for five days at 18 ± 2°C temperature and 80 ± 5% RH.	The authors reported no significant changes in luminosity (L*), redness (a*), and yellowness (b*) of the ozone-treated samples as compared to control samples.	[55]

TABLE 5.2 *(Continued)*

Commodity	Type	Ozone Treatment Parameters	Storage and Holding Parameters	Effect on Color	References
–	Edamame (soybeans)	An ozone generator was used to prepare an aqueous ozone solution of 6 ppm/L. The ozonation time to prepare ozone solution varied from 0 to 15 min.	The treatment time varied from 0 to 24 min.	Luminosity (L*), redness (a*), and yellowness (b*) analysis showed no statistical difference in control and ozone-treated samples.	[56]
–	Green pepper	A lab-scale ozone generator was used to generate a 2-ppm ozone solution.	Pepper and control samples were treated with ozone water at 8 ± 1°C for 10 and 45 min.	Ozone-treated samples showcased a higher L* value as compared to control samples. Ozone-treated peppers had a less intense green color.	[57]
–	*Toona sinensis*	Ozone gas was generated using a laboratory corona discharge generator. The ozone treatment was carried out in a 2 m^3 chamber at a concentration of 4.28 mg/m^3 for 30 min and samples were sealed in a polyethylene bag.	The ozone-treated samples in polythene bags were stored at 1 ± 0.5°C and 90 ± 5% RH for 30 days. Samples were observed after 0, 10, 20, and 30 days.	Ozone and polyethylene treatments resulted in maintaining the green color of *Toona sinensis.*	[58]
–	Romaine lettuce	Ozone was generated using an air-cooled corona discharger. The concentration of ozone solution was 5 ppm.	The samples were treated with ozone solution for 0, 3, 6, 9, 12, and 15 min.	Samples treated with ozone lost their green color and appeared translucent.	[59]
Meat	Chicken	An ozone generator provided ozone gas with concentrations of 2, 5, and 10 mg/L.	Samples were exposed to various concentrations of ozone gas for 1 hour, sealed in a polyethylene/polyamide bag, and tested at 0, 2, 4, 6, 8, 10, 12, 14, and 16 days of storage.	No significant differences in control and ozone-treated samples were recorded in L*, a*, and b* values.	[60]

TABLE 5.2 *(Continued)*

Commodity	Type	Ozone Treatment Parameters	Storage and Holding Parameters	Effect on Color	References
–	Ground beef	Ozone solution (1%) was generated into tap water using an ozone generator. Ozone treatments were used in combination with 0.5% w/v cetylpyridinium chloride solution and 5% v/v acetic acid solution.	Samples were placed in ozone solution for 15 min at 7.2°C and given I min to drip dry, followed by placing the samples into a clean meat container containing 400 mL of cetylpyridinium chloride or acetic acid.	The authors reported that samples treated with ozone had lighter colors than the control samples. The change in color differences may be attributed to the lower pH of treated samples as compared to control samples.	[61]
–	Duck breast	An ozone generator was used to produce ozone gas in a chamber with a continuous flow of ozone ($10*10^{-6}$ kg ozone/m^3/hour).	A 20 g sample was placed in the ozone chamber and was set on for 15 min and off for 105 min, and this cycle continued for a 3-day storage period.	The ozone-treated samples showed a reduced redness. Ozone facilitated an increasing degree of L*, b*, and h? (hue-angle) value on the duck surface during storage.	[25]
–	Fish (Tilapia)	Aqueous ozone solution of 6 ppm was prepared using an ozone generator.	The samples were ozone treated for 1 hour, and then the samples were stored in water (20°C) before refrigeration at 0 and 5°C. The samples were analyzed after 0, 4, 8, 11, 14, 18, 22, 25, and 30 days.	The sample results were compared to control samples. In control samples, eye color was dull due to storage, whereas in ozone-treated fish, eye color was as fresh as that of a live fish.	[62]
Eggs	Whole shell eggs	Ozone gas (9.7% w/w ozone in oxygen) was generated using a corona discharge from very dry compressed clean oxygen.	Whole shell eggs were heated prior to ozone treatment at 57°C in a water bath and transferred in a chamber filled with ozone at 10–12 psig(pounds per square inch gauge) for 40 min.	Authors reported no color change when ozone-treated eggs were compared with control samples.	[63]

TABLE 5.2 *(Continued)*

Commodity	Type	Ozone Treatment Parameters	Storage and Holding Parameters	Effect on Color	References
–	Egg yolk gels	An ozone generator generated ozone gas at a rate of 1 g/h.	100 mL of liquid egg yolk was treated with ozone gas by direct contact, and the egg yolk was stirred at low speed. The sample treatment times were 0, 10, 20, 30, and 40 min.	L* and b* values of ozone-treated egg yolk gels were more significant than the control sample. On the other hand, a* values of ozone-treated egg yolk gels were slightly less than the control sample.	[64]
Milk	Milk and whey concentrate.	An ozone generator was used to produce ozone gas and was connected to a flask containing the sample. The ozone generator supplied the flask with 3.5 g ozone/h.	Ozone treatment was applied to skim milk and whey concentrate for 5-, 10-, 15-, 30-, or 60-min. Samples that were not treated served as controls. The samples were maintained at 20°C during the treatments.	For milk concentrate samples, ozone treatments reduced the samples' yellowness and greenness. Higher lightness and lower yellowness can be attributed to carotenoid denaturation because of ozone treatments. On the other hand, ozone treatments increased lightness and greenness in whey concentrate samples.	[65]
–	Whey powder	Ozone gas was generated using an ozone generator and fed into a flask containing whey powder samples. An ozone generator with 3.5 g ozone/g capacity was used in the experiment.	Whey concentrate samples were treated under a continuous flow of ozone gas for 30 min varying up to 120 min at 20°C. Untreated whey powders served as controls. Following ozonation, whey concentrate samples were spray-dried to get whey powders. A chroma meter was used to determine the effect of ozone treatments on whey powder samples.	Authors reported an increased L* causing higher whiteness compared to control in 120 min treated samples, whereas L* remained comparable in samples treated for 30–90 mins. Similar trends were observed for a* values where ozone treatments resulted in lower a* value for 120 min than control samples. During treatment, ozonation caused carotenoid degradation leading to a higher L* value and lower a* value in the samples.	[43]

provides a summary of the effects of ozone treatments on different food products. In these experiments, researchers primarily used ozone in gaseous and aqueous forms to control microbial spoilage of food commodities. The researchers examined the effect of ozone on color profile and other physiochemical properties, in addition to studying microbial spoilage.

5.6 CHALLENGES AND FUTURE OPPORTUNITIES ASSOCIATED WITH OZONE TECHNOLOGY

Ozone is hazardous to humans with long exposure at concentrations above 4 ppm (parts per million) [66]. The federally mandated eight-hour workday ozone gas exposure limit in workplaces is 0.1 ppm [67]. The National Institute of Occupational Safety and Health (NIOSH) advises against ever exceeding the upper limit of 0.1 ppm. To ensure worker comfort and safety, each system design should be thoroughly examined [66]. Personal protective equipment such as gloves, goggles, face shields, fire-resistant lab coats, closed-toed shoes, and long pants must always be worn when handling the equipment. Since the gas is highly explosive at higher concentrations, heat, flame, or combustible materials should not be used near the ozone generator [68]. The gas has a pungent and disagreeable odor at concentrations of 1 ppm and irritates the human eyes and throat [66].

Ozone gas is highly corrosive to equipment. In the presence of 0.1 ppm of dissolved ozone in water, the effectiveness of two corrosion inhibitors, I-hydroxyethylidene diphosphonate and tolyltriazole, decreased in both soft and hard water, leading to pitting on the carbon steel [69]. Varga & Szigeti [70] reported that particular care must be taken to protect equipment when pulsed ozonation is applied to dairy chilling water systems that contain copper or carbon steel parts. The corrosion resistance properties of stainless steel Grade 304 and 316 to ozone are good and excellent, respectively [71]. Ozone damages rubber and other polymers used for gaskets and O-rings [72].

Ozone is very unstable in water and breaks down quickly to oxygen. The low stability of ozone makes it difficult and impractical to maintain a sufficient amount of dissolved ozone for microbiological disinfection under remote ozone production conditions. However, the low stability of ozone is considered advantageous for its use as a disinfectant [66].

The half-life of ozone is extremely short (< 1 minute) in processing water that contains suspended soil and organic matter. In systems where water is recycled or the source water is of poor quality, pre-conditioning or filtration of the water before ozonation is necessary [67]. In systems with adequate filtration, a 3 to 4 log reduction in microbial load can be expected with short treatment exposure [66].

Phytotoxicity symptoms or limited product injury in the form of fruit, rachis, or calyx damage or desiccation have been observed in ozonated air in table grapes and tomatoes [66]. The cuticle of some apple varieties turned sticky and varnish-like when treated with ozone [67]. It is possible that the sensitivity of the produce to ozone may be crop or variety specific and thus may require experimentation before the technique is adopted.

Penetration of ozone in fruit containers is another major issue. Ozone entry in ordinary fruit packages is poor, but satisfactory gas entry occurs in packing cases with large vent holes or open tops. Plastic bags or fiberboard cartons allowed 0.1 ppm or less penetration compared to 0.6 ppm into returnable plastic containers (RPCs) with large vents [73]. More suppression of fungal growth on product surfaces in RPCs was recorded than in packages with more restrictive air-exchange openings [66].

Ozone reacts rapidly with exposed oxidizable plant materials but does not efficiently penetrate cell walls, natural openings, or wounds. Since infections typically occur within or beneath the plant surface, it is ineffective in reducing microbial decay once an infection has set in [66].

The activity of ozone decreases in dry air and at higher temperatures. Lie & Wang [74] found that 200 ppm of ozone was required to kill spores of molds and sour rot of citrus in humid air at a relative humidity (RH) of about 95% at 5°C, whereas a 5–10 times higher dose of ozone was needed in dry air holding 35% RH. Temperature also affects the action of ozone. High temperature (40°C) tends to be less suited since the solubility of ozone is low (0.112 liter per liter of water), as compared to the high solubility of 0.64 liter per liter of water at 0°C [75]. The effectiveness of ozone is also influenced by other factors, such as the culture's physiological state, the medium's pH, and the presence of additives [76].

Ozone applications have an adverse effect on the color of some fruits and vegetables, such as iceberg lettuce [77], carrots [78], apple, and blackberry juice [45, 47, 79]. However, generally no adverse effect has been noticed in the majority of crops, such as broccoli until day 9 [80, 81], lettuce [82, 83], potato strips [84], cantaloupe [85], pineapple [86], etc. Degradation of the color parameter has also been reported in some meat products [25, 87], though most other meat products do not have associated problems. Prabha & co-workers [88] have reported that applying ozone at large concentrations for efficient disinfection may alter the sensory qualities of food products. Some effects of ozone on the color parameters of certain foods are shown in Table 5.2.

In the early stages of mass food production and processing, food quality and safety were the most significant aspects. The impact of thermal or nonthermal processing techniques on sensory and physical properties was secondary. With the improvement of food production and processing techniques over time and with stricter standards for quality assurance, sensory and physical properties of food have become among the most influential factors in product sales [89]. Looking ahead, ozone processing for food products needs to address consumers' concerns regarding the oxidation of lipids, including phospholipids and cholesterol in the foods, which can result in color loss, production of off-flavors, and damage to the nutritional value of food.

5.7 SUMMARY

Ozone applications have been found to have adverse effects on the color attributes of some food products, but overall, the advantages of the technology outweigh the disadvantages. The limitations of the technique in preserving the natural color of fruits, vegetables, or other foods may be overcome by using a lower dose or TT of ozone gas or by implementing further refinements in technology.

KEYWORDS

- **aqueous ozone**
- **betalains**
- **carotenoids**
- **gaseous ozone**
- **natural color pigments**
- **nonthermal processing**
- **ozonation**
- **post-harvest practices**

REFERENCES

1. Sandoval, J. R. M., Rosas, M. E. M., Sandoval, E. M., & De Ávila, H. C., (2018). Colorimetry and image processing. In: Travieso-Gonzalez, C. M., (ed.), *Colorimetry and Image Processing* (pp. 61–78). In Tech Publishing: Rijeka, Croatia.
2. Mielby, L. A., Wang, Q. J., Jensen, S., Bertelsen, A. S., Kidmose, U., Spence, C., & Byrne, D. V., (2018). See, feel, taste: The influence of receptacle color and weight on the evaluation of flavored carbonated beverages. *Foods*, *7*(8). https://doi.org/10.3390/FOODS7080119.
3. Aghajanzadeh, S., Ziaiifar, A. M., & Verkerk, R., (2021). *Effect of Thermal and Non-Thermal Treatments on the Color of Citrus Juice: A Review*. https://doi.org/10.1080/87559129.2021.2012799.
4. Nowacka, M., Dadan, M., Janowicz, M., Wiktor, A., Witrowa-Rajchert, D., Mandal, R., Pratap-Singh, A., & Janiszewska-Turak, E., (2021). Effect of nonthermal treatments on selected natural food pigments and color changes in plant material. *Compr. Rev. Food Sci. Food Saf.*, *20*(5), 5097–5144. https://doi.org/10.1111/1541–4337.12824.
5. Delgado-Vargas, F., Jiménez, A. R., Paredes-López, O., & Francis, F. J., (2010). *Natural Pigments: Carotenoids, Anthocyanins, and Betalains — Characteristics, Biosynthesis, Processing, and Stability,* *40*(3), 173–289. https://doi.org/10.1080/10408690091189257.
6. Jadhav, H. B., Annapure, U. S., & Deshmukh, R. R., (2021). Non-thermal technologies for food processing. *Front. Nutr.*, *8*, 248. https://doi.org/10.3389/FNUT.2021.657090/BIBTEX.
7. Pereira, R. N., & Vicente, A. A., (2010). Environmental impact of novel thermal and non-thermal technologies in food processing. *Food Res. Int.*, *43*(7), 1936–1943. https://doi.org/10.1016/J.FOODRES.2009.09.013.
8. Chacha, J. S., Zhang, L., Ofoedu, C. E., Suleiman, R. A., Dotto, J. M., Roobab, U., Agunbiade, A. O., Duguma, H. T., Mkojera, B. T., Hossaini, S. M., et al., (2021). Revisiting non-thermal food processing and preservation methods—action mechanisms, pros and cons: A technological update (2016–2021). *Foods*, *10*(6), 1430. https://doi.org/10.3390/FOODS10061430.
9. Choudhary, R., & Bandla, S., (2012). Ultraviolet pasteurization for food industry. *Int. J. Food Sci. Nutr. Eng.*, *2*(1), 12–15. https://doi.org/10.5923/J.FOOD.20120201.03.
10. Thirumdas, R., Sarangapani, C., & Annapure, U. S., (2015). Cold plasma: A novel non-thermal technology for food processing. *Food Biophys.*, *10*(1), 1–11. https://doi.org/10.1007/S11483-014-9382-Z/FIGURES/3.
11. Rubin, M. B., (2001). The history of ozone. The Schönbein period, 1839–1868. *Bull. Hist. Chem.*, *26*, 40–56.
12. McHugh, T., (2015). *Ozone Processing of Foods and Beverages.* Food Technology Magazine.
13. Miller, F. A., Silva, C. L. M., & Brandão, T. R. S., (2013). A review on ozone-based treatments for fruit and vegetables preservation. *Food Eng. Rev.*, *5*(2), 77–106. https://doi.org/10.1007/s12393-013-9064-5.

14. Tzortzakis, N., & Chrysargyris, A., (2017). Postharvest ozone application for the preservation of fruits and vegetables. *Food Rev. Int.*, *33*(3), 270–315. https://doi.org/10.1080/87559129.2016.1175015.
15. Guzel-Seydim, Z. B., Greene, A. K., & Seydim, A. C., (2004). Use of ozone in the food industry. *LWT – Food Sci. Technol.*, *37*(4), 453–460. https://doi.org/10.1016/j.lwt.2003.10.014.
16. Brodowska, A. J., Nowak, A., & Śmigielski, K., (2017). *Ozone in the Food Industry: Principles of Ozone Treatment, Mechanisms of Action, and Applications: An Overview, 58*(13), 2176–2201. https://doi.org/10.1080/10408398.2017.1308313.
17. Niveditha, A., Pandiselvam, R., Prasath, V. A., Singh, S. K., Gul, K., & Kothakota, A., (2021). Application of cold plasma and ozone technology for decontamination of *Escherichia coli* in foods: A Review. *Food Control*, *130*, 108338. https://doi.org/10.1016/J.FOODCONT.2021.108338.
18. Brueggemann, N., Puehmeier, T., Fiekens, R., Richardt, F. J., & Salvermoser, M., (2017). Cooling conditions of ozone generators. *Ozone Sci. & Eng.*, *39*(3), 196–201.
19. Victorin, K., (1992). Review of the genotoxicity of ozone. *Mutat. Res. Genet. Toxicol.*, *277*(3), 221–238. https://doi.org/10.1016/0165-1110(92)90045-B.
20. Yeoh, W. K., Ali, A., & Forney, C. F., (2014). Effects of ozone on major antioxidants and microbial populations of fresh-cut papaya. *Postharvest Biol. Technol.*, *89*, 56–58. https://doi.org/10.1016/J.POSTHARVBIO.2013.11.006.
21. Kim, J. G., Yousef, A. E., & Dave, S., (1997). Application of ozone for enhancing the microbiological safety and quality of foods: A review. *J. Food Prot.*, *62*(9), 1071–1087. https://doi.org/10.4315/0362-028X-62.9.1071.
22. Sarron, E., Gadonna-Widehem, P., & Aussenac, T., (2021). Ozone treatments for preserving fresh vegetables quality: A critical review. *Foods*, *10*(3). https://doi.org/10.3390/FOODS10030605.
23. Botondi, R., Barone, M., & Grasso, C., (2021). A review into the effectiveness of ozone technology for improving the safety and preserving the quality of fresh-cut fruits and vegetables. *Foods*, *10*(4), 748.
24. Karaca, H., & Velioglu, Y. S., (2007). *Ozone Applications in Fruit and Vegetable Processing, 23*(1), 91–106. https://doi.org/10.1080/87559120600998221.
25. Muhlisin, M., Utama, D. T., Lee, J. H., Choi, J. H., & Lee, S. K., (2016). Effects of gaseous ozone exposure on bacterial counts and oxidative properties in chicken and duck breast meat. *Korean J. Food Sci. Anim. Resour.*, *36*(3), 405.
26. Cano, C., Meneses, Y., & Chaves, B. D., (2019). Ozone-based interventions to improve the microbiological safety and quality of poultry carcasses and parts: A review. *J. Food Prot.*, *82*(6), 940–947. https://doi.org/10.4315/0362-028X.JFP-18-489.
27. Fontes, B., Cattani, H. A. M., De Souza, B. G., Costa, S. F., Van, D. H. I. M., Levin, A. S., & Rasslan, S., (2012). Effect of low-dose gaseous ozone on pathogenic bacteria. *BMC Infect. Dis.*, *12*(1), 1–6. https://doi.org/10.1186/1471-2334-12-358/TABLES/2.
28. Rangel, K., Cabral, F. O., Lechuga, G. C., Carvalh, J. P. R. S., Villas-Bôas, M. H. S., Midlej, V., & De-Simone, S. G., (2022). Detrimental effect of ozone on pathogenic bacteria. *Microorganisms*, *10*(1). https://doi.org/10.3390/MICROORGANISMS10010040.
29. Toby, S., (1984). Chemiluminescence in the reactions of ozone. *Chem. Rev.*, *84*(3), 277–285. https://doi.org/10.1021/CR00061A003.

30. Rice, R. G., Robson, C. M., Miller, G. W., & Hill, A. G., (1981). Uses of ozone in drinking water treatment. *J. Am. Water Works Assoc.*, *73*(1), 44–57. https://doi.org/10.1002/J.1551-8833.1981.TB04637.X.
31. Koike, K., Nifuku, M., Izumi, K., Nakamura, S., Fujiwara, S., & Horiguchi, S., (2005). Explosion properties of highly concentrated ozone gas. *J. Loss Prev. Process Ind.*, *18*(4–6), 465–468. https://doi.org/10.1016/J.JLP.2005.07.020.
32. Jamil, A., Farooq, S., & Hashmi, I., (2017). *Ozone Disinfection Efficiency for Indicator Microorganisms at Different pH Values and Temperatures, 39*(6), 407–416. https://doi.org/10.1080/01919512.2017.1322489.
33. Patil, S., Valdramidis, V. P., Cullen, P. J., Frias, J., & Bourke, P., (2010). Inactivation of *Escherichia coli* by ozone treatment of apple juice at different PH levels. *Food Microbiol.*, *27*(6), 835–840. https://doi.org/10.1016/J.FM.2010.05.002.
34. Marino, M., Maifreni, M., Baggio, A., & Innocente, N., (2018). Inactivation of foodborne bacteria biofilms by aqueous and gaseous ozone. *Front. Microbiol.*, *9*, 2024. https://doi.org/10.3389/FMICB.2018.02024/BIBTEX.
35. Kim, J. G., Yousef, A. E., & Chism, G. W., (2007). Use of ozone to inactivate microorganisms on lettuce. *J. Food Saf.*, *19*(1), 17–34. https://doi.org/10.1111/J.1745-4565.1999.TB00231.X.
36. Pironti, C., Moccia, G., Motta, O., Boccia, G., Franci, G., Santoro, E., Capunzo, M., De Caro, F., (2021). The influence of microclimate conditions on ozone disinfection efficacy in working places. *Environ. Sci. Pollut. Res. Int.*, *28*(45), 64687. https://doi.org/10.1007/S11356-021-15457-2.
37. Rodriguez-Amaya, D. B., (2016). Natural food pigments and colorants. *Curr. Opin. Food Sci.*, *7*, 20–26. https://doi.org/10.1016/J.COFS.2015.08.004.
38. Misawa, N., (2010). Carotenoids. *Compr. Nat. Prod. II*, 733–753. https://doi.org/10.1016/B978-008045382-8.00009-5.
39. Rodriguez-Amaya, D. B., (2019). Update on natural food pigments: A mini-review on carotenoids, anthocyanins, and betalains. *Food Res. Int.*, *124*, 200–205. https://doi.org/10.1016/J.FOODRES.2018.05.028.
40. Saini, R. K., Nile, S. H., & Park, S. W., (2015). Carotenoids from fruits and vegetables: Chemistry, analysis, occurrence, bioavailability, and biological activities. *Food Res. Int.*, *76*, 735–750. https://doi.org/10.1016/J.FOODRES.2015.07.047.
41. Zhou, Y., & Singh, B. R., (2004). Effect of light on anthocyanin levels in submerged, harvested cranberry fruit. *J. Biomed. Biotechnol.*, *2004*(5), 259–263. https://doi.org/10.1155/S1110724304403027.
42. Segat, A., Misra, N. N., Fabbro, A., Buchini, F., Lippe, G., Cullen, P. J., & Innocente, N., (2014). Effects of ozone processing on chemical, structural and functional properties of whey protein isolate. *Food Res. Int.*, *66*, 365–372. https://doi.org/10.1016/J.FOODRES.2014.10.002.
43. Sert, D., & Mercan, E., (2022). The impact of ozone treatment on whey concentrate on the flow behavior, functional and microbiological characteristics of whey powder. *Int. Dairy J.*, *134*, 105447. https://doi.org/10.1016/J.IDAIRYJ.2022.105447.
44. Okpala, C., (2014). Changes of physico-chemical properties of sequential minimal ozonetreated ice-stored pacific white shrimp (*Litopenaeus vannamei*). *Curr. Nutr. Food Sci.*, *10*(3), 218–227. https://doi.org/10.2174/1573401310666140306003557.

45. Torres, B., Tiwari, B. K., Patras, A., Wijngaard, H. H., Brunton, N., Cullen, P. J., O'Donnell, C. P., (2011). Effect of ozone processing on the cooler, rheological properties and phenolic content of apple juice. *Food Chem.*, *124*(3), 721–726. https://doi.org/10.1016/j.foodchem.2010.06.050.
46. Tiwari, B. K., O'Donnell, C. P., Muthukumarappan, K., & Cullen, P. J., (2009). Anthocyanin and color degradation in ozone treated blackberry juice. *Innov. Food Sci. Emerg. Technol.*, *10*(1), 70–75. https://doi.org/10.1016/J.IFSET.2008.08.002.
47. Tiwari, B. K., O'donnell, C. P., Patras, A., Brunton, N., & Cullen, P. J., (2009). Effect of ozone processing on anthocyanins and ascorbic acid degradation of strawberry juice. *Food Chem.*, *113*(4), 1119–1126.
48. C, K., Nur, M., Sabrina, D. T., & Al-Baarri, A. N., (2022). The effect of ozone on the physical and chemical qualities of red chili pepper (*Capsicum anuum* L.). *J. Food Nutr. Res.*, *10*(4), 287–292. https://doi.org/10.12691/JFNR-10-4-4.
49. Öztürk, G. F., Özdestan, O., & Ünal, M. K., (2020). Effect of ozone treatment on the physical, microbiological, and sensorial properties of Spanish-style table olives. *Grasas y Aceites*, *71*(1), e348–e348. https://doi.org/10.3989/GYA.0103191.
50. Karaca, H., & Velioglu, Y. S., (2014). Effects of ozone treatments on microbial quality and some chemical properties of lettuce, spinach, and parsley. *Postharvest Biol. Technol.*, *88*, 46–53. https://doi.org/https://doi.org/10.1016/j.postharvbio.2013.09.003.
51. Onopiuk, A., Półtorak, A., Moczkowska, M., Szpicer, A., & Wierzbicka, A., (2017). The impact of ozone on health-promoting, microbiological, and color properties of *Rubus ideaus* raspberries. *15*(4), 563–573. https://doi.org/10.1080/19476337.2017.1317669; http://mc.manuscriptcentral.com/tcyt (accessed on 23 January 2024).
52. Juhnevica-Radenkova, K., Radenkovs, V., Kundzins, K., & Seglina, D., (2019). Effect of ozone treatment on the microstructure, chemical composition and sensory quality of apple fruits. *Food Sci. Technol. Int.*, *25*(3), 252–267. https://doi.org/10.1177/1082013218815285.
53. Tiwari, B. K., O'Donnell, C. P., Patras, A., Brunton, N., & Cullen, P. J., (2009). Anthocyanins and color degradation in ozonated grape juice. *Food Chem. Toxicol.*, *47*(11), 2824–2829. https://doi.org/10.1016/J.FCT.2009.09.001.
54. Skog, L. J., & Chu, C. L., (2001). Effect of ozone on qualities of fruits and vegetables in cold storage. *Can. J. Plant Sci.*, *81*(4), 773–778. https://doi.org/10.4141/P00-110.
55. De Souza, L. P., Faroni, L. R. D., Heleno, F. F., Cecon, P. R., Gonçalves, T. D. C., Da Silva, G. J., & Prates, L. H. F., (2018). Effects of ozone treatment on postharvest carrot quality. *LWT*, *90*, 53–60.
56. Yudiastuti, S. O. N., Wijaya, R., & Budiati, T., (2021). The effect of ozonation time and contact time of edamame washing on color changes using the continuous type ozone washing method. *IOP Conf. Ser. Earth Environ. Sci.*, *672*(1), 012066. https://doi.org/10.1088/1755-1315/672/1/012066.
57. Özen, T., Koyuncu, M. A., & Erbaş, D., (2021). Effect of ozone treatments on the removal of pesticide residues and postharvest quality in green pepper. *J. Food Sci. Technol.*, *58*(6), 2186–2196. https://doi.org/10.1007/S13197-020-04729-3.
58. Lin, S., Chen, C., Luo, H., Xu, W., Zhang, H., Tian, J. J., Ju, R., & Wang, L., (2019). The combined effect of ozone treatment and polyethylene packaging on postharvest

quality and biodiversity of toona sinensis (A. Juss.) M. Roem. *Postharvest Biol. Technol.*, *154*, 1–10. https://doi.org/10.1016/J.POSTHARVBIO.2019.04.010.

59. Bermúdez-Aguirre, D., & Barbosa-Cánovas, G. V., (2013). Disinfection of selected vegetables under nonthermal treatments: Chlorine, acid citric, ultraviolet light, and ozone. *Food Control*, *29*(1), 82–90. https://doi.org/10.1016/J.FOODCONT.2012.05.073.
60. Gertzou, I. N., Karabagias, I. K., Drosos, P. E., & Riganakos, K. A., (2017). Effect of combination of ozonation and vacuum packaging on shelf-life extension of fresh chicken legs during storage under refrigeration. *J. Food Eng.*, *213*, 18–26. https://doi.org/10.1016/J.JFOODENG.2017.06.026.
61. Pohlman, F. W., Stivarius, M. R., McElyea, K. S., Johnson, Z. B., & Johnson, M. G., (2002). The effects of ozone, chlorine dioxide, cetylpyridinium chloride and trisodium phosphate as multiple antimicrobial interventions on microbiological, instrumental color, and sensory color and odor characteristics of ground beef. *Meat Sci.*, *61*(3), 307–313. https://doi.org/10.1016/S0309–1740(01)00198-X.
62. Gelman, A., Sachs, O., Khanin, Y., Drabkin, V., & Glatman, L., (2005). Effect of ozone pretreatment on fish storage life at low temperatures. *J. Food Prot.*, *68*(4), 778–784. https://doi.org/10.4315/0362-028X-68.4.778.
63. Kamotani, S., Hooker, N., Smith, S., & Lee, K., (2010). Consumer acceptance of ozone-treated whole shell eggs. *J. Food Sci.*, *75*(2). https://doi.org/10.1111/J.1750-3841.2009.01468.X.
64. Sun, Y., Wang, Q., Jin, H., Li, Z., & Sheng, L., (2021). Impact of ozone-induced oxidation on the textural, moisture, micro-rheology, and structural properties of egg yolk gels. *Food Chem.*, *361*, 130075. https://doi.org/10.1016/J.FOODCHEM.2021.130075.
65. Sert, D., & Mercan, E., (2021). Effects of ozone treatment to milk and whey concentrates on degradation of antibiotics and aflatoxin and physicochemical and microbiological characteristics. *LWT*, *144*, 111226. https://doi.org/10.1016/J.LWT.2021.111226.
66. Suslow, T., (2004). *Ozone Applications for Postharvest Disinfection of Edible Horticultural Crops*. UCANR Publications.
67. Smilanick, J. L., (2003). Use of ozone in storage and packing facilities. In: *Washington Tree Fruit Postharvest Conference* (pp. 1–10).
68. *Standard Operating Procedure Using the Ozone Generator*, (2010). UCLA Chem. BioChem.
69. Gan, F., Chin, D. T., & Meitz, A. K., (1996). *Effect of Ozone on Corrosion of Carbon Steel and Copper in the Presence of HEDP and TTA in Cooling Water*. Journal ID: ISSN 0001-2505.
70. Varga, L., & Szigeti, J., (2016). Use of ozone in the dairy industry: A review. *Int. J. Dairy Technol.*, *69*(2), 157–168.
71. *Ozone Resistant Materials*. https://www.oxidationtech.com/ozone/ozone-resistant-materials.html (accessed on 23 January 2024).
72. *Ozonation in Food Applications*. https://www.gov.mb.ca/agriculture/food-safety/at-the-food-processor/ozonation.html (accessed on 23 January 2024).
73. Palou, L., Smilanick, J. L., Crisosto, C. H., Mansour, M., & Plaza, P., (2003). Ozone gas penetration and control of the sporulation of *Penicillium digitatum* and

Penicillium italicum within commercial packages of oranges during cold storage. *Crop Prot.*, *22*(9), 1131–1134.
74. Li, C. S., & Wang, Y. C., (2003). Surface germicidal effects of ozone for microorganisms. *AIHA J.*, *64*(4), 533–537.
75. Chuwa, C., Vaidya, D., Kathuria, D., Gautam, S., Sharma, S., & Sharma, B., (2020). Ozone O_3 an emerging technology in the food industry. *Food Nutr. J.* https://doi.org/10.29011/2575-7091.100124.
76. Nath, A., Mukhim, K., Swer, T., Dutta, D., Verma, N., Deka, B. C., & Gangwar, B., (2014). A Review on application of ozone in the food processing and packaging. *J. Food Prod. Dev. Packag.*, 7–21.
77. Koseki, S., & Isobe, S., (2006). Effect of ozonated water treatment on microbial control and on browning of iceberg lettuce (*Lactuca sativa* L.). *J. Food Prot.*, *69*(1), 154–160.
78. Liew, C. L., & Prange, R. K., (1994). Effect of ozone and storage temperature on postharvest diseases and physiology of carrots (*Daucus Carota* L.). *J. Am. Soc. Hortic. Sci.*, *119*(3), 563–567. https://doi.org/10.21273/jashs.119.3.563.
79. Shah, N., Sulaiman, A., Sidek, N. S. M., & Supian, N. A. M., (2019). Quality assessment of ozone-treated citrus fruit juices. *Int. Food Res. J.*, *26*(5), 1405–1415.
80. Das, B. K., & Kim, J. G., (2010). Microbial quality and safety of fresh-cut broccoli with different sanitizers and contact times. *J. Microbiol. Biotechnol.*, *20*(2), 363–369.
81. Renumarn, P., Srilaong, V., Uthairatanakij, A., Kanlayanarat, S., & Jitareerat, P., (2014). The effects of immersion methods and concentration of ozonated water on the microbial counts and the quality and sensory attributes of fresh-cut broccoli. *Int. Food Res. J.*, *21*(2).
82. Beltrán, D., Selma, M. V., Marin, A., & Gil, M. I., (2005). Ozonated water extends the shelf life of fresh-cut lettuce. *J. Agric. Food Chem.*, *53*(14), 5654–5663.
83. Wang, J., Wang, S., Sun, Y., Li, C., Li, Y., Zhang, Q., & Wu, Z., (2019). Reduction of *Escherichia coli*O157:H7 and naturally present microbes on fresh-cut lettuce using lactic acid and aqueous ozone. *RSC Adv.*, *9*(39), 22636–22643. https://doi.org/10.1039/c9ra03544c.
84. Oner, M. E., Walker, P. N., & Demirci, A., (2011). Effect of in-package gaseous ozone treatment on shelf life of blanched potato strips during refrigerated storage. *Int. J. Food Sci. & Technol.*, *46*(2), 406–412.
85. Toti, M., Carboni, C., & Botondi, R., (2018). Postharvest gaseous ozone treatment enhances quality parameters and delays softening in cantaloupe melon during storage at 6° C. *J. Sci. Food Agric.*, *98*(2), 487–494.
86. Nur, A. M. P., Hairiyah, M., Wan, M. R. W. H., & Nur, I. M., (2010). Effect of ozonated water wash on quality of fresh-cut 'josephine' pineapple during storage. In: *VII International Pineapple Symposium* (Vol. 902, pp. 487–492).
87. Ayranci, U. G., Ozunlu, O., Ergezer, H., & Karaca, H., (2020). Effects of ozone treatment on microbiological quality and physicochemical properties of turkey breast meat. *Ozone Sci. & Eng.*, *42*(1), 95–103.
88. Prabha, V., Barma, R. D., Singh, R., & Madan, A., (2015). Ozone technology in food processing: A review. *Trends Biosci.*, *8*, 4031–4047.
89. Garber, Jr. L. L., Hyatt, E. M., Starr, Jr. R. G., (2000). The effects of food color on perceived flavor. *J. Mark. Theory Pract.*, *8*(4), 59–72.

CHAPTER 6

COLD PLASMA PROCESSING AND ITS IMPACT ON THE COLOR PROFILE OF FOOD PRODUCTS

SANGEETHA KARUNANITHI, U. GNANA MOORTHY ESWARAN, PROSHANTA GUHA, AND PREM PRAKASH SRIVASTAV

Department of Agricultural and Food Engineering, Indian Institute of Technology, Kharagpur, West Bengal, India

ABSTRACT

Cold plasma (CP) is the most emerging technology among the food industry, which also influences the food quality parameters, particularly color, after processing. Therefore, it is crucial to gain knowledge about the impact of CP on food color. This chapter focuses on the application of CP technology and its influences on the color profile of food products. It covers the basics of CP technology in food processing, types of CP, principles and equipment, factors affecting CP efficiency on color, and the application of CP in the food industry, including cereals, fruits, vegetables, meat, milk, spices, and liquid food processing. The chapter also discusses the principle mechanism of CP, which alters the color of foods during processing. The major mechanisms affecting the color of food, such as etching, oxidation, extraction, etc., are briefly explored. This information may be helpful for researchers, food processors, and policy makers in the development of processed foods using CP techniques.

Nonthermal Food Processing Technologies: Impact on Color Profile.
Kaavya Rathnakumar and R. Pandiselvam (Eds.)

6.1 INTRODUCTION

Food color is a noticeable factor that has a significant impact on consumer impression and, as a result, product success. The existence of pigments (synthetic or natural) or chemical reactions is the main sources of food color (enzymatic or non-enzymatic). The certain unfavorable color alteration in food products because of treatment conditions applied in the line of manufacturing leads to less desirability of consumer acceptability. Color is among the most crucial components with respect to vegetables and fruits, apart from other organoleptic characters like taste and texture, for food product satisfaction. Table 6.1 describes the natural color pigments responsible for the color of various food products, mainly fruits and vegetables. The presence of natural and manmade colors, as well as observed chemical reactions while processing, are important contributors to the distinctive color look of products [14]. Although color plays a significant characteristic parameter of nutritive in addition to functional alterations in food through handling, it is mostly employed as a quality control parameter by industries and universities as any color change offers information about the influence of processing parameters. The process involved high temperature is extensively utilized for preserving food through limiting pathogenic as well as contaminating microorganisms, despite a lot of drawbacks such as overcooking, textural damage, alternating sensory and flavor characteristics, nutritional value loss due to thermal exposure, etc. [75]. The food manufacturing industry is seeking innovative approaches to address the demand for safer, more nutritious, and "fresh-like" food quality parameters due to increasing consumer knowledge. An alternative emerging technology is required to potentially improve the quality of food while keeping technical expenses within reasonable limits [12]. This has created a significant need for creative and innovative food processing research. Over the last few decades, researchers have explored several innovative processing technologies to create safer and longer-lasting food with enhanced quality and nutritional content. Some of these technologies include high pressure processing (HPP), dielectric heating (microwave and radiofrequency heating), pulsed electric field (PEF), ohmic heating, pulsed light (PL), ultrasound, infrared heating, ozone processing, cold plasma (CP), and other alternative technologies that are being investigated with or without temperature involvement. In addition to that, the innovative methods should operate near ambient conditions, eliminating one or more of the negative consequences associated with traditional thermal processing. This benefit is driving the food industry's increasing implementation of these technologies. Cold plasma (CP) seems to be an innovative processing method employed in the food sector to assure food quality and safety among all developing nonthermal techniques. This technique has been increasingly applied in other sectors, such as decontaminating the surface of medical devices and preserving thermo-liable components in the electronics industry [44]. Researchers have been attracted to this technique due to its shorter processing times (caused by high ion temperature) and minimal thermal damage (caused by sudden cooling process following exposure) [15]. The reaction between reactive radicals produced by plasma and the bioactive compounds in sample tissue is an interesting phenomenon to be explored. However, the application of CP on food products produces both positive and slightly negative impacts on certain quality aspects, mainly color, which will be the focus of this particular chapter.

TABLE 6.1 Pigments Responsible for Food Color

Pigments	Color	Source
Carotenoids	Orange, yellow to brown, light red	Carrots, fruits, and vegetables
Chlorophyll (a, b, c, d)	Green	Leafy vegetable, herbs
Lycopene and lutein	Red	Tomato, corn, green vegetables
Phenolic: Anthocyanin, catechins, tannins	Yellow, red, blue, orange	Pomegranate, cocoa, tea, fruits
Betelains	Yellow, orange, red, violet	Beetroot, flower, fruits

6.2 COLOR MEASUREMENT

The CIE Lab (International Commission on Illumination Laboratory) scale is used to quantify color changes that occur in processed food samples by the L*, a*, and b* variables. In this scale, L* represents brightness or luminosity, a* represents the green (–) to red (+) scale, and b* indicates the blue (–) to yellow (+) scale, as shown in Figure 6.1. The variation of the color value before treatment (L*) and after treatment (Lo*) is necessary for calculating the overall color change (ΔE), whiteness index (WI),

Chroma (C*), as well as hue (H*), which are determined using Equations (1)–(4), respectively [39].

$$\Delta E = \sqrt{(L^* - L_o^*)^2 + (a^* - a_o{}^*)^2 + (b^* - b_o^*)^2} \quad (1)$$

$$WI = 100 - \sqrt{(100 - L^*)^2 + (a^*)^2 + (b^*)^2} \quad (2)$$

$$C^* = 100 - \sqrt{(a^*)^2 + (b^*)^2} \quad (3)$$

$$H^* = \tan^{-1}\left(\frac{b^*}{a^*}\right) \quad (4)$$

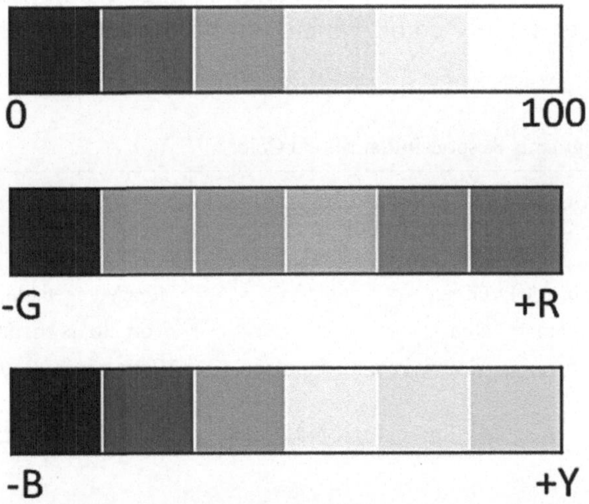

FIGURE 6.1 CIE color sale used in food industry.

6.3 COLD PLASMA (CP)

The commonly known states of matter are solid, liquid, and gas. In addition to these, plasma is considered the fourth state of matter. Solid, liquid, gas, and plasma are the four primary states of matter that can be formed by increasing the energy level. Plasma is defined as a "partially or completely ionized gas containing a plethora of highly reactive species including electrons, negative ions, positive ions, free radicals, excited or non-excited atoms, and photons at near ambient temperature" [13]. The term "plasma" was coined by Langmuir to describe the region where ions and electrons have a balanced charge. Electrons play a crucial role in the complex chemistry of plasma, involving the excitation and disintegration of atoms, development of free radicals, and maintenance of non-equilibrium states to keep the plasma active [38]. The gas used for the production of CP remains at ambient temperature because the electrical energy is focused solely on increasing the temperature of electrons rather than raising the temperature of the gas. Electricity, heat, radio waves, and microwaves (MWs) can all provide adequate energy for the ionization of gas molecules [50]. Despite the fact that plasma is generated by ionizing gas molecules, every species of plasma consists of both positive and negative charges, resulting in a quasi-neutral state. Natural phenomena such as stars, thunderstorms, as well as synthetic sources such as neon fluorescent lights, televisions working with plasma, and so on, all fall under the category of plasma. Plasma can be classified in a variety of ways, depending on whether it occurs naturally or is purposefully manufactured. The classification of plasma is shown in Table 6.2 [50, 69, 71].

TABLE 6.2 Classification of Plasma

Plasma		
Thermal/Equilibrium Plasma	**Nonthermal/Nonequilibrium Plasma**	
Hot plasma	Atmospheric pressure cold plasma	Vacuum or low-pressure cold plasma
Gas molecules, electrons as well as ions are thermodynamically equilibrium.	Gas molecules, electrons as well as ions are thermodynamically out of equilibrium.	
Temperatures: 20,000 K (100–150°C).	<60°C.	
Degree of ionization: High	Degree of ionization: Low	
$T_i \geq T_e$ where, T_e: Electron temperature, T_i: ion temperature.	$T_i < T_e$	

6.3.1 PRINCIPLE OF COLD PLASMA (CP)

Several types of actions occur concurrently or successively when CP species come in contact with a food surface. It produces various reactions such as surface etching, depolymerization, functionalization, deposition, structural modification, etc., to impart numerous unique properties to solid food products.

6.3.1.1 ETCHING

The main principle involved in the plasma process is surface etching, also called plasma etching. This plasma etching process is utilized for roughening the surface at a microscopic scale. The reactive plasma species bombard the surface, causing etching on exposed surface atoms and removing them into the gas phase, which cracks the surface of the materials [21]. The etching phenomena of plasma species on the surface of the material are shown in Figure 6.2. Plasma etching works by two mechanisms: (i) ion bombardment at the surface, hence removing the materials required to form cracks/trenches, and (ii) the gas monomers present in the reactor can interact among substrate molecules, forming volatile byproducts that will be removed [83]. This is the process of creating active sites over the substrate's surface (ablation of the surface) and volatile species that diffuse into the gas phase by combining reactive gas phase species within atoms or groups of atoms on the substrate.

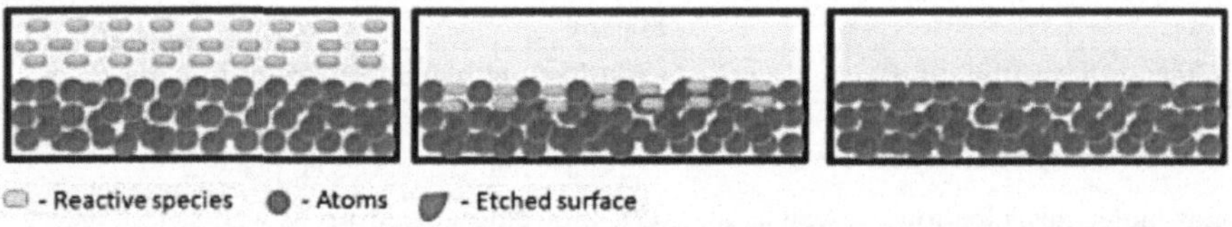

FIGURE 6.2 Mechanisms of etching by plasma species.

6.3.1.2 CROSS LINKING/POLYMERIZATION

Crosslinking/polymerization of starch creates polymeric structured chains through the interaction of high-energy electrons and reactive free radicals produced by plasma. Crosslinking occurs when a water molecule (H_2O) is removed, causing fragmentation and the formation of new linkages or chains (C-O-C) among the polymeric chains [37]. The glycosidic linkages are cleaved by hydroxyl radicals (*OH) generated from the breakdown of water molecules and oxygen present in the plasma generation [52].

6.3.1.3 DEPOLYMERIZATION

Ionization has been the most significant factor in the depolymerization mechanism, which results in starch modification. Plasma ion bombardment promotes the depolymerization of amylose and amylopectin starch structured chains into tiny particles. After depolymerization, maltose, maltotriose, and maltotetrose are prevalent final products that modify the functional properties of the treated sample [28].

6.3.2 PLASMA GENERATION SYSTEM

The changes of state of materials from solid to liquid to gas are well familiarized. However, increasing the energy level further in gas causes ionization with different gas molecules and develops another state of matter called plasma, as shown in Figure 6.3. The most common plasma generating systems are dielectric barrier discharges (DBD), corona discharges (CDs), plasma jets (PJs), and MW discharges, which are utilized in various industrial sectors. However, the design and configuration of the plasma system influence the effectiveness of plasma species on food substrates, as depicted in Figure 6.4.

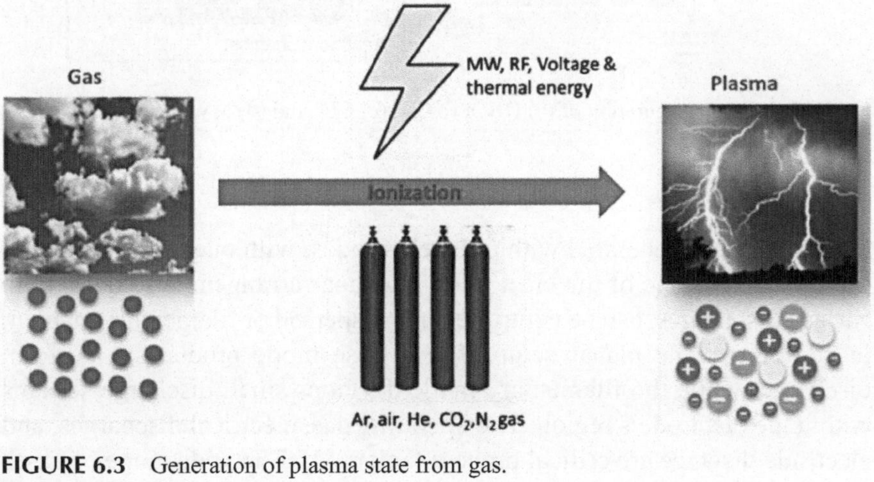

FIGURE 6.3 Generation of plasma state from gas.

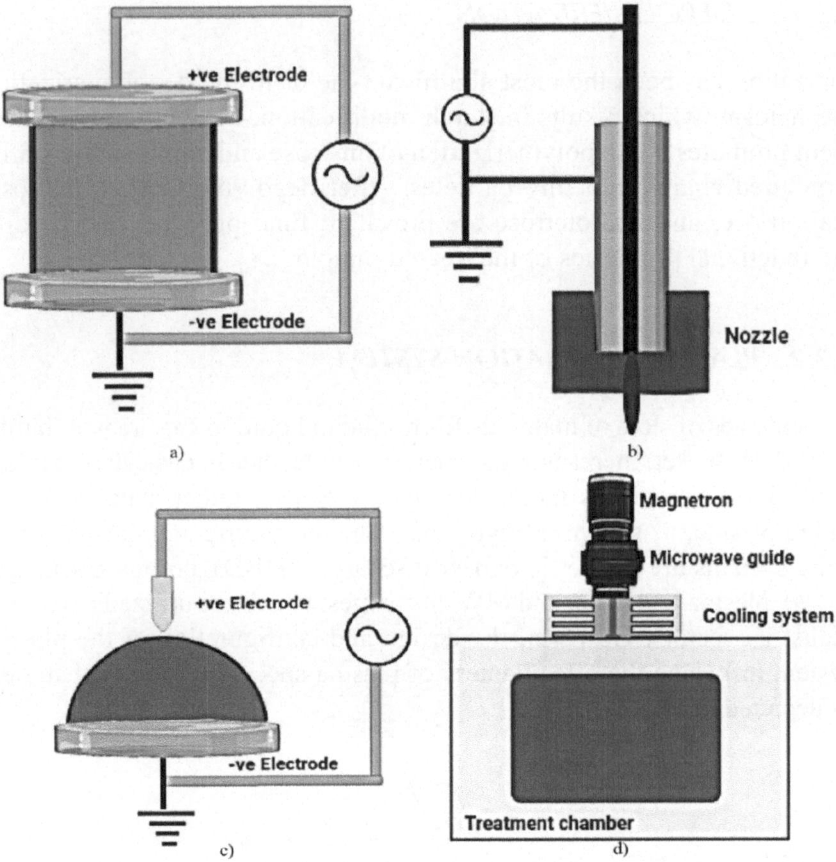

FIGURE 6.4 Configuration of (a) DBD; (b) PJ; (c) CD; and (d) MW plasma systems.

6.3.2.1 DBD PLASMA SYSTEM

Plasma is often generated within the electrodes, with one dielectric material attached to one of the electrodes. Coplanar arrangements, along with surface discharges, can be controlled in a dispersed or filamentary manner, in addition to the planar setup. The diffuse mode produces a uniform discharge, while the filamentary mode develops small discharge streams within the electrode's region. The operating gas, electrical discharges, and electrode distance are critical process factors [4]. The wide range of gases that can be used at relatively low gas flow requirements, the ability to produce a uniform discharge over long distances, and the versatility of DBDs are significant advantages of DBD over other systems. However, a relatively higher ignition voltage (>10 kV) is required based on the electrode gap, which demands specific precautions or isolation of the DBD system.

6.3.2.2 PJ PLASMA SYSTEM

Small "plasma flames" are produced by PJs with RF energy. PJs usually consist of two electrodes (ring and needle shape). However, recently PJs have been designed with a single electrode, which improves their applicability in the food system. The distance between electrodes varies at the millimeter level, and the working gas (noble gases) is ignited under a voltage of about 100 V. PJs have several benefits, including their small size, targeted area exposure, and potential to penetrate through confined areas [56].

6.3.2.3 CD PLASMA SYSTEM

Corona discharges (CDs) form at pointed electrode shapes whenever the electric field is strong enough to drive randomly emitted electrons until the gas attains ionization. Point-to-plane (sharp curve to flat shape electrode) as well as cylindrical shapes are typical patterns used in CDs. The pointed electrode can have a negative or positive potential, and the configurations can be performed in direct current or pulsed manner [81]. CDs are typically produced by extremely simple devices with low initial and operating costs. As a result, they are a good model for studying the microbicide capabilities of electric discharges in the lab. However, they are only useful for non-uniform treatments of very small regions [19].

6.3.2.4 MW PLASMA SYSTEM

The MW discharge system produces a plasma region without electrodes. MW is produced using a magnetron operating at a frequency of 2.45 GHz, which transmits towards the operating system through a wave conductor/coaxial cable. The gas, consisting of electrons, absorbs MW energy, causing a rise in kinetic energy and leading to ionizing processes through inelastic collisions [79]. The neutral gas temperatures increase (from room temperature to >1,000 K (Kelvin)) based on the MW power utilized. The main advantage of MW-driven discharges is the ability to generate plasma in an air environment without electrodes. The gas usage is within a reasonable range of standard liters per minute. A significant number of reactive species (e.g., N_xO_y, O_3, O_2, and OH) can be generated based on the discharge gas [80]. The drawback of MW-driven plasma is the regional constraint. An array of discharges has to be applied to decontaminate large sample sizes.

6.4 APPLICATION OF CP IN FOOD INDUSTRY

The CP technique has a wide application in various food sectors, including cereals, fruits, vegetables, dairy, beverages, and meat processing industries [8, 13]. Figure 6.5 illustrates the application of CP in different food sectors, highlighting its significant functional characteristics. CP is a fundamental technique that can be utilized to eliminate microorganisms and modify the structure of food to enhance its appearance and flavor. CP interacts with the outer layer of functional and bioactive compounds such as macromolecules, proteins, amino acids, and carbohydrates by disrupting the surface structure [7, 84]. This method has been extensively studied for 30 years and is now considered a viable approach for inactivating microorganisms, deactivating enzymes, and stabilizing bioactives. It offers advantages such as reduced energy consumption, faster response times, and lower costs compared to traditional techniques. However, due to delays in legislative clearance, its commercial applicability is currently limited [7]. CP, in combination with various innovative techniques, has made a significant impact in the food industry. The Plasma method, also known as "cold plasma" and "nonthermal approach," is particularly suitable for the manufacturing of thermosensitive food products. Additionally, CP processing is environmentally friendly as it does not leave behind chemical residues and utilizes ambient air as an energy source [54, 69].

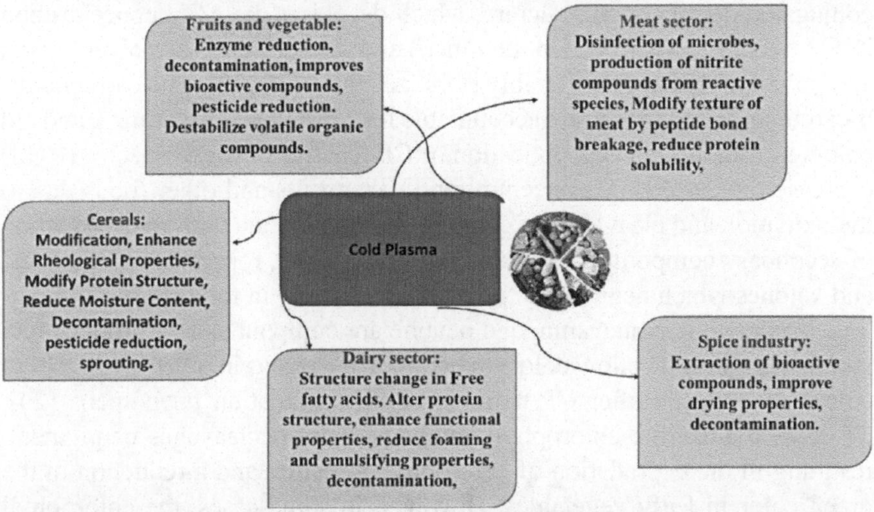

FIGURE 6.5 Application cold plasma in various food industry.

6.5 COLD PLASMA (CP) MECHANISM FOR COLOR CHANGES

Bioactive components in tissues destroyed by reactive species consist of RNS and ROS (OH, O, O_3, NO) and cause color-producing pigments to change the color of the material. The CP reactive species are the strongest oxidizing agents based on oxidation potential, such as 2.7, 2.1, 1.8, 2.4, and 2.1 for OH•, O_3, H_2O_2, O•, ONOO–, respectively [28]. The CP reactive species come into contact with the surface of food materials and cause various mechanisms shown in Figure 6.6, resulting in an alteration of natural color. Overall, chlorophylls, anthocyanin (ACN), carotenoids, myoglobin, and protein interact with CP reactive species and cause enhancement or degradation of pigment compounds based on several factors discussed in upcoming sections. The reddish-blue hue of plant products is due to the presence of ACNs, which are found inside the cellular vacuoles in plant cells. The plasma species create etching or fission on the cell membrane, which disturbs and degrades the cellular membrane and increases the mass transfer rate. As a result, a higher mass of ACN is leached out from the vacuole, leading to an increase in ACN concentration [23]. However, ACNs are constructed in agglomerates/co-pigmented forms, which are vulnerable to reactive species like O_2, O_3, etc., and are responsible for dissociation. Plasma species interact with ACN and cause oxidation of pigments or chromophores by breaking the conjugated double bond structure, which decreases the ACN concentration in CP-treated foods [39]. On the other hand, the presence of polyacylation and polyglycylation considerably stabilizes the ACN [32]. The carotenoids, β-carotene, and lycopene are accountable for causing yellow, orange, and red colors in food products. Application of CP reactive nitrogen species (RNS) decrease the carotenoid concentration in kiwifruits and other foods due to the oxidation and cleavages of conjugated double bond with the production of secondary compounds like carbonyl compounds, roperoxides, epoxides, and ketones which degrade the carotenoid structure in foods [47, 63]. Betaline, betaxanthin, betacyanin, and betaine are compounds that produce red, violet, orange, and yellow colors in beetroot, as well as in other foods. These colors are reduced after CP treatment due to oxidation phenomena [24]. CP ROS oxidize the chlorophyll and cause type II cleavages of pigment, resulting in the degradation of chlorophyll structure and a reduction in the green color in leafy vegetables. However, in some cases, the chlorophyll content is retained due to the denaturation of dechelatase and chlorophyllase catabolic enzymes, which prevent the degradation of type II cleavages and stabilize the chlorophyll [57].

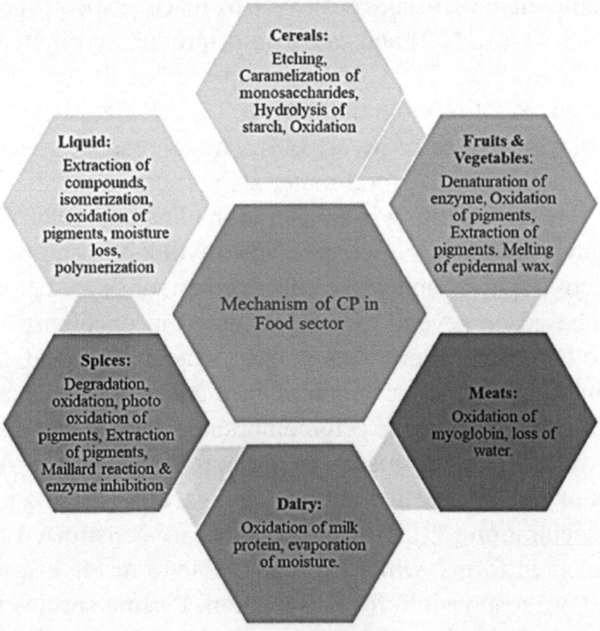

FIGURE 6.6 Mechanism involved in color changes of CP treated foods.

6.6 FACTOR AFFECTING COLOR PROPERTIES OF FOOD

The production of reactive species is influenced by plasma operating conditions such as the generating system, voltage, gas flow rate, type of gas, and exposure time [32]. Naturally, the concentration of radical species in plasma varies with distance, which affects interactions [79]. Furthermore, the interactions of bioactive compounds and pigments (carotenoid, chlorophyll, anthocyanin) with plasma are influenced by the reactive species and the distance between the sample and electrode, as well as the physical conditions of the sample. The pigments and color of plant materials are stabilized or destroyed based on these characteristics. Overall, the CP significantly influences the color change depending on variables such as the product category (solid, liquid, whole or sliced material), plasma producing parameters (voltage, treatment time (TT), generating power, gas used, DBETS (distance between electrode to sample)), and storage conditions.

Sample matrix – surface pigments may be destroyed in solids, and plasma can rupture tissues and cause excess pigment to flow out, resulting in higher ACN in blueberries treated with CP. In the presence of liquids, the treatments allow pigments and coloring compounds to freely react with reactive intermediates, regardless of volume, resulting in the reduction of ACN in blueberry juice treated with CP-PJ due to the degradation of ACN [57]. This shows the importance of sample matrix in color compounds. TT: The pomegranate juice color values are significantly reduced in L* (23.36–21.83), a* (45.94–44.59), b* (32.06–30.95), C* (56.04–54.28), H* (34.94-34.67) for long exposure time (3 to 7 min). However, there is no significant difference in total color value, which is less than 2.33 (ΔE – 2.24–2.33) up to 7 min [39]. The color value might vary based on prolonged TT and types of food products due to oxidation of pigmented compounds. In other cases, increasing the treatment duration produces desirable changes in the color of cereal products, like the WI value increased as the TT increased, which gives a high acceptance rate among consumers [74]. Sample volume: As the sample volume increases from 3 to 5 cm^3, the color values are significantly reduced for L* (23.36–21.70), a* (45.94–44.56), b* (32.06–30.87), C* (56.04–54.2), H* (34.94–34.71) [39]. However, there is no significant difference in total color value, which is less than 2.47 up to 5 cm^3, making it difficult to distinguish by visual inspection [68]. Therefore, the total color difference is insignificant based on sample volume variation, but it does increase when compared with the untreated sample. This is because the penetration depth is not a significant factor for the liquid food system, as the plasma reactive species act on the bulk liquid independent of penetration depth.

Gas flow rate: When the flow rate of gaseous molecules increases from 0.75 to 1.25 Dm^3/min (cubic decimeter per minute), the color values are significantly reduced for L* (23.36–22.45), a* (45.94–45.01), b* (32.06–31.14), C* (56.04–54.74), H* (34.94–34.67). However, there is a significant difference in the total color value of 3.02 up to 1.25 Dm^3/min. In terms of the gas flow effect, PJ offers lighter color as the flow rate of gas increases. However, the L* value stabilizes up to 1 Dm^3/min, except for H*, which exhibits significant variation even with a slight modification of the flow rate. The flow rate of gas and color are indirectly proportional to each other. As the gas flow rate increases (0.75–1.25 L/min), the color reduction was more when treated with AP CP jet system with the 2.5 kV at 2 kHz [39]. Also, the noticeable reduction in the value of L*, a*, and b*. Gag types: The strawberry, packed with O_2 and N_2, was exposed to CP. It was found that the O_2-packed strawberry exhibited a more attractive color compared to the N_2-packed products [3]. Additionally, the tomato red value increased when N_2-based CP treatment was applied. The use of Ar or Ar combined with O_2 as an effective gas source has been shown to achieve color preservation in fresh produce [51]. This suggests that plasma created using an inert gas source may be advantageous in maintaining color.

Voltage: The difference in a* was insignificant between 20 and 50 W. However, a noticeable change in color, leading to darkening, was observed at a power of 60 W, which was substantially higher than the control. Similarly, applying a voltage beyond 80 V to milk resulted in a reduction of lightness over a period of 120 s [82]. Distance: Placing the plasma source too close to the product, along with prolonged exposure time, caused significant surface degradation and resulted in a high alteration in color [62]. Mode: All of these plasma devices can be used both directly and indirectly. The direct mode of exposure to food products is beneficial for causing etching due to the wide variety of reactive species with short shelf life (up to ms). In the indirect mode of application, reactive species such as nitric oxide (NO) or ozone, which have a longer shelf life, can penetrate into the products. Remote plasmas have the benefit of being able to use considerably larger ionization energies, results in higher degree of ionization and therefore a comparable production of heat. The indirect plasma treatment had a lesser influence on cucumbers, as well as the presence of green-colored apples with chlorophyll, whereas it produced a significant influence on carrots, as well as the presence of red-colored tomatoes with carotenoid compounds. Storage period: The duration of storage had a considerable impact on b* (with a 2-unit rise) compared to the control post-treatment sample (0 s + 5-day storage). L* was greatly enhanced by extending the treatment duration to 180 seconds, but b* was significantly lowered by increasing TTs up to 300 seconds [90]. However, there were no significant changes in the stored products compared to the immediately treated products, which indicates that the changes occur more after treatment rather than during storage.

6.7 EFFECT OF COLD PLASMA (CP) ON SOLID FOOD'S COLOR IN DIFFERENT FOOD INDUSTRY

6.7.1 CEREAL PROCESSING INDUSTRY

CP processing has also been observed to provide certain desirable color variations in a few food items. Consumers prefer rice with increased lightness and decreased yellowness. The plasma treatment increased the brightness and WI (whiteness index) of brown rice [66]. Similarly, when CP was applied at 250 W, 15 kHz for 5 to 20 minutes on brown rice, the L* value increased while the a* and b* values decreased after plasma treatment [41]. For the CP treated samples, lightness levels were higher, but yellowness and redness were slightly reduced. The WI (56.89–58.28) improved with prolonged exposure time. The gelatinization and parboiling processes are closely related to the lightness value. The study concludes that for longer treatment duration, the color of basmati rice improved to a desirable level. The plasma treatment enhanced the whiteness of rice samples by etching the bran layer, which reveals the internal white cores to the outer layer [73]. The levels of L*, a*, and b* were mildly significant when pearl millet was processed with CP (40–45 kV up to 15 minutes) [43]. However, as the voltage level and treatment time increased, the color intensity (ΔE) also increased. The color changes in the grains are due to the oxidation of color pigments on the grain's surface by plasma reactive species, particularly O_3, as well as the caramelization of monosaccharides in the sample [89]. When starch molecules are exposed to high-energy electron species, they undergo hydrolysis into smaller sugar fragments. Sugar acids are another mechanism for color change.

6.7.2 FRUITS AND VEGETABLE PROCESSING INDUSTRY

Endogenous pigments have a massive impact on the color of fruits and vegetables. The oxidizing capacity of CP species damages the structure of pigments during plasma treatment, resulting in tissue damage and subsequent color alterations. Plasma pretreatment enhances the extraction of colorful ACNs, as well as other substances such as organic acids, flavonoids, amino acids, alkaloids, and polysaccharides, which may cause copigmentation when the PJ system is employed [39]. Copigmentation helps ACNs retain their stability while also causing color variations due to changes in the chromophore group's spectrum characteristics. These effects have been found to make ACN compounds appear bluer and more intense in coloration in pomegranate juice [39]. As a result, copigmentation mostly reduces the levels of L* and b*. The color changes might be due to RNS and ROS, which oxidize pigments and starch granules into small fragments [43]. Fruit darkening is also attributed to moisture loss along with the melting of epidermal wax on the fruit's surface [67]. The authors, Sudheesh & Sunooj [70], discovered that chlorophyll-rich green apples and cucumbers are more susceptible to damage compared to carotenoid-rich foods like carrots. The loss of pigment may result in discoloration; however, the ACNs in plasma-treated blueberries decreased noticeably, while the appearance became darker and bluer. Furthermore, the retention of pigment compounds is due to the destruction of enzymes responsible for discoloration. The breakdown of ACNs by CP was believed to be affected by variations in PPO and α-glucosidase activity. The minor alteration of chlorophyll with respect to CP exposure occurs through destruction of catabolic enzymes such as magnesium dechelatas and chlorophyllase. Because of this, the retention of chlorophyll was high in kiwifruit treated with CP [88]. Insignificant changes in color were noticed in tomato and carrot when an argon-based CP needle was applied for up to 10 minutes with a voltage of 12.83 kV [47]. Mandarins showed minimal color change, which may be attributed to their thick peel that resists plasma exposure [80]. However, a considerable color alteration in lettuce was observed when exposed to the same settings for longer than 7 minutes [10, 53, 63, 72, 92]. Apple, strawberry, cherry tomatoes, kiwifruit, carrots, and lettuce treated with CP did not experience significant color loss [10, 53, 63, 72, 92]. Several other authors reported significant color degradation with extreme exposure time [40, 67]. Overall, our findings indicate that CP exposure only had a minor impact on food color at shorter treatment durations.

6.7.3 MEAT PROCESSING INDUSTRY

The beef patties treated with CP strongly influenced and reduced the L* value as the voltage increased up to 70 kV. However, the L* value does not produce significant changes on storage up to 7 days [77]. The change in meat color after plasma treatment is more pronounced than in stored meat, and it depends on the raw beef color. The red (a*) and yellowish color (b*) of the CP-exposed chicken cutlets did not vary significantly over the duration of treatment, but lightness (L*) continued to rise (3.1 to 3.6) as the time increased from 60 to 300 s [77]. The decrease in L* value is proportional to the voltage. The Plasma jet had no effect on the b* of chicken breasts, beef loins, and pig shoulder. However, the red (a*) color of pork loin, beef patties, herring fillets, and sliced dry beef significantly decreased. These findings suggest that a high voltage DBD-CP application might have a negative impact on the raw beef redness. The myoglobin present in meat develops choleglobin by interacting with H_2O_2 and O_2 plasma reactive species, which decreases the red color of the meat. Fresh meat slices have a delightful bright red color due to OMb on their surface. The redness of the skin color was reduced when Mb was oxidized to brown MMb. The changes in L* value also depend on the loss of moisture content. Plasma reactive species induce the oxidation of deoxy and oxy myoglobin (OMb) and produce met myoglobin (MMb), which reduces the red color (a* = 4.54 from 7.71 at 70 kV) and develops a gray-brown color in meat [9]. ROS and RNS in plasma (O_3, OH, and H_2O_2) are responsible for developing MMb through oxidation, which also results in the cleavage of the porphyrin ring. This produces nitrimyoglobin, causing a green color in meat [18, 86]. The plasma treatment on pork and beef induces insignificant changes in b* values, which are attributed to the development of MMb [33]. However, a study found that CP can be used as an alternative to chemical (sodium nitrite) for processing pork jerky, as it retains the red color of the meat with optimum plasma processing conditions [34]. The study shows that the redness increases as the processing duration extends due to the interaction of plasma reactive species (NO, RNS, and ROS). Nitrosylmyoglobin (NO interacts with Mb) is responsible for bright red coloration, while nitrosometmyoglobin (NO binds with MMb) is responsible for a dark red color by increasing the a* value of meats [87]. These investigations broaden the present study field for the creation of new CP-based products that are natural and devoid of chemical additives.

6.7.4 MILK PRODUCTS PROCESSING INDUSTRY

Color is an essential sensory quality parameter in milk analysis. Buffalo milk appears white, whereas cow milk appears as a creamy yellow tint. Cattle breed, feed mixture, lactation stages, milking frequency, and seasonal calving all influence milk color, which ranges from yellowish white to nearly white [64]. The decrease in color values of L* and b* (lightness and yellow) in dairy foods is noticed when processed with CP over prolonged periods. Milk treated with a high voltage level for extended periods of time (after 80 V and 120 s) decreased the lightness [82]. However, the ΔE of the CP treated sample (4.03) had less change in color compared to UHT (36.02) and pasteurized sample (9.02), which is acceptable to the consumer [82]. When nonfat dried milk was processed by CP with a brief duration, there were no noticeable color changes [17]. However, little alterations were found for samples treated with a long duration. The development of brown color in CP treated samples was observed with a reduction of L* and b* values while an increase in a* value. Exposure of N_2-based CP with a short duration on milk powder up to 120 s does not produce significant changes in color ($\Delta E = 1.5$) and falls within the acceptable color range (0.5–1.5) [17]. The yellowish color of whey protein isolate (WPI) increases as the duration of CP exposure extends, due to the production of carotenoids that contribute to the yellow color. The protein from milk undergoes oxidation by CP species, resulting in a slight yellowing of the dairy color, as determined by a colorimeter. Although the instrumental method showed a significant decrease in L* and an increase in b*, there was no noticeable difference in the color of the plasma-treated cheese products [29]. When DBD plasma was applied to cheese for an extended period, the L* values decreased (from 69.48 to 66.57–68.64), and the b* values increased significantly (from 34.28 to 35.84–37.75). However, there was no significant difference in the overall color difference value between the control and treated cheese. The decrease in the L* value may have been caused by water evaporation during the APP process. A significant decrease in L* followed by an increase in a* was observed when sliced cheddar cheese was treated with CP at 15 kHz for an extended period of 10 minutes. A higher b* value (e.g., >6) indicates non-enzymatic changes in milk products [20]. To minimize variations in the color of dairy food, the treatment period should be kept under 5 minutes and in low oxygen environments to reduce protein and lipid oxidation, which leads to a more intense yellow color.

6.7.5 *SPICE AND HERBS PROCESSING INDUSTRY*

The color of nuts and spices also plays a crucial role in quality analysis. With longer plasma treatment periods of 120s, the values of L*, b*, and C* declined significantly, while the a* value increased dramatically [48]. In conclusion, significant color changes to a dark pistachio color have been observed due to prolonged TT, resulting from Maillard reactions and degradation of chlorophyll and carotenoids [30]. The development of degraded compounds like amino and carbonyl compounds, resulting from the breakdown of glycosidic and peptide bonds under radiation, triggers the Maillard reactions and eventually generates colorful chemicals [6]. Similar results have been found for walnuts and almonds, with color reduction after plasma exposure for 20 minutes at 40 W [2]. Saffron is a crucial compound that produces natural flavor and acts as a color agent in food manufacturing, making color an essential quality characteristic. Therefore, studying the effects of LPCP-produced species on flavor and color change is essential. The saffron are exposed for 5 min at 110 W produced greatest L* value (23.30) and a* value with reduced b* value, which is much more brightness than control sample. The overall color difference (ΔE) in LPCP-processed products increased compared to untreated sample. This phenomenon is beneficial to saffron. The higher color value in CP-treated saffron is due to the presence of higher pigment compounds extracted in solvent due to the etching phenomenon, which discharges compounds into the surrounding liquid [31]. Saffron's principal pigments belong to the carotenoid family. The formation of conjugated double bonds analyzed to provide a noticeable change in the color of these compounds. The greater the absorption, the redder the color intensity of the product by increasing the amount of conjugated double bonds. The color changes occur due to the oxidation of pigments by reactive O_2 species, which break and scratch the surface of the saffron [22]. When peppermint is processed with CP at 60 W power, the a* value changes to -4.18 for the CP-treated sample compared to -6.37 in the control sample, which represents the decline of green color and the decrease of L* value indicates the development of dark green of peppermint from light green color. Overall, the color changes in chlorophyll leaves are due to the photooxidation of pigments by ROS and light of CP [35]. However, the green color losses increase as the voltage rises. The interaction of plant cells with O_2 molecules results in the Millard reaction as well as enzymatic browning through polyphenol oxidase, producing fast darkening. Spirulina flour experienced discoloration towards green and blue color after air-based CP exposure [60]. The effect of CP technology on solid products' color has been tabulated (Table 6.3).

6.8 EFFECT OF COLD PLASMA (CP) ON THE LIQUID FOODS PROCESSING INDUSTRY

Several studies show that exposure to cold plasma (CP) has diverse impacts on different foods. In plasma-treated acerola juices, color became lighter, brightness and hue improved, while chroma indicated insignificant alteration [27]. The improvement in color is attributed to the presence of more carotenoids. In N_2-based CP treatment, the b* values differed significantly, while the values of a* and L* remained stable compared to air-based plasma exposure. Upon exposure to CP, apple juice underwent a considerable color change towards a more yellow and bright hue. This alteration was linked to pigments isomerization as well as oxidation processes [78]. When Chinese bayberries were exposed to plasma activated water (PAW), the color changed to dark violet due to an increase in ACN concentration, resulting in increased color index of red grapes (CIRG) values compared to the control [45]. The overall color change of tomato juice significantly increased with prolonged exposure to plasma, due to the destruction of carotenoids by plasma reactive species [5]. The samples began to become somewhat yellowish, as seen by variations in the color angle. Color measurements in orange juice, pumpkin puree, and coconut water all changed significantly after being treated with CP

TABLE 6.3 Effect of Cold Plasma on Color of Solid Foods

Sl. No.	Sample	CP System	Treatment Conditions	Purposed	Effect	Inference	References
1.	Cherry tomatoes	DBD-F: 50 Hz, V:60 kV.	V:30 kV, TT: 30–300 s.	Decontamination	L*: ↑, a*: ↓, b*: ↑, C*: ↓, H*: ↓, ΔE: ↑ (2.88). OC: 30 s.	Oxidation of carotenoid pigments.	[92]
2.	Blueberries	DBD-V: 230 V, F: 50 Hz	TT: 2–5 min, V: 60–80 kV.	Degraded pesticide compounds.	L*: ↓, a*: ↑, b*: ↓, OC: 80 kV, 5 min.	Reduction of moisture and cutaneous wax melting causes dark color.	[67]
3.	Button mushrooms.	Ar/O_2-APPJ-PAW, F: 10 kHz.	S: 400 g, V: 18 kV, DBETS: 10 mm, PAW: 500 mL at 20 min, TT: 5–15 min.	Decontamination and quality retention.	L*: ↓, a*: ↑, b*: ↑, ΔE: ↓ (1.6). OC: 10 min.	Microbial reduction and white color retention is due to inactivation of enzymes (tyrosinase) causes browning reaction by oxidization, bleach ability, and anti-browning capacity of PAW.	[85]
4.	Pumpkin puree	CD-CP, V: 220 V, DBETS: 1 cm.	S: 25 g, TT: 5–20 min.	To inactivate *E. coli.*	L*: ↓, a*: ↓, b*: ↑, OC: 5 min	Oxidation of carotenoid.	[65]
5.	Wheat flour	DBD-V: 230 V, F: 50 Hz.	TT: 5–30 min, V:80 kV.	Functionality improvement.	L*: ↑, a*: ↓, b*: ↓, WI: ↑.	Oxidation of carotenoids and disintegration of conjugated double bonds.	[16]
6.	Spirulina algae	DBD-F: 10 kHz.	P: 10–20 W, TT: 5 min.	Spore inactivation	L*: ↑, a*: ↓, b*: ↓, ΔE: ↑ (2.3).	Plasma causes less color change in chlorophyll due to O_2 gas.	[11]
7.	Raw chicken breast meat.	DBD-CP, V: 30 kV	V: 60–80 kV, TT: 60–180s, Gas %: O_2 (35–95)/ CO_2(30–60)/N_2(5).	Decontamination and quality retention.	L*: ↑, a*: ↑, b*: ↓, OC: 35+ 60+5% of O_2+ CO_2+ N_2, 60 Kv, 60 s.	O_3 interact with meat and cause color reduction.	[91]

TABLE 6.3 *(Continued)*

Sl. No.	Sample	CP System	Treatment Conditions	Purposed	Effect	Inference	References
8.	Pepper seeds, crushed oregano, and paprika powder.	MW CP torch-P:1.2 kW, F: 2.45 GHz.	TT: 7 s, break time: 5 s.	Decontamination and quality retention.	L*: ↑, a*: ↓, b*: ↓, ΔE: ↑ (46.9) for oregano and paprika powder. ΔE*: ↓ (pepper seeds).	Oxidation of pigments (carotenoids).	[36]
9.	Sliced cheese.	He-DBD-F: 50 kHz, DBETS: 5 mm.	V: 3.5 kV, TT: 1–15 min.	Inactivate endogenous enzyme and quality retention.	L*: ↓, a*: ↑, b*: ↑.	Moisture evaporation by oxidation.	[1]
10.	Bolti fish (*Tilapia nilotica*)	DBD-AP, V: 220 V, F:50 Hz.	DBETS: 48 mm, V: 40–60 kV, TT: 1–4 min.	Inactivate endogenous enzyme and quality retention.	L*: ↓, a*: ↓, b*: ↑, During storage a*: ↑.	Colors loss: pigment and protein oxidation, damage of fish tissue.	[55]
11.	Non-fat milk powder.	N_2-fluidized bed plasma-P: 480 W.	V: 4.4 KV, TT: 20–120 s.	Pathogen inactivation.	L*: ↓, a*: ↑, b*: ↓. ΔE: ↑ (1.14). Development of light brown color.	Less TT and N_2 gas contribute to less changes in color.	[17]

[61, 65, 84]. In contrast to the previous findings, no significant color change was observed after CP treatment in experiments involving cherry tomatoes, siriguela, blueberry, mandarins, and strawberries juices [32, 58, 93]. CP, produced using Ar gas, influenced the darkness of pomegranate juice, resulting in an enhancement of its blueness in CP-exposed fruit juices [39]. These variations have been attributed to possible chemical alterations in the constituent colors, specifically the polymerization of phenolic chemicals. Insignificant modifications in the color of orange juice were found when exposed to air-based CP for up to 60 seconds at 70 kV voltage [84]. Similarly, insignificant alterations in the color of raw milk were observed following CP treatment for 15 minutes. The total color variation (ΔE) of CP (9 kV for 9 minutes) treated milk was 0.25, while prolonged exposure to plasma (20 minutes) resulted in slightly greater color differences (ΔE of 0.52) [25]. However, insignificant variations in the total color deviation value were observed after exposure times of 5 and 10 minutes (0.72 and 0.5 of ΔE) for CP-treated milk. It was reasoned that a larger L^* value may be linked to a higher proportion of fat globules capable of diffracting light more efficiently [20]. The interaction of the aldehyde group of lactose from milk protein with the residue of the amino acid lysine or lysyl results in an increase in the b^* value due to the Maillard's reaction [49]. The effect of CP technology on the color of solid products has been tabulated (Table 6.4).

6.9 SUMMARY

- Cold Plasma (CP) is considered the latest innovative approach in ensuring food safety and quality, particularly among nonthermal techniques. Plasma exists in a "quasi-neutral" state, where it is composed mostly of ions, atoms, photons, and free electrons, all of which have a "net neutral charge." By increasing the energy level in the gas, ionization occurs, resulting in the development of a plasma state. CP treatments are environmentally friendly as they do not leave behind any chemical residues and utilize ambient air as an energy source.
- When plasma species come into contact with a food surface, they produce various reactions. These reactions include surface etching (creating pores and causing tissue damage), depolymerization (breaking down starch to form new compounds), and cross-linking

TABLE 6.4 Effect of Cold Plasma on Color of Liquid Foods

Sl. No.	Sample	CP System	Treatment Conditions	Purposed	Effect	Inference	References
1.	Orange juice	DBD-V: 20 V, F: 60 Hz.	V: 90 kV, TT: 30–120 s.	Microbial inactivation	ΔE: ↑ (1.08), OC: 120 s.	Less change in color because of short TT and He gas.	[84]
2.	White grape juice.	DBD-V: 230 V, F: 60 Hz.	V: 80 kV, TT: 1 to 4 min.	Sterilization	L*: ↔, a*: ↑, b*: ↑, C*: ↑, OC: 80 kV, 2 min.	Degradation of ascorbic acid caused non-enzymatic browning.	[59]
3.	Apple juice	Air DBD	P: 30–50 W, TT: 0–40 s, DBETS: 2 mm.	Inactivation of *E. coli.*	L*: ↓, b*: ↓, a*: ↑, ΔE: ↑.	Phenolic compounds polymerization.	[42]
4.	Pomegranate juice.	Ar PJ: V: 2.5 kV, F: 25 kHz, C: 3 mA.	DBETS: 1.5 cm. P: 4 W, TT: 3–7 min.	Inactivation	L*: ↑, a*: ↓, b*: ↓, ΔE: ↑ (1.5–3.0).	Copigmentation/↑ of anthocyanin concentration/ chromophore group changes.	[39]
5.	Carrot juice	Air DBD-V: 0–80 kV.	S: 250 mL DBETS: 30 mm, V: 60–80 kV, TT: 3–4 min.	Sterilization	L*: ↓, a*: ↓, b*: ↑, OC: 70 kV, 3 min.	Less activation of enzymes causes enzymatic browning.	[76]
6.	Apple juice	DBD–DBETS: 8 cm.	F: 50, 200, 400, 600, and 900 Hz, P: 20 kV, TT: 15 min S: 20 mL.	Inactivate enzymes	L*: ↓, a*: ↑, b*: ↑, ΔE: ↑ (> 10).	Less reduction of PPO enzyme activity.	[26]
7.	Tender coconut water:	DBD-V: 120 V, 60 Hz, 5 cm.	90 Kv, 120 s.	Microbial destruction: L. monocytogenes and *E. coli.*	L*: ↔, a*: ↓, b*: ↓, ΔE: ↔ (0.37–1.7).	Oxidation/polymerization (phenols, proteins)/ degradation (ascorbic acid).	[46]
8.	Milk	Air-DBD: F: 15 kHz, P: 250 W.	S: 10 mL, TT: 5–10 min.	Inactivation of MO's and ↑ food quality and safety.	L*: ↑, b*: ↑, a*: ↓	↑ L*: ↑ quantity of fat globules.	[82]

(cleaving C-OH polymeric chains to add new C-O-C bonds by eliminating water). These reactions result in the development of unique properties in solid and liquid food products.

- Food color is a noticeable factor that has a significant impact on consumer impression. The CIE LAB scale is used to quantify color changes that occur in processed food samples using the L*, a*, and b* variables.
- The DBD plasma is generated between two electrodes containing insulating material by electrical discharge. Plasma jet systems are capacitive-coupled devices where gas is passed between two coaxial electrodes with high velocity, producing miniature "plasma fires" that typically originate in the radio frequency (RF) band. The CD system consists of point-to-plate geometries (a strongly curved electrode positioned in the opposite direction of a flat electrode) and cylindrical arrangements. In MW, the electrons absorb the microwave (MW) energy, resulting in an increase in kinetic energy and inelastic collisions that discharge energy in terms of UV photons as well as visible light, which leads to ionization processes.
- Color changes can be induced by CP due to various mechanisms such as oxidation, degradation, extraction, etching, polymerization, melting of surface wax, and moisture loss. The extent of these changes depends on the type of products.
- The effect of cold plasma (CP) on the color profile of food depends on several factors. Longer treatment times (TT) result in a darker color in pomegranate juice, while enhancing the whiteness value in brown rice. The total color difference is insignificant when considering variations in sample volume. To preserve the color of fresh produce, Ar or Ar combined with O_2 can be used as an effective gas source. Indirect plasma treatment has a minimal influence on raw chlorophyll. CP exposure only has a minor impact on food color during shorter treatment durations. CP can be used as an alternative to chemical (sodium nitrite) for processing pork jerky, although it may slightly decrease the redness of the meat in some products. The development of a brown color has been observed in CP-treated milk powder, while no significant changes were found in milk products like cheese. CP enhances the darkness of nuts and the lightness of green leafy vegetables, resulting in

a reduction in green color. Chinese bayberries exposed to PAW undergo a color change to dark violet due to an increase in anthocyanin (ACN) concentration, while the color of raw milk remains largely unchanged.

- The food business operators can adopt CP for functionalization based on desirability without causing significant changes in color profile under optimized conditions.

KEYWORDS

- **cold plasma**
- **color profile**
- **corona discharges**
- **etching**
- **extraction**
- **liquid foods processing industry**
- **oxidation**
- **pigments**
- **solid and liquid foods**

REFERENCES

1. Montaño, J. A., et al., (2022). Effect of the application of cold plasma energy on the inactivation of microorganisms, proteins, and lipids deterioration in adobera cheese. *J. Food Qual., 2022*.
2. Ahangari, M., Ramezan, Y., & Khani, M. R., (2021). Effect of low-pressure cold plasma treatment on microbial decontamination and physicochemical properties of dried walnut kernels (*Juglans regia* L.). *J. Food Process Eng., 44*(1), e13593.
3. Ahmadnia, M., Sadeghi, M., Abbaszadeh, R., & Ghomi, M. H. R., (2021). Decontamination of whole strawberry via dielectric barrier discharge cold plasma and effects on quality attributes. *J. Food Process. Preserv., 45*(1), e15019.
4. Albertos, I., Martin-Diana, A., Cullen, P. J., Tiwari, B. K., Ojha, S. K., et al., (2017). Effects of dielectric barrier discharge (DBD) generated plasma on microbial reduction and quality parameters of fresh mackerel (*Scomberscombrus*) fillets. *Innov. Food Sci. & Emerg. Technol., 44*, 117–122.
5. Ali, M., Cheng, J. H., & Sun, D. W., (2021). Effects of dielectric barrier discharge cold plasma treatments on degradation of anilazine fungicide and quality of tomato (*Lycopersicon esculentum* Mill) juice. *Int. J. Food Sci. & Technol., 56*(1), 69–75.

6. Alinezhad, M., Hojjati, M., Barzegar, H., Shahbazi, S., & Askari, H., (2021). Effect of gamma irradiation on the physicochemical properties of pistachio (*Pistacia vera* L.) nuts. *J. Food Meas. Charact., 15*(1), 199–209.
7. Alves, F. E. G., De Brito, E. S., & Rodrigues, S., (2020). Effects of cold plasma processing in food components. In: *Advances in Cold Plasma Applications for Food Safety and Preservation* (pp. 253–268). Elsevier.
8. Attri, P., Venkatesu, P., Kaushik, N., Han, Y. G., Nam, C. J., et al., (2012). Effects of atmospheric-pressure non-thermal plasma jets on enzyme solutions. *J. Korean Phys. Soc., 60*(6), 959–964.
9. Bauer, A., Ni, Y., Bauer, S., Paulsen, P., Modic, M., et al., (2017). The effects of atmospheric pressure cold plasma treatment on microbiological, physical-chemical, and sensory characteristics of vacuum packaged beef loin. *Meat Sci., 128*, 77–87.
10. Bermúdez-Aguirre, D., Wemlinger, E., Pedrow, P., Barbosa-Cánovas, G., & Garcia-Perez, M., (2013). Effect of atmospheric pressure cold plasma (APCP) on the inactivation of *Escherichia coli* in fresh produce. *Food Control., 34*(1), 149–157.
11. Beyrer, M., Pina-Perez, M. C., Martinet, D., & Andlauer, W., (2020). Cold plasma processing of powdered *Spirulina* algae for spore inactivation and preservation of bioactive compounds. *Food Control., 118*, 107378.
12. Blikra, M. J., Altintzoglou, T., Løvdal, T., Rognså, G., Skipnes, D., et al., (2021). Seaweed products for the future: Using current tools to develop a sustainable food industry. *Trends Food Sci. & Technol., 118*, 765–776.
13. Bourke, P., Ziuzina, D., Boehm, D., Cullen, P. J., & Keener, K., (2018). The potential of cold plasma for safe and sustainable food production. *Trends Biotechnol., 36*(6), 615–626.
14. Carrillo, C., Nieto, G., Martinez-Zamora, L., Ros, G., Kamiloglu, S., et al., (2022). Novel approaches for the recovery of natural pigments with potential health effects. *J. Agric. Food Chem.*
15. Chacha, J. S., Zhang, L., Ofoedu, C. E., Suleiman, R. A., Dotto, J. M., et al., (2021). Revisiting non-thermal food processing and preservation methods—Action mechanisms, pros and cons: A technological update (2016--2021). *Foods, 10*(6), 1430.
16. Chaple, S., Sarangapani, C., Jones, J., Carey, E., Causeret, L., et al., (2020). Effect of atmospheric cold plasma on the functional properties of whole wheat (*Triticum aestivum* L.) grain and wheat flour. *Innov. Food Sci. & Emerg. Technol., 66*, 102529.
17. Chen, D., Peng, P., Zhou, N., Cheng, Y., Min, M., et al., (2019). Evaluation of *Cronobacter sakazakii* inactivation and physicochemical property changes of non-fat dry milk powder by cold atmospheric plasma. *Food Chem., 290*, 270–276.
18. Cheng, J. H., Chen, Y. Q., & Sun, D. W., (2021). Effects of plasma activated solution on the color and structure of metmyoglobin and oxymyoglobin. *Food Chem., 353*, 129433.
19. Choi, S., Puligundla, P., & Mok, C., (2016). Microbial decontamination of dried Alaska pollock shreds using corona discharge plasma jet: Effects on physicochemical and sensory characteristics. *J. Food Sci., 81*(4), M952–M957.
20. Coutinho, N. M., Silveira, M. R., Rocha, R. S., Moraes, J., Ferreira, M. V. S., et al., (2018). Cold plasma processing of milk and dairy products. *Trends Food Sci. & Technol., 74*, 56–68.

21. Dai, Z., Chen, S., Xie, X., & Zhou, L., (2019). Investigation of grinding and lapping surface damage evolution of fused silica by inductively coupled plasma etching. *Int. J. Precis. Eng. Manuf., 20*(8), 1311–1323.
22. Darvish, H., Ramezan, Y., Khani, M. R., & Kamkari, A., (2022). Effect of low-pressure cold plasma processing on decontamination and quality attributes of saffron (*Crocus sativus* L.). *Food Sci. & Nutr., 10*(6), 20, 82–90.
23. Dong, X. Y., & Yang, Y. L., (2019). A novel approach to enhance blueberry quality during storage using cold plasma at atmospheric air pressure. *Food Bioprocess Technol., 12*(8), 1409–1421.
24. Dzimitrowicz, A., Jamroz, P., Cyganowski, P., Bielawska-Pohl, A., Klimczak, A., & Pohl, P., (2021). Application of cold atmospheric pressure plasmas for high-throughput production of safe-to-consume beetroot juice with improved nutritional quality. *Food Chem., 336*, 127635.
25. Fajar, M. I., Amalda, E. C., Syah, M. A. A., et al., (2020). Design of a switching flyback transformer as high voltage and frequency power supply for raw milk cold plasma sterilization. *Int. J. Sci. Technol. & Manag., 1*(3), 230–236.
26. Farias, T. R. B., Rodrigues, S., & Fernandes, F. A. N., (2020). Effect of dielectric barrier discharge plasma excitation frequency on the enzymatic activity, antioxidant capacity and phenolic content of apple cubes and apple juice. *Food Res. Int., 136*, 109617.
27. Fernandes, F. A. N., Santos, V. O., & Rodrigues, S., (2019). Effects of glow plasma technology on some bioactive compounds of acerola juice. *Food Res. Int., 115*, 16–22.
28. Ge, X., Shen, H., Su, C., Zhang, B., Zhang, Q., et al., (2021). The improving effects of cold plasma on multi-scale structure, physicochemical and digestive properties of dry heated red adzuki bean starch. *Food Chem., 349*, 129159.
29. Heo, Y. S., Yim, D. G., Baek, K. H., Kang, T., Lee, Y. E., et al., (2021). Effect of inkjet-printed flexible dielectric barrier discharge plasma on reduction of pathogen and quality changes on sliced cheese. *LWT, 143*, 111128.
30. Hertwig, C., Leslie, A., Meneses, N., Reineke, K., Rauh, C., & Schlüter, O., (2017). Inactivation of *Salmonella enteritidis* PT30 on the surface of unpeeled almonds by cold plasma. *Innov. Food Sci. & Emerg. Technol., 44*, 242–248.
31. Hosseini, S. I., Farrokhi, N., Shokri, K., Khani, M. R., & Shokri, B., (2018). Cold low pressure O_2 plasma treatment of *Crocus sativus*: An efficient way to eliminate toxicogenic fungi with minor effect on molecular and cellular properties of saffron. *Food Chem., 257*, 310–315.
32. Hou, Y., Wang, R., Gan, Z., Shao, T., Zhang, X., et al., (2019). Effect of cold plasma on blueberry juice quality. *Food Chem., 290*, 79–86.
33. Huang, M., Wang, J., Zhuang, H., Yan, W., Zhao, J., & Zhang, J., (2019). Effect of in-package high voltage dielectric barrier discharge on microbiological, color and oxidation properties of pork in modified atmosphere packaging during storage. *Meat Sci., 149*, 107–113.
34. Inguglia, E. S., Oliveira, M., Burgess, C. M., Kerry, J. P., & Tiwari, B. K., (2020). Plasma-activated water as an alternative nitrite source for the curing of beef jerky: Influence on quality and inactivation of *Listeria innocua*. *Innov. Food Sci. & Emerg. Technol., 59*, 102276.

35. Kashfi, A. S., Ramezan, Y., & Khani, M. R., (2020). Simultaneous study of the antioxidant activity, microbial decontamination, and color of dried peppermint (*Mentha piperita* L.) using low pressure cold plasma. *LWT, 123*, 109121.
36. Kim, J. E., Choi, H. S., Lee, D. U., & Min, S. C., (2017). Effects of processing parameters on the inactivation of *Bacillus cereus* spores on red pepper (*Capsicum annum* L.) flakes by microwave-combined cold plasma treatment. *Int. J. Food Microbiol., 263*, 61–66.
37. Kopuk, B., Gunes, R., & Palabiyik, I., (2022). Cold plasma modification of food macromolecules and effects on related products. *Food Chem., 382*, 132356.
38. Kourtzanidis, K., Dufour, G., & Rogier, F., (2020). Self-consistent modeling of a surface AC dielectric barrier discharge actuator: In-depth analysis of positive and negative phases. *J. Phys. D.Appl. Phys., 54*(4), 45203.
39. Kovačević, D. B., Putnik, P., Dragović-Uzelac, V., Pedisić, S., Jambrak, A. R., & Herceg, Z., (2016). Effects of cold atmospheric gas phase plasma on anthocyanins and color in pomegranate juice. *Food Chem., 190*, 317–323.
40. Lacombe, A., Niemira, B. A., Gurtler, J. B., Fan, X., Sites, J., et al., (2015). Atmospheric cold plasma inactivation of aerobic microorganisms on blueberries and effects on quality attributes. *Food Microbiol., 46*, 479–484.
41. Lee, K. H., Kim, H. J., Woo, K. S., Jo, C., Kim, J. K., et al., (2016). Evaluation of cold plasma treatments for improved microbial and physicochemical qualities of brown rice. *LWT., 73*, 442–447.
42. Liao, X., Li, J., Muhammad, A. I., Suo, Y., Chen, S., et al., (2018). Application of a dielectric barrier discharge atmospheric cold plasma (Dbd-Acp) for *Eshcerichia coli* inactivation in apple juice. *J. Food Sci., 83*(2), 401–408.
43. Lokeswari, R., Sharanyakanth, P. S., Jaspin, S., & Mahendran, R., (2021). Cold plasma effects on changes in physical, nutritional, hydration, and pasting properties of pearl millet (*Pennisetum glaucum*). *IEEE Trans. Plasma Sci., 49*(5), 1745–1751.
44. Ma, C., Nikiforov, A., De Geyter, N., Morent, R., & Ostrikov, K. K., (2022). Plasma for biomedical decontamination: From plasma-engineered to plasma-active antimicrobial surfaces. *Curr. Opin. Chem. Eng., 36*, 100764.
45. Ma, R., Yu, S., Tian, Y., Wang, K., Sun, C., et al., (2016). Effect of non-thermal plasma-activated water on fruit decay and quality in postharvest Chinese bayberries. *Food Bioprocess Technol., 9*(11), 1825–1834.
46. Mahnot, N. K., Mahanta, C. L., Farkas, B. E., Keener, K. M., & Misra, N. N., (2019). Atmospheric cold plasma inactivation of *Escherichia coli* and *Listeria monocytogenes* in tender coconut water: Inoculation and accelerated shelf-life studies. *Food Control., 106*, 106678.
47. Mahnot, N. K., Siyu, L. P., Wan, Z., Keener, K. M., & Misra, N. N., (2020). In-package cold plasma decontamination of fresh-cut carrots: Microbial and quality aspects. *J. Phys. D. Appl. Phys., 53*(15), 154002.
48. Makari, M., Hojjati, M., Shahbazi, S., & Askari, H., (2021). Elimination of *Aspergillus flavus* from pistachio nuts with dielectric barrier discharge (DBD) cold plasma and its impacts on biochemical indices. *J. Food Qual., 2021*.
49. Mazumder, M. A. R., Hongsprabhas, P., & Thottiam, V. R., (2019). In vitro and *in vivo* inhibition of Maillard reaction products using amino acids, modified proteins, vitamins, and genistein: A review. *J. Food Biochem., 43*(12), e13089.

50. Mishra, R., Bhatia, S., Pal, R., Visen, A., & Trivedi, H., (2016). Cold plasma: Emerging as the new standard in food safety. *Res. Inv. Int. J. Eng. Sci., 6*(2), 15–20.
51. Misra, N. N., (2016). Quality of cold plasma treated plant foods. In: *Cold Plasma in Food and Agriculture* (pp. 253–271). Elsevier.
52. Misra, N. N., Pankaj, S. K., Segat, A., & Ishikawa, K., (2016). Cold plasma interactions with enzymes in foods and model systems. *Trends Food Sci. & Technol., 55*, 39–47.
53. Misra, N. N., Patil, S., Moiseev, T., Bourke, P., Mosnier, J. P., et al., (2014). In-package atmospheric pressure cold plasma treatment of strawberries. *J. Food Eng., 125*, 131–138.
54. Misra, N. N., Schlüter, O., & Cullen, P. J., (2016). *Cold Plasma in Food and Agriculture: Fundamentals and Applications*. Academic Press.
55. Mohamed, E. E., Younis, E. R., & Mohamed, E. A., (2021). Impact of atmospheric cold plasma (ACP) on maintaining bolti fish (*Tilapia nilotica*) freshness and quality criteria during cold storing. *J. Food Process. Preserv., 45*(5), e15442.
56. Nishime, T. M. C., Borges, A. C., Koga-Ito, C. Y., Machida, M., Hein, L. R. O., & Kostov, K. G., (2017). Non-thermal atmospheric pressure plasma jet applied to inactivation of different microorganisms. *Surf. Coatings Technol., 312*, 19–24.
57. Nowacka, M., Dadan, M., Janowicz, M., Wiktor, A., Witrowa-Rajchert, D., et al., (2021). Effect of nonthermal treatments on selected natural food pigments and color changes in plant material. *Compr. Rev. Food Sci. Food Saf., 20*(5), 5097–5144.
58. Paixão, L., Fonteles, T. V., Oliveira, V. S., Fernandes, F. A. N., & Rodrigues, S., (2019). Cold plasma effects on functional compounds of siriguela juice. *Food Bioprocess Technol., 12*(1), 110–121.
59. Pankaj, S. K., Wan, Z., Colonna, W., & Keener, K. M., (2017). Effect of high voltage atmospheric cold plasma on white grape juice quality. *J. Sci. Food Agric., 97*(12), 4016–4021.
60. Pina-Pérez, M. C., Úbeda-Manzanaro, M., Beyrer, M., Martinez, A., & Rodrigo, D., (2021). *In vivo* Assessment of cold atmospheric pressure plasma technology on the bioactivity of *spirulina*. *Front. Microbiol., 12*.
61. Porto, E., Alves, F. E. G., Silva, L. M. A., Fonteles, T. V., Do Nascimento, R. B. R., et al., (2020). Ozone and plasma processing effect on green coconut water. *Food Res. Int., 131*, 109000.
62. Ramanan, K. R., Sarumathi, R., & Mahendran, R., (2018). Influence of cold plasma on mortality rate of different life stages of *Tribolium castaneum* on refined wheat flour. *J. Stored Prod. Res., 77*, 126–134.
63. Ramazzina, I., Berardinelli, A., Rizzi, F., Tappi, S., Ragni, L., et al., (2015). Effect of cold plasma treatment on physico-chemical parameters and antioxidant activity of minimally processed kiwifruit. *Postharvest Biol. Technol., 107*, 55–65.
64. Rathod, N. B., Kahar, S. P., Ranveer, R. C., & Annapure, U. S., (2021). Cold plasma an emerging nonthermal technology for milk and milk products: A review. *Int. J. Dairy Technol., 74*(4), 615–626.
65. Santos, Jr. L. C. O., Cubas, A. L., V., Moecke, E. H. S., Ribeiro, D. H. B., & Amante, E. R., (2018). Use of cold plasma to inactivate *Escherichia coli* and physicochemical evaluation in pumpkin puree. *J. Food Prot., 81*(11), 1897–1905.

66. Sarangapani, C., Devi, R. Y., Thirumdas, R., Trimukhe, A. M., Deshmukh, R. R., & Annapure, U. S., (2017). Physico-chemical properties of low-pressure plasma treated black gram. *LWT-Food Sci. Technol., 79*, 102–110.
67. Sarangapani, C., O'Toole, G., Cullen, P. J., & Bourke, P., (2017). Atmospheric cold plasma dissipation efficiency of agrochemicals on blueberries. *Innov. Food Sci. & Emerg. Technol., 44*, 235–241.
68. Smet, C., Baka, M., Steen, L., Fraeye, I., Walsh, J. L., et al., (2019). Combined effect of cold atmospheric plasma, intrinsic and extrinsic factors on the microbial behavior in/on (food) model systems during storage. *Innov. Food Sci. & Emerg. Technol., 53*, 3–17.
69. Stoica, M., Alexe, P., & Mihalcea, L., (2014). Atmospheric cold plasma as new strategy for foods processing-an overview. *Innov. Rom. Food Biotechnol., 15*.
70. Sudheesh, C., & Sunooj, K. V., (2020). Cold plasma processing of fresh-cut fruits and vegetables. In: *Fresh-Cut Fruits and Vegetables* (pp. 339–356). Elsevier.
71. Surowsky, B., Schlüter, O., & Knorr, D., (2015). Interactions of non-thermal atmospheric pressure plasma with solid and liquid food systems: A review. *Food Eng. Rev., 7*(2), 82–108.
72. Tappi, S., Berardinelli, A., Ragni, L., Dalla, R. M., Guarnieri, A., & Rocculi, P., (2014). Atmospheric gas plasma treatment of fresh-cut apples. *Innov. Food Sci. & Emerg. Technol., 21*, 114–122.
73. Thirumdas, R., Saragapani, C., Ajinkya, M. T., Deshmukh, R. R., & Annapure, U. S., (2016). Influence of low-pressure cold plasma on cooking and textural properties of brown rice. *Innov. Food Sci. & Emerg. Technol., 37*, 53–60.
74. Thirumdas, R., Trimukhe, A., Deshmukh, R. R., & Annapure, U. S., (2017). Functional and rheological properties of cold plasma treated rice starch. *Carbohydr. Polym., 157*, 1723–1731.
75. Ucar, Y., Ceylan, Z., Durmus, M., Tomar, O., & Cetinkaya, T., (2021). Application of cold plasma technology in the food industry and its combination with other emerging technologies. *Trends Food Sci. & Technol., 114*, 355–371.
76. Umair, M., Jabbar, S., Nasiru, M. M., Sultana, T., Senan, A. M., et al., (2019). Exploring the potential of high-voltage electric field cold plasma (HVCP) using a dielectric barrier discharge (DBD) as a plasma source on the quality parameters of carrot juice. *Antibiotics., 8*(4), 235.
77. Wang, X., Wang, Z., Zhuang, H., Nasiru, M. M., Yuan, Y., et al., (2021). Changes in color, myoglobin, and lipid oxidation in beef patties treated by dielectric barrier discharge cold plasma during storage. *Meat Sci., 176*, 108456.
78. Wang, Y., Wang, Z., Zhu, X., Yuan, Y., Gao, Z., & Yue, T., (2020). Application of electrical discharge plasma on the inactivation of *Zygosaccharomyces rouxii* in apple juice. *LWT, 121*, 108974.
79. Wiktor, A., Hrycak, B., Jasiński, M., Rybak, K., Kieliszek, M., et al., (2020). Impact of atmospheric pressure microwave plasma treatment on quality of selected spices. *Appl. Sci., 10*(19), 6815.
80. Won, M. Y., Lee, S. J., & Min, S. C., (2017). Mandarin preservation by microwave-powered cold plasma treatment. *Innov. Food Sci. & Emerg. Technol., 39*, 25–32.

81. Wu, T. Y., Sun, N. N., & Chau, C. F., (2018). Application of corona electrical discharge plasma on modifying the physicochemical properties of banana starch indigenous to Taiwan. *J. Food Drug Anal., 26*(1), 244–251.
82. Wu, X., Luo, Y., Zhao, F., & Mu, G., (2021). Influence of dielectric barrier discharge cold plasma on physicochemical property of milk for sterilization. *Plasma Process. Polym., 18*(1), 1900219.
83. Xiao, D., Ruan, Q., Bao, D. L., Luo, Y., Huang, C., et al., (2020). Effects of ion energy and density on the plasma etching-induced surface area, edge electrical field, and multivacancies in $MoSe_2$ nanosheets for enhancement of the hydrogen evolution reaction. *Small., 16*(25), 2001470.
84. Xu, L., Garner, A. L., Tao, B., & Keener, K. M., (2017). Microbial inactivation and quality changes in orange juice treated by high voltage atmospheric cold plasma. *Food Bioprocess Technol., 10*(10), 1778–1791.
85. Xu, Y., Tian, Y., Ma, R., Liu, Q., & Zhang, J., (2016). Effect of plasma activated water on the postharvest quality of button mushrooms, *Agaricus bisporus*. *Food Chem., 197*, 436–444.
86. Yong, H. I., Han, M., Kim, H. J., Suh, J. Y., & Jo, C., (2018). Mechanism underlying green discoloration of myoglobin induced by atmospheric pressure plasma. *Sci. Rep., 8*(1), 1–9.
87. Yong, H. I., Lee, S. H., Kim, S. Y., Park, S., Park, J., et al., (2019). Color development, physiochemical properties, and microbiological safety of pork jerky processed with atmospheric pressure plasma. *Innov. Food Sci. & Emerg. Technol., 53*, 78–84.
88. Zhao, Y., Chen, R., Liu, D., Wang, W., Niu, J., et al., (2019). Effect of nonthermal plasma-activated water on quality and antioxidant activity of fresh-cut kiwifruit. *IEEE Trans. Plasma Sci., 47*(11), 4811–4817.
89. Zhu, F., (2018). Effect of ozone treatment on the quality of grain products. *Food Chem., 264*, 358–366.
90. Zhuang, H., Rothrock, Jr. M. J., Hiett, K. L., Lawrence, K. C., Gamble, G. R., et al., (2019). In-package air cold plasma treatment of chicken breast meat: Treatment time effect. *J. Food Qual., 2019*.
91. Zhuang, H., Rothrock, Jr. M. J., Line, J. E., Lawrence, K. C., Gamble, G. R., et al., (2020). Optimization of in-package cold plasma treatment conditions for raw chicken breast meat with response surface methodology. *Innov. Food Sci. & Emerg. Technol., 66*, 102477.
92. Ziuzina, D., Misra, N. N., Cullen, P. J., Keener, K. M., Mosnier, J. P., et al., (2016). Demonstrating the potential of industrial scale in-package atmospheric cold plasma for decontamination of cherry tomatoes. *Plasma Med., 6*(3, 4).
93. Ziuzina, D., Misra, N. N., Han, L., Cullen, P. J., Moiseev, T., et al., (2020). Investigation of a large gap cold plasma reactor for continuous in-package decontamination of fresh strawberries and spinach. *Innov. Food Sci. & Emerg. Technol., 59*, 102229.

CHAPTER 7

EFFECT OF OHMIC HEATING ON COLOR PROFILE OF LIQUID FOOD

PRATIKSHA SHRESTHA,[1] AMIRHOSSEIN ABEDINI,[2,3] PANKAJ PARMAR,[4] KAAVYA RATHNAKUMAR,[5] and RAJIB LOCHAN POUDYAL[6]

[1]*Department of Food Technology and Quality Control (DFTQC), Kathmandu, Nepal*

[2]*Department of Environmental Health, Food Safety Division, Faculty of Public Health, Tehran University of Medical Sciences, Tehran, Iran*

[3]*Students' Scientific Research Center, Tehran University of Medical Sciences, Tehran, Iran*

[4]*Parul Institute of Technology, Parul University, Faculty of Agriculture, Dairy Technology Department, Gujarat, India*

[5]*Department of Food Science, University of Wisconsin, Madison, USA*

[6]*Japan Ginger Company 780-0821 Sakuraicho 1-3-12, Koch-Shi, Kochi, Japan*

ABSTRACT

Ohmic heating (OH) can also be referred to as Joule heating and is commonly used in the food processing industry to prevent contamination of food by pathogenic microorganisms and maintain the sensory, functional, and nutritional properties of food, thereby increasing the overall

Nonthermal Food Processing Technologies: Impact on Color Profile.
Kaavya Rathnakumar and R. Pandiselvam (Eds.)

shelf life of the food. This chapter provides a comprehensive review of ohmic heating, its basic principles, and its application in the processing of liquid food. Recently, industries have been utilizing ohmic heating for the processing of milk and milk beverages, alcoholic beverages, soft drinks, purees, syrups, fruit juice, and similar fluid foods. One of the major challenges during food processing is preserving the natural bioactive compounds present in the food. This chapter covers the retention or degradation of carotenoids, chlorophyll, xanthophyll, riboflavin, anthocyanin (ACN), and other bioactive color compounds during ohmic-assisted processing of various liquid foods. The challenges to overcome for the commercial use of ohmic heaters in food processing and recent advances in this field are also discussed at the end of this chapter.

7.1 INTRODUCTION

For the processing and preservation of liquid food products in the food industry, heating is the most common and essential unit operation being performed at one stage or another, with different purposes such as flavor and color development, shelf-life enhancement, destruction, and inactivation of undesirable microorganisms and enzymes. Heat transfer mainly occurs via conduction or convection, or by a combination of both, through the steady and non-steady state in conventional heat treatment (HT). In these processes, heat energy needs to be generated outside the food material and subsequently passed through it. Low viscous products like fruit juices, milk, nectars, buttermilk, whey, etc., are usually pasteurized or sterilized, while high viscous food products such as fruit purees, honey, *Basundi,* evaporated milk, sweetened condensed milk, etc., resist rapid heat transfer and require longer exposure time. One of the essential requirements that every dairy processor looks for in an improved heating method is the rapidity and effectiveness of that technique within narrow temperature and exposure time gradients, so that the final product can be safeguarded against both over and under HT. OH is a novel heating method that can be applied to obtain uniform heating of the food due to the inherent electrical resistivity of the food itself. When compared to other new heating technologies like microwave (MW) heating, induction heating, and radio frequency (RF) heating, ohmic heating possesses many desired characteristics for food processing [1]. Ohmic heating technology can be successfully used for various food production processes, such as pasteurization, enzyme inactivation, thawing, extraction, blanching, sterilization, and many others [2]. However, this method of heating is best suited for liquid or semi-solid food processing. Unlike conventional heating, ohmic heating can preserve the nutritional properties of food and therefore can be applied for the processing of baby food [3], where the retention of nutrients is a major concern. Commercial products like fruit and vegetable juice, pulp, puree, etc., have been the prime interest for most of the research and studies done on OH processing [4].

The principle of OH is the transformation of electrical energy to thermal energy, resulting in internal energy generation and a subsequent temperature rise [5]. When electricity passes through an electrically conductive material, charged particles move within the system, causing the atoms in the system to become agitated and leading to a rise in temperature. In food materials, the charged molecules of minerals and proteins migrate towards the electrode of opposite polarity. This movement of polar molecules results in an increase in temperature. OH is recommended for processing a wide range of food products that are rich in water, polar compounds such as proteins and minerals, and so on. The primary requirement for food processing using OH is that the food must be electrically conductive. It is challenging to process foods with low water and ionic content using OH. Milk is an ideal food for OH processing due to its high water content, presence of ionic molecules, and electrical conductivity between 4 and 5.5 $mScm^{-1}$[6].

Organoleptic and physicochemical properties of food products are considered essential for their preservation and quality assessment, which can be greatly affected by the intensity and nature of the treatment [5]. The intensity of OH processing can be determined by certain indicators, which are also used for conventionally processed foods. These indicators include HMF (hydroxy methyl furfural), furosine, and lactulose content, color indices such as L*, a*, b*, ΔE, C* and h*, FAST Index, protein denaturation, and protein (β-lactoglobulin) aggregation, among others [7].

Color is defined as the light that is absorbed, reflected, or transmitted through a food medium and is perceived by the eyes in the form of wavelength [8, 9]. It is an essential property of food that influences customer buying behavior, making it the most widely used parameter in the food industry for assessing food quality. Natural colors in food have been recognized for their health benefits and safety due to their anti-inflammatory action, antioxidant properties, and anti-carcinogenic effects, as well as various other biological functions, in comparison to artificial colors.

Color is expressed through qualitative and quantitative analysis. Qualitative analysis is conducted through visual inspection (sensory analysis) by sensory panels, comparing various samples. However, in some cases, the sensory method may not detect small changes or provide exact results due to human influences [10]. Traditionally, this analysis was done by sensory experts, but due to its subjective nature, results can vary from person to person. Therefore, color is now measured quantitatively using instrumental techniques [11]. Quantitative measurement is performed using instruments that evaluate samples and convert the data into an easily analyzable and interpretable form [9]. This method is based on the absorption, reflection, and illumination of light, as well as the wavelength and amplitude emitted from the light source, using instruments such as colorimeters, spectrophotometers, and spectroradiometers.

The stability of natural color components in food depends on various factors such as processing methods, time, temperature, and storage conditions. Ohmic heating preserves the color and nutritional value of food with higher yield compared to the conventional process. Short processing time and uniform temperature are specific advantages of Ohmic heating over convection and conduction mechanisms [12]. Ohmic heating is more effective in liquid food (beverages) due to its good electrical conductivity compared to solids. The degradation kinetics of color and its application in predicting experimental color changes using empirical models for food products is a complex phenomenon [13]. Total color retention can be determined to assess the extent of deterioration caused by thermal treatment. In process design, kinetic modeling is considered an important parameter to derive reaction kinetic information, calculate the reaction rate as a function of experimental variables, and predict changes during processing and storage [14]. To optimize thermal processing for improving product color, the reaction rate and activation energy of a reaction need to be determined. Non-enzymatic browning usually follows zero or first-order reaction kinetics [15].

Ohmic heating can also be used in conjunction with other processing technologies. It has been found to be more effective when combined with other thermal processing technologies. The electrical conductivity (EC) of foods determines how efficiently ohmic heaters should be designed [2], making it a key parameter to be quantified as heat generation depends on EC. The applied voltage gradient, frequency, temperature, moisture content, particle size, concentration, and concentration of electrolytes in the food greatly influence its EC [16]. EC increases linearly with increasing concentration of ionic constituents, temperature, and voltage gradient. Conversely, when particle size and frequency are increased, the rate of heating decreases [17].

7.2 POSITIVE AND NEGATIVE ASPECTS OF OH

In OH, food is heated due to its own electrical resistance, resulting in a fast heat transfer process compared to conduction or convection. This high heat transfer rate leads to shorter processing times, allowing for microbial safety without any negative impact on the food product [18]. Additionally, ohmic heating ensures uniform temperature increase throughout the food due to volumetric heating. For foods with similar electrical conductivity (EC), uniform heating of different particles in liquid food can be achieved simultaneously, preserving the nutritional quality and preventing cold point formation [19]. The positive and negative aspects of ohmic heating in fluid food processing are summarized in Table 7.1 and should be considered when selecting ohmic heaters for different purposes [20].

Ohmic heating technology has been successfully applied in laboratory-scale thermal processing of food [21]. The principles can be utilized to develop commercially applicable ohmic heaters for household use. Ohmic heating has shown great potential in replacing conventional thermal processing for a variety of food products, including cooking, roasting, drying, pasteurization, sterilization, and microbiological inhibition. Comparative studies have been conducted to understand the microbial inactivation principles of ohmic heating compared to conventional heating, and it has been found that inactivation primarily occurs due to heat [22]. Nonthermal inactivation of microorganisms can also be achieved by employing high current and high frequency during ohmic heating [4]. However, there are differing opinions among researchers regarding the effectiveness of this nonthermal method for microbial inactivation [23].

Slow heat transfer during thermal processing of particulate food in cans is the major reason for substantial quality degradation of food products [24]. This problem can be solved by using ohmic heating, as it can uniformly heat the liquid food faster than with conduction or convection.

Pasteurization with ohmic heating is a very good alternative for protecting epoxy xanthophylls and other xanthophylls [25], which are otherwise difficult to protect during processing.

Despite the mentioned advantages, the major problem with ohmic heating is that the food product may not be heated evenly if it predominantly contains non-conductive constituents like fat [26]. Therefore, ohmic heating treatment cannot be performed for fatty food and low moisture content, as electric current cannot flow through them. Another important negative aspect to consider is the electrolysis of the food product due to electrochemical reactions [22]. Since ohmic heating uses electrodes for passing electrical current through and is kept in direct contact with food, metal ions can migrate at the interface between the electrode and food. This eventually causes the electrode to corrode and invites fouling problems [27] in long-term use. Moreover, ohmic heating requires highly skilled technical manpower and an accurate process control mechanism.

TABLE 7.1 Summarizing Positive and Negative Aspects of Ohmic Heating

Positive Aspects	Negative Aspects
• Required temperature is achieved very quickly. • Rapid uniform heating of liquid food. • Eliminates the problem of surface fouling. • Easy shut down of the system that takes no time. • Shutting off the current completely stops heat transfer. • Environment friendly system. • Minimal maintenance cost.	• Narrow frequency band. • Complicated coupling of electrical field distribution and temperature. • Difficult to monitor and control. • Lack of generalized information.

7.3 PRINCIPLE OF OH FOR LIQUID FOOD

Humans have used heating as one of the oldest methods of storing and producing food throughout history. While conventional processing technologies can ensure product safety, they often result in damage to the product and a reduction in its nutritional value [28, 29]. Surveys of food buyers indicate a preference for products that are high in nutritional value, organic, and natural [30]. Among the various scientific technologies available, heating technology has made significant advancements, including induction heating, dielectric heating, and ohmic heating [31, 32]. These methods generate high energy within the food itself, allowing for the production of nutritious and safe organoleptic foods. Ohmic heating (OH) was initially used for milk pasteurization [33].

Today, OH is widely employed in countries such as the United States, Japan, and European countries. It is primarily used for pasteurizing liquid and pumpable foods, including fruit pieces in syrup, fruit juices, sauces, purees, dairy products, and eggs. The usage of OH has been authorized by the *FDA* (United States Food and Drug Administration) [17, 34]. OH, also known as Joule heating, electric heating, and EC heating, is a method that generates heat within the food by passing alternating current through it [35]. Unlike direct heating, OH converts electrical energy into heat within the food, which is justified by the Joule effect. The food is pumped into a vertical tube with a set of electrodes that act as a heater, raising the temperature to the desired level. These chambers are made of polytetrafluoroethylene and are connected by stainless steel interface pipes with an insulating layer on the inner side. The insulation material used is polyether and polyvinylidene fluoride or glass [22].

An air drain valve is installed on top of the heater to ensure that the column is constantly filled. To prevent boiling of the processed products in the column, it is necessary to apply a relatively high return pressure on the system. For acidic products that are processed at 90°C–95°C, pressure is applied twice and for low acid products that are processed at 140°C–120°C, pressure is applied 4 times. Generally, food contains both water and ionic salts and therefore is electrically conductive, as well as resistant at the same time. When electric current is applied, heat is generated due to the resistance of food to electricity. A food's rate of heating is significantly influenced by its electrical resistance. A multimeter that is in contact with a conductivity container is used to test the electrical resistance of food.

In examining the principle of OH, it should be noted that many things are important. These include the type of electrodes, voltage, current, and power applied, EC, flow properties, the structure of the food, the electrical energy, and the temperature [36–38]. An overview of OH components includes an energy source to power the system, a variable power supply (variac) to regulate voltage, current, and voltage units, an ohmic heating cell with electrodes, a temperature measuring system, and a microcomputer for data entry. Each component has a specific task [35], as shown in Figure 7.1.

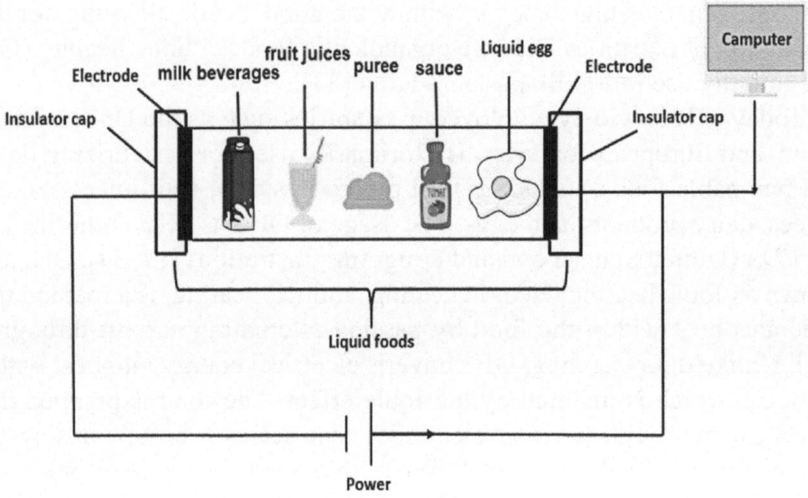

FIGURE 7.1 The basis of ohmic heating performance.

The food positioned between the positive and negative electrodes acts as both a conductor and a resistor in the circuit. The inherent resistance of food compounds to the passage of electricity, due to the state and structure of the food, leads to heat generation throughout the food. This indicates that the conversion of energy from electrical to thermal occurs within the matter itself. Ohm's law forms the basis of OH, which itself represents the relationship between current, voltage, and resistance (Table 7.2 and Eqn. (1)).

$$I = V/R\Delta\nabla$$

According to Eqn. (2) of Table 7.2, The relationship between Maxwell's equations and Ohm's law can be used to investigate the voltage distribution in OH. Electrical conductivity (σ) is a factor that indicates the quality of the passage of electrical current through food.

$$\nabla(\sigma \nabla V) + \partial\rho c/\partial t = 0$$

Actually, the relationship between current density and electric field intensity determines EC. According to Equation (4) of Table 7.2, heat is generated in the food when a constant voltage is flowing through the food medium. This heat generation is characterized by the volumetric heat generation rate.

$$U = |\nabla V|^2 \sigma$$

In examining the speed at which liquid food transfers heat in the single-phase state, which itself is affected by EC, we can refer to Eqn. (5) from Table 7.2.

$$\nabla(K\nabla T) + u = \rho Cp \partial T/\partial t$$

TABLE 7.2 Equations Used in Ohmic Heating

1. Ohm's law	$I = V/R$	I = current (unit: A). V = potential difference (unit: V). R = resistance (Unit: Ω).
2. Voltage distribution (Maxwell's equations, Ohm's law, and continuity equation).	$\nabla(\sigma \nabla V) + \partial \rho c/\partial t = 0$	σ = electrical conductivity (unit: S/m). V = potential difference (unit: V). ρc = current density (unit: A/m³). t = time (unit: s).
3. Steady state for voltage distribution.	$\nabla(\sigma \nabla V) = 0$	
4. Volumetric heat generation rate.	$U = \|\nabla V\|^2 \sigma$	U = generation rate (unit: W/m³)
5. Heat transfer.	$\nabla(K\nabla T) + u = \rho Cp\partial T/\partial t$	K = thermal conductivity (unit: W/mK). P = density (unit: kg/m³). Cp = specific heat capacity (unit: J/kgK).

In examining the advantages and disadvantages of all alternative methods for heating food, it should be considered that no serious damage is done to the food. This is because, in this case, the reference of this method will be ambiguous. OH has been designed to have very little effect on food. On the other hand, considering the important effects of environmental pollution on the world, the existing methods in this regard must also have the necessary approvals. OH, given that it uses electrical energy, can have this positive factor that does not harm the environment [35, 39]. It also reduces fuel consumption. In the study of safe product production, it can be mentioned that this method can affect bacteria and mycotoxins and cause inactivation of microorganisms and enzymes [20, 40, 41]. The initial cost of using OH is high, but over time it will save costs. It should be noted that this method is a desirable method for use in liquid foods. These benefits can be examined in terms of nutrient preservation and color [42, 43]. Disadvantages include the complexity of the system, lack of sufficient information, readjustments according to the nature of food and liquids, and complex monitoring [17, 42] (Figure 7.2).

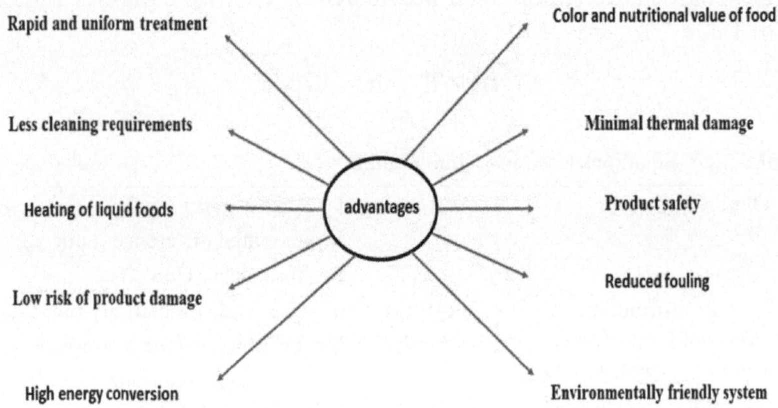

FIGURE 7.2 Advantages of ohmic heating.

7.4 OH EFFECT ON COLOR ATTRIBUTE OF FLUID FOOD

Foods come in many forms, one of which is liquid. In the group of liquid foods, we can mention fruit juices, oil, milk, milk beverages, syrups, sauces, purees, and liquid eggs. It should be noted that when we talk about liquid foods, we have a liquid diet based on water, such as clear broths, sports drinks, fruit juices, carbonated drinks, tea, and coffee [42]. For this reason, and due to the high consumption of these products, the sensory properties of these foods are very important, including features such as color and taste [42].

Lovibond, Munsell, CIELAB, Hunter Lab, and CIE are available methods of expressing color. Among them, CIELAB is the most preferred method in the laboratory of food processing industries due to its uniform color distribution and its ability to provide results that are approximately close to what the human eye perceives. CIELAB represents color using the calorimetric coordinates L*, a*, b*. L* represents lightness on a scale of 0 to 100, a* represents redness (+a*) to greenness (–a*), and b* represents yellowness (+b*) to blueness (–b) [8]. Each color is represented by a color dot with three numbers (L*, a*, b*) in the color space. These three numbers represent the color coordinates of the color point. From these values, we can calculate chroma (C*), hue angle (h), and color difference (ΔE) using the following formula:

Chroma (C*) = $(a^{*2} + b^{*2})^{1/2}$

Hue angle (h) = $\tan^{-1}$(b*value/a*value)?

Total color difference (ΔE) = $(\Delta L^{*2} + \Delta a^{*2} + \Delta b^{*2})^{1/2}$, where; ΔL, Δa, and Δb represents color parameter difference from initial to final (Figure 7.3).

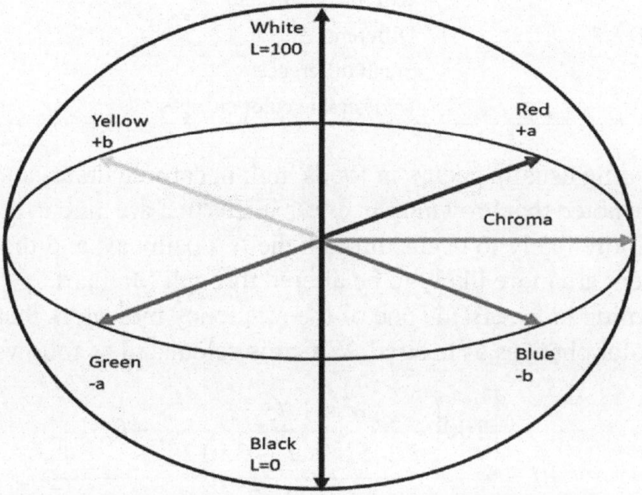

FIGURE 7.3 The CIELAB color space diagram.

The definitions of chromaticity and hue angle are very important because humans recognize colors by their brightness and chromaticity. In fact, chromaticity is obtained from the combination of hue and saturation (chroma), and the hue angle examines colors at the same brightness level. The qualitative expression of color is based on identifying differences in the absorption of different wavelengths, with one color being compared to gray. This analysis is conducted using a 360-degree grid consisting of 0 degrees, 90 degrees, 180 degrees, and 270 degrees, where each point represents bluish-red, yellow, green (RGB), and blue, respectively. TCD is an important indicator for distinguishing colors in food with each other and consists of ΔL*, Δa*, and Δb*. And is determined by the subsequent

Eqn. (3) in Table 7.2. The TCD value indicates the extent of variation between two comparative sample as shown in Table 7.3.

$$\sqrt{\left(\text{ÄL}^*\right)^2 + \left(\text{Äa}^*\right)^2 + \left(\text{Äb}^*\right)^2}$$

TABLE 7.3 TCD Features

TCD	Different
TCD > 3	Extremely different
1.5 < TCD < 3	Different
TCD < 1.5	Small difference
TCD < 3	Undesirable color changes

Browning usually occurs in foods high in carbohydrates and protein. It should be noted that browning in citrus juices that are high in ascorbic acid (AA) is more likely to occur through the AA pathway, and those lower in AA content are more likely to be altered through Maillard.

Browning index (BI) is one of the efficiency indicators that can report brown color changes as desired. Which is calculated as follows:

$$BI = \frac{\left[100\left(\frac{a^* + 1.75 \times L^*}{5.645L^* + a^* - 3.012b^*} - 0.31\right)\right]}{0.17}$$

Food color parameters can change during production and storage, which can affect the sensory properties of the product. The color is the first feature that consumers see, and it can greatly influence their decision to purchase or not [42]. Various processes, including physical, bacterial, thermal, and nonthermal methods, can impact the color of liquid food. Thermal methods like OH and MW heating may have less harmful effects compared to direct (traditional) heating methods, as they do not generate heat from inside the food. The objective of these techniques is to ensure that the sensory and nutritional characteristics of liquid meals remain unchanged, while indirect heating helps ensure their safety. Liquid materials provide favorable conditions for the growth of microorganisms due to their high nutrient accessibility, so it is important to ensure the shelf life of these products. The structure and ingredients of liquid foods can be affected by OH, which is an important and influential factor. Additionally, time, temperature, and voltage play crucial roles in OH applications, and by programming these parameters, the detrimental effects of this method on food color can be minimized. This chapter will examine the impacts of OH on the color of milk and milk beverages, fruit juices, liquid eggs, sauces, and purees.

7.4.1 MILK AND MILK BEVERAGES

Less consideration is given to milk and its products, particularly liquid milk-based beverages [44]. Color is a crucial quality characteristic for all food products, including liquid milk and milk beverages. The appearance is the first thing that comes to the customer's mind, and if it is unusual or unappealing, the customer will not purchase it. During dairy processing, color is largely influenced by chemical reactions such as Maillard and caramelization reactions. While some products like *Basundi, Rabri, Kheer*, etc., desire a certain color, it is not desirable in others. There have been a few studies on the impact of OH on the color attributes of milk and milk beverages.

Researchers have explored the potential of OH for pasteurization of milk and lactose-free milk. The milk samples were preheated to 50°C, then homogenized using 2-stage homogenizers, followed by pasteurization using OH and the conventional method of steam at 63°C for 30 minutes. The milk was then cooled to 4°C, and one portion was hydrolyzed by lactase while the other served as a control sample. The samples were analyzed for various physicochemical attributes, including color profile. It was observed that the L*, a*, and b* values were significantly lower compared to the control samples [45]. This could be due to the Maillard reaction products resulting from intense heating.

Parmaret et al. [46] assessed the potential of OH for milk concentration. They applied OH to three types of milk: cow milk, buffalo milk, and mixed milk (1:1). The voltage applied was 13.33 V/cm for 75 minutes of processing in a batch-type ohmic heater. They studied various physicochemical properties, including color attributes. They observed a significant increase in redness (a*) and yellowness (b*) values, which may be attributed to the electrochemical reaction between stainless steel electrodes and milk ingredients, followed by electrode corrosion. On the other hand, whiteness (L*) values decreased significantly, which may be due to the increased concentration of HMF and other Maillard reaction products over time [46].

Parmar et al. [47] examined the effect of OH treatment on the shelf-life of Basundi (a milk dessert). They prepared *Basundi* using buffalo milk and 5% sucrose (w/w of milk) at 13.33 V/cm. The prepared product was stored at two different temperatures, namely 30 ± 1°C and 4 ± 1°C, using two different packaging materials: polystyrene cups and self-stand Retort Pouches. They analyzed the stored product for various physicochemical, sensory, and microbiological tests. It was observed that lightness values (L*) significantly decreased, while redness (a*) and yellowness (b*) values increased significantly over time [47]. The decrease in L* values could be attributed to Maillard reaction products, while the increase in a* and b* values could be due to electrochemical reactions and their products. The carotenoid content in cow milk is higher than that in buffalo milk, which distinguishes them from each other in terms of color [48]. Furthermore, the release of carotenoids during vigorous heating in un-homogenized milk partly contributes to color variations.

7.4.2 FRUIT AND VEGETABLE JUICES

Juice is obtained by applying extraction methods to fruits. Juices have natural colors such as chlorophyll, carotenoids, and flavonoids [49, 50]. The different color compounds found in various fruit sources are shown in Figure 7.4.

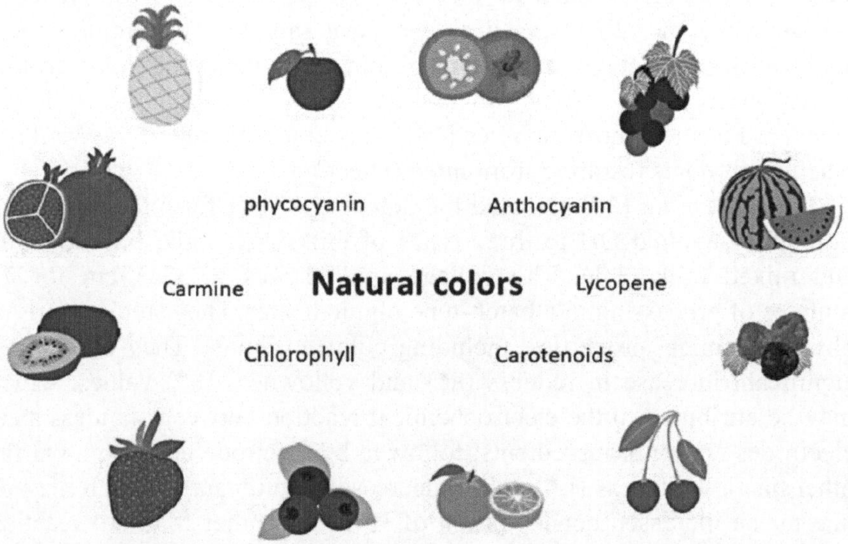

FIGURE 7.4 Fruit as source of natural colors.

These natural dyes have healing properties for humans. They exhibit health effects such as antioxidant activity, anticancer properties, delaying aging, and anti-inflammation [51]. Fruit juices also contain dietary fiber, which can promote health by improving gut health, reducing the risk of obesity, cardiovascular disease, type 2 diabetes, and certain types of cancer, as well as lowering LDL cholesterol levels [52, 53]. Fruit juices are rich in vitamins (A, E, C, thiamin, riboflavin, niacin, pyridoxine, and folate) and minerals (phosphorus, calcium, zinc, iron, magnesium, and potassium), according to research on vitamins and minerals. Due to these health properties, various guidelines recommend high consumption of natural fruits and juices. For instance, the 2020–2025 dietary guidelines suggest replacing half of your fruit intake with fruit juice [53–55].

Due to high consumption and the existence of health properties in different parts of this food product, especially colors, it is very important to maintain their nutritional and sensory properties. When examining the preservation of nutritional value and sensory properties, care should be taken to use the correct methods for extracting fruit juice from fruits that have the least impact on its appearance and also ensure the stability of the product during storage. Fruit juice deterioration is greatly facilitated by the presence of enzymes. The enzymes peroxidase (POD), pectin methyl esterase (PME), and polyphenol oxidase (PPO) are the three important enzymes that contribute to juice spoilage [56]. OH technology is one of the methods that can inactivate these enzymes [57]. When this approach has no negative impact on the product's color, these beneficial benefits are important. Because the first parameter to attract customers is the color of the product. In this section, we will examine the influence of OH on the color of fruit juices.

In 2017, a study by Ishita et al. examined the color and pH changes during OH treatment of watermelon juice. They found that the voltage gradient had a significant impact on the color of the product. The color parameter L* showed a direct relationship with the voltage gradient, treatment time, and storage duration. Specifically, at voltage gradients of 23.33 V/cm and treatment times of 1 min, 3 min, and 5 min, the L* values were 33.06, 33.03, and 33.01, respectively. These values did not show significant changes compared to the results obtained at 10 V/cm [58]. The decrease in color intensity in fruit juice is primarily attributed to the activity of oxidative enzymes, which can be associated with a decrease in ascorbic acid (AA) content. The enzyme activity in this category is influenced by the thermal processing and storage conditions of the fruit juice. Another study conducted in 2016 by Makroo et al. investigated the enzyme inactivation during OH treatment of tomato juice and its impact on the physical and chemical characteristics of tomato paste. The reduction in AA content observed in the previous study was found to be significant and increased with time. Additionally, the hue angle index, which is an important indicator of color change and browning in fruit juice, showed a significant decrease with increasing OH treatment time from 0 to 60 seconds [59]. These studies suggest that color changes in fruit juice are the result of a combination of factors, including enzymatic activity. This point has been noted in the study of Saxena et al. [86]. When the voltage is 24 V/cm lightness increases with increasing temperature (similar to redness and opposite to yellowness), which is related to the degradation of chlorophyll pigment. Lightness will decrease when we reach 32 and 48 V/cm and decrease in polyphenol oxidase activity is evinced [4, 59].

Li et al. in 2019 [60] noted that OH can stimulate POD activity initially, but then rapidly deactivate it. Similarly, ohmic HT at a deactivation temperature of 35°C can significantly decrease polyphenol oxidase activity [60]. Brochier et al. noted in 2019 and 2016 that OH can induce POD inactivation at 80°C [61]. Consistent with previous studies, the Makroo study of pineapple juice in 2022 reported that polyphenol oxidase activity decreased at 80 ± 2°C and 16–36 V/cm. Other enzymes, such as pectin methyl esterase (PME) [62], are also involved in the dye and can affect the OH process.

Non-enzymatic browning can also be a significant factor in affecting the color (loss of color/brown pigments) of fruit juice. Browning is a major concern for fruit juice manufacturing industries. The BI can be used as a reference tool for assessing overall color change in samples.

Sometimes, when different parameters are changed, there is no change in color. In a study conducted by Brochier and colleagues [61], it was observed that the color (yellowness – b* parameter) of the food sample was not affected by processing time, applied voltage gradient, and frequency [61]. In a new application of OH drying for liquid foods such as sour orange juice, Torshizi and colleagues [63] examined the effect of OH drying on several quality parameters of sour orange juice. In this study, the voltage variable was examined at 8.33, 10.83, and 13.33 V/cm with weight reductions of 10%, 20%, and 30% respectively. The efficiency of this method in reducing the weight of the product increases with increasing voltage, showing a linear relationship. However, at 13.33 V/cm, the highest red color was observed. The reason for this can be explained as follows: at lower voltage gradients, the ohmic process requires more time to remove water from the fruit juice sample, resulting in the need for more energy. This increase in time and energy leads to a decrease in useful energy and a decrease in energy efficiency [63]. Another parameter that can be considered in the study of the effects of OH on the color of fruit juices is the effect of pH.

In examining the differences between OH and conventional heating methods, Rodríguez et al. Studied carrot juice in 2021. Both methods were able to induce POD inactivation (80°C – 7 min) up to 99% [18] and similar result were obtained where 62.8 to 72.6% enzyme inactivation was achieved after 25 min of HT [61]. In the study of conventional heating, sensory tests could not be successful compared to OH, especially in the parameter L*, which had a significant difference with the control sample. This study stated that OH can be used instead of HT to pasteurize carrot juice. It should be noted that both methods were safe in terms of microbial properties. These findings align with the results of thermal processing of sour cherry juice [64]. Conventional heating can have a significant impact on phenolic compounds and color of food. However, carotenoids and vitamin C compounds were well preserved when ohmic heating was applied at pasteurization temperature for thermal processing of vegetable purees and fruit juice [65, 66]. Similarly, Rodrigues et al. in 2021 [89] studied sugarcane juice and observed that phenolic compounds did not change significantly in terms of 0, 2, 10, and 20 minutes and 3.9 and 20.5 V/mc. In general, sugarcane juice treated ohmically and conventionally were yellower (higher b*) and presented lighter coloration (higher luminosity, L*) than fresh juice [89]. OH can be used in various methods and modes, such as atmospheric condition (OHAC) and vacuum condition (OHVC). Fadavi et al. in 2018 studied these two methods on tomato juice and found that the acidity, turbidity, lycopene, and pH mean percentage changes for OHVC were 24.03, 30.77, 7.99, and 0.47, respectively. These values were lower than those for OHAC and CVH, indicating that the ohmic heating treatment carried out in the application of OHVC preserves juice quality [67].

The reason for the red color of tomato juice is lycopene. The more the methods used are programmed, the less impact they can have on the color. If the OH time increases, the oxidizing agents may increase, thus affecting lycopene and its sensory properties. Several compounds can be affected by the OH process, such as citric acid (CA), ascorbic acid (AA), limonin, naringin, and overall total polyphenol content (TPC). In 2021, Doan et al. investigated the stability of these compounds in pomelo juice due to OH. These compounds themselves can affect the color of the final product. The applied electric field strength dramatically reduced antioxidant activity and TPC at a modest electric field strength of 20 V/cm. Similar outcomes were obtained when ohmic heating was applied at 30 V/cm and 60 Hz [68]. In another study, Tuçe et al. in 2022 [69] examined OH-assisted vacuum evaporation of pekmez samples using voltage gradients of 17.5 V/cm, 20 V/cm, 22.5 V/cm, and 25 V/cm. As a result, lower h* values were observed for the samples, and the color shifted from yellow to dark brown-red. It was found that conventional vacuum evaporation and OH-assisted vacuum evaporation at relatively low voltage gradients (17.5 and 20 V/cm) had a more red tone than vacuum evaporation using ohmic heating and large voltage gradients (22.5 and 25 V/cm) [69]. In reviewing all the studies, we conclude that OH has beneficial applications in the fruit juice industry, such as inactivating enzymes and microorganisms, and may be used in the future as a useful method for pasteurization. It is of utmost importance to pay attention to the parameters affecting OH. Parameters such as voltage, temperature, and other factors play a significant role. Additionally, the characteristics of the juice used are also very important, including the amount of enzymatic and antioxidant activity, acidity, pH, and physicochemical properties.

Rascón et al. [70] reported that there is no color degradation in the Aguamiel beverage after applying ohmic heating compared to conventional pasteurization in terms of luminosity L*, hue, and chromaticity, thanks to the shorter processing time. Achir et al. [25] analyzed the carotenoid profiles in grapefruit juice and blood orange and found that epoxy xanthophyll losses were minimized from 70% to 30%, and hydroepoxyxanthophyll losses were minimized from 40% to 20% by ohmic heating compared to conventional heating. However, carotene species (lycopene and beta-carotene) remained stable. Makroo et al. [59] observed similar results in watermelon juice during ohmic heating.

7.4.3 LIQUID EGG

The egg is composed of various parts, including the yolk, white, and shell. Eggs are known for their high protein content and are considered an excellent source of protein in terms of quantity and quality. They also contain water, protein, fat, and carbohydrates. Eggs are rich in essential amino acids such as histidine, arginine, lysine, leucine, isoleucine, valine, cysteine, methionine, threonine, tryptophan, and phenylalanine. Additionally, they provide important minerals like phosphorus, calcium, magnesium, sodium, potassium, zinc, and iron. Eggs are also a valuable source of vitamins, including fat-soluble vitamins (A, D, E, and K) and water-soluble vitamins B1, B2, B3, B6, B9, and B12 [71, 72]. Furthermore, eggs contain both monounsaturated and polyunsaturated fatty acids. Due to their high nutritional value and health benefits, eggs have a wide range of applications for people of all ages. They can be used in various food products, such as meringues, pasta, bakery items, and even in healthy and sports nutrition products [73, 74]. However, it is crucial to handle eggs properly to preserve their nutritional and technological properties. Liquid eggs, in particular, can undergo quality deterioration even with mild heat treatment. To prevent such undesirable changes, Ohmic heating can be a better alternative for homogeneous and rapid heating of liquid whole eggs [21]. Alamprese et al. [75] investigated the effects of Ohmic heating on the technological properties of whole eggs, including lab parameters, under different conditions. They observed changes in the L* parameter in all conditions, but the change was not similar to the a* parameter. The highest change was reported in the b* parameter at 65.5°C for 3 minutes. The study concluded that color changes in liquid eggs due to thermal conditions are always possible. However, the least color change was observed in continuously Ohmic-heated samples. These effects and changes are influenced by temperature and time, and with proper planning, undesirable alterations in the color of liquid eggs can be minimized [75]. In a separate study conducted by Wang et al. in 2021, two approaches of Ohmic heating were examined to assess the color changes in liquid eggs. Method one involved rapid Ohmic heating, while method two included OH with total 10 min heating time. The studied parameters were in accordance with the previous study of L*a*b*. In applying these two different methods, color parameters had significant changes. For example, in the case of the parameter Δa* in the case of two methods and in the whole egg, the method of one number is 0.82 ± 0.18 and the method of two 1.28 ± 0.42 showed that these changes were significant. According to Table 7.3, a significant change occurred in the parameter Δb*. This indicates that time is a crucial factor in the effect of OH on the color of whole liquid eggs. The study also discovered that variations in proteins under different temperature and time conditions could impact egg color [76]. When comparing OH and conventional heating methods, we can refer to the study conducted by Darvishi et al. in 2012 [77]. The parameters examined were based on Table 7.3 and two previous studies. For this particular study, they considered the conditions as 30 V/cm, 60 Hz, and analyzed the egg white, yolk, and the whole egg. In the comparison of the two methods, the values of parameter L* and h* did not show significant differences. However, the results for greenness (a*) and yellowness (b*) of the ohmic heated egg samples were notably different from those obtained via conventional heating. Furthermore, the a* and b* values for convectional heating were the highest compared to the results of the ohmic heated egg samples. One of the key parameters in OH is EC, and different materials have different EC. In this study, the conductivity parameter increased linearly with temperature for whole yolks, whites, and eggs. Of course, the egg white had a higher EC. Ultimately, they concluded that OH had a more significant impact on properties such as the color of the liquid egg compared to convective heat [77]. Llave et al. [78] examined the impact of OH on the color of whole liquid egg. They used a computer vision system (CVS) to assess color changes. Using this method showed that, due to the use of OH according to Table 7.4, the uniform color gradually changed from a plain orange to a vivid yellow. In general, it can be said that OH has various parameters that should be considered for use in liquid eggs. These parameters include temperature, time, EC, and voltage. By effectively planning these parameters, the efficiency of this method can be increased and the harmful effects on the color of liquid eggs can be reduced.

7.4.4 PUREE AND PULP

Puree is suitable for infants, children, and the elderly due to its easier consumption [79, 80]. This food industry product is primarily derived from plants such as rice, potatoes, and barley. Therefore, careful attention should be given to the processing of these materials before they are turned into puree, as the presence of toxins and enzymatic activity can significantly impact the final product and potentially harm the consumer's health [81]. Enzymatic activity can also affect the color and sensory properties of the product. Blanching is an important method that, when performed correctly, can prevent adverse effects on the product. Additionally, blanching can deactivate enzymes, preserve sensory properties, and extend the shelf life of the product [82]. Ohmic heating can be used as an alternative technique for effective blanching of vegetable purees, resulting in improved color retention [35].

It is important to note that, like all other foods, the properties of the puree material, such as pH, temperature, time, and voltage, are crucial factors influencing the ohmic heating method [1, 56]. These factors can directly impact the changes in color parameters. In this section, we will explore the effects of ohmic heating on the color characteristics of puree. Numerous studies have investigated the main characteristics of ohmic heating and its impact on puree. The main factors include voltage gradient and ohmic holding time, which can affect the properties of the puree. Enzymatic activity is also a significant factor in increasing the amount of beta-carotene, both during puree storage and the ohmic heating process. During storage, enzymatic activity increases, leading to an increase in beta-carotene levels. It should be noted that ohmic heating at temperatures above 70°C can enhance enzyme activity, potentially resulting in higher levels of beta-carotene.

Mango puree of TSS 20 °Brix, 22 °Brix, and 24 °Brix, and acid content of 0.50 g/100 g, 0.56 g/100 g, and 0.62 g/100 g were treated with ohmic heating (20 to 40 V/c and heating rate of 1.71 and 4.70 °C/s for acid content 0.50 and 0.62 g/100 g, respectively). It was confirmed that an increase in TSS and acid content can cause an increase in heating rate and water activity of mango puree, respectively [83]. Water activity and heating rate can influence non-enzymatic reactions, eventually affecting the color of the puree.

In an experiment conducted by Massimiliano & co-workers [84], it was demonstrated that better quality apple puree (EC of apple puree $\sigma = 0.17$ Sm^{-1} at 25°C) was obtained by ohmic treatment (107°C for flow rate of 2,000 L/h) than conventional HT. However, brightness (L* value) was significantly reduced in both ohmic and conventional heat processing, conceivably due to the Maillard reaction and caramelization. In general, ohmic heating can cause changes in the color of puree. To maintain physicochemical and sensory properties similar to fresh produce, important considerations should be taken into account for parameters such as voltage gradient, holding time, and temperature of ohmic heaters [45].

TABLE 7.4 Impact of OH on Color Properties of Different Liquid Foods

Author/Year/ County	Food Matrix	Process Condition	a*	b*	L*	TCD	BI	Chroma	Hue	References
Ishita/2017/India	Watermelon juice.	10, 13.33, 16.66, 20, and 23.33 V/cm at 50 Hz. 95°C for 1 min, 3 min, and 5 mins.	Significant change	Significant change	23.33 V/cm: 33.06, 33.03, and 33.01 at 1 min, 3 min and 5 min, respectively, 10 V/cm for 1 min, 3 min and 5 min showed the L-value as 33.11, 33.08 and 33.09, respectively.	–	–	–	–	[58]
Makroo/2016/ India	Tomato juice	0, 15, 30, 45, 60 s	0: 6.77 15: 5.02 30: 3.84 45: 3.77 60: 3.05	0: 4.7 15: 4.47 30: 3.67 45: 3.60 60: 3.20	0: 28.77 15: 27.26 30: 25.77 45: 26.30 60: 27.18	–	–	–	0: 34.76 15: 41.68 30: 43.70 45: 43.68 60: 46.37	[85]
Saxena/2016/India	Sugarcane juice	24, 32, and 48 V/cm. 60, 70, 80, 90°C	90°C (48 V/cm): 2.329E-02	90°C (48 V/cm): 4.44E-02	90°C (48 V/cm): 7.06 E-04	–	–	–	–	[86]
Makroo/2016/ India	Watermelon juice	90°C 0, 15, 30, 45, 60 s	60s: 4.19	60s: 2.56	60s: 24.23	–	–	–	–	[59]
Li/2019/China	Water chestnut juice	A:36.7 V/cm, 50 Hz B:22 V/cm, 400 Hz C:36.7 V/cm, 400 Hz	A: 1.08 ± 0.03 B: 1.23 ± 0.02 C: 1.26 ± 0.01	A: 1.62 ± 0.09 B: 1.63 ± 0.02 C: 1.58 ± 0.01	A: 4.15 ± 0.21 B: 5.45 ± 0.11 C: 5.54 ± 0.06	–	–	–	–	[60]
Brochier/2019/ Brazil	Sugarcane juice	60 and 80°C 0–16.7 V/cm 0–10^5 Hz	0.64 ± 0.04	1.3 ± 0.1	31.8 ± 0.4	–	–	–	–	[61]
Torshizi/2020/Iran	Sour orange juice	8.33, 10.83, 13.33 V/cm	–	–	–	–	–	–	–	[63]

TABLE 7.4 *(Continued)*

Author/Year/ County	Food Matrix	Process Condition	a*	b*	L*	TCD	BI	Chroma	Hue	References
Shao/2021/China	Apple juice	45 Hz to 500 Hz. 0 V to 320 V	–0.22 ± 0.05	–0.15 ± 0.09	30.41 ± 0.35	–	–	–	–	[18]
Rodríguez/2021/ Argentina	Carrot juice	60 kHz	18.96 ± 0.46	30.44 ± 0.59	41.73 ± 0.21	–	148.6 ± 3.8	35.87 ± 0.73	58.14 ± 0.26	[18]
Makroo/2022/ India	Pineapple juice	80 ± 2°C (held for 1 min) 16–36 V/cm	16 V: –0.68 ± 0.00 20 V: –0.68 ± 0.01 24 V: –0.68 ± 0.01 28 V: –0.69 ± 0.00 32 V: –0.71 ± 0.00 36 V: –0.72 ± 0.00	16 V: 0.54 ± 0.03 20 V: 0.56 ± 0.03 24 V: 0.57 ± 0.03 28 V: 0.58 ± 0.03 32 V: 0.58 ± 0.00 36 V: 1.17 ± 0.01	16 V: 24.07 ± 0.72 20 V: 24.04 ± 0.57 24 V: 24.01 ± 0.93 28 V: 24.19 ± 1.22 32 V: 24.29 ± 0.47 36 V: 25.95 ± 0.17	–	–	16 V: 0.87 ± 0.01 20 V: 0.88 ± 0.02 24 V: 0.89 ± 0.02 28 V: 0.90 ± 0.02 32 V: 0.92 ± 0.02 36 V: 1.37 ± 0.00	16 V: 38.60 ± 0.06 20 V: 39.72 ± 1.05 24 V: 39.90 ± 1.17 28 V: 40.19 ± 1.02 32 V: 39.27 ± 1.61 36 V: 58.37 ± 0.28	[62]
Kim/2017/Korea	Tomato juice	0.06, 0.2, 0.5, 1 kHz	0.06: 0.55 ± 0.39 0.2: 0.49 ± 0.18 0.5: 0.52 ± 0.22 1: 0.52 ± 0.31	0.06: 3.52 ± 0.44 0.2: 3.49 ± 0.05 0.5: 3.41 ± 0.08 1: 3.54 ± 0.38	0.06: 34.42 ± 0.50 0.2: 34.26 ± 0.27 0.5: 34.47 ± 0.71 1: 34.26 ± 0.55	–	–	–	–	[87]
Kim/2017/Korea	Apple juice	26.7 V/cm; 25 kHz 85–100°C for 30–90 s	85°C (90s): 3.0 ± 0.02 90°C (90s): 3.0 ± 0.03 95°C (95): 3.0 ± 0.1 100°C (90s): 3.0 ± 0.04	85°C (90s): 29.2 ± 0.6 90°C (90s): 29.7 ± 0.3 95°C (95): 29.2 ± 0.3 100°C (90s): 27.1 ± 0.3	85°C (90s): 66.3 ± 0.3 90°C (90s): 66.4 ± 0.04 95°C (95): 66.1 ± 0.2 100°C (90s): 65.4 ± 0.1	–	–	–	–	[88]
Norouzi/2021/Iran	Sour cherry juice	The range of S/m for 30–50 V was 0.70–2.52.	30 V: 7.72 35 V: 1.31 40 V: 1.24 45 V: 37.7 50 V: 2.13	30 V: 0.04 35 V: 0.13 40 V: 0.18 45 V: 4.19 50 V: 1.08	30 V: 3.02 35 V: 3.96 40 V: 0.36 45 V: 1.44 50 V: 0.97	–	–	–	–	[64]

TABLE 7.4 *(Continued)*

Author/Year/ County	Food Matrix	Process Condition	a*	b*	L*	TCD	BI	Chroma	Hue	References
Rodrigues/2021/ Brazil	Sugarcane juice	30 V/cm, 60 Hz 20 min.	0.31 ± 0.02	1.60 ± 0.05	32.10 ± 0.12	–	–	–	–	[89]
Brochier/2018/ Brazil	Sugarcane juice	10–10^5 Hz. 3.9 and 20.5 V/cm 25 min.	3.9 V/cm: 0.83 ± 0.04 20.5 V/cm: 0.55 ± 0.07	3.9 V/cm: 2.61 ± 0.07 20.5 V/cm: 1.6 ± 0.2	3.9 V/cm: 32.3 ± 0.3 20.5 V/cm: 32.4 ± 0.1	–	–	–	–	[61]
Darvishi/2020/ Iran	Mulberry juice	15, 20, 25, 30 V/cm, 50 Hz.	–	–	–	–	–	–	–	[90]
Fadavi/2018	Tomato juice	Atmospheric condition (OHAC) and vacuum condition (OHVC).	–	–	–	–	–	–	–	[67]
Doan/2021/Viet Nam	Pomelo juice	20–80°C 60 Hz 30 V/cm	–	–	–	–	–	–	–	[68]
Abedelmaksoud/ 2018/Egypt	Apple juice	40 V/cm, 80°C	–3.26 ± 0.06	1.57 ± 0.09	38.85 ± 0.17	–	–	–	–	[91]
Bhat/2017/India	Bottle gourd (*Lagenaria siceraria*) juice.	60–90°C and time 1–5 min	–	–	–	–	60°C, 5 min: 174.64 70°C, 5 min: 156.00 80°C, 5 min: 86.40 90°C, 4 min: 86.41	–	–	[36]

TABLE 7.4 *(Continued)*

Author/Year/ County	Food Matrix	Process Condition	a*	b*	L*	TCD	BI	Chroma	Hue	References
Darvishi/2020/ Iran	Grape juice	15, 20, 25, and 30 V/cm at 50 Hz	–	–	–	–	–	–	–	[92]
Park/2017/Korea	Apple juice	30, 40, 50, and 60 V/cm	30 V/cm, ?Brix 72, 60s: 0.38 ± 0.02 60 V/cm, ?Brix 72, 60s: 0.36 ± 0.07	30 V/cm, ?Brix 72, 60s: 4.09 ± 0.05 60 V/cm, ?Brix 72, 60s: 4.19 ± 0.08	30 V/cm, ?Brix 72, 60s: 26.44 ± 0.08 60 V/cm, ?Brix 72, 60s: 26.03 ± 0.02	–	–	–	–	[93]
Tunc/2021/Turkey	Grape pekmez	17.5, 20, 22.5, and 25 V/cm 57, 28.5, 32, 39, and 50 minutes	17.5 V/cm: 1.73_0.20 20 V/cm: 1.73_0.16 22.5 V/cm: 1.99_0.30 25 V/cm: 1.98_0.10	17.5 V/cm: 0.43_0.12 20 V/cm: 0.41_0.11 22.5 V/cm: 0.63_0.36 25 V/cm: 0.68_0.18	17.5 V/cm: 25.89_1.06 20 V/cm: 25.53_0.90 22.5 V/cm: 25.15_0.99 25 V/cm: 25.66_0.99	–	–	17.5 V/cm: 1.78_0.19 20 V/cm: 1.78_0.15 22.5 V/cm: 2.09_0.29 25 V/cm: 2.09_0.10	17.5 V/cm: 14.07_1.57 20 V/cm: 13.40_1.20 22.5 V/cm: 17.80_2.54 25 V/cm: 18.98_0.89	[69]
Brochier/2016/ Brazil	Sugarcane juice	2.0, 3.0, 3.5, and 4.0 min for 60, 70, 75 and 80°C. 3.57 to 4.39 V/cm	–	–	–	–	–	–	–	[94]
Alamprese/ 2019/ Italy	Liquid whole egg	65.5°C×3 min (coded 65.5–3), 70°C×1 min (coded 70–1), and 67°C×4.5 min (coded 67–4.5)	Batch 1: n.d. Batch 2: 2.2 ± 0.4 Batch 3: 3.0 ± 0.4 65.5°C×3 min: 2.6 ± 0.7 70°C×1 min: 2.9 ± 0.7 67°C×4.5 min: 3.1 ± 0.7	Batch 1: n.d. Batch 2: 42.2 ± 0.2 Batch 3: 40.8 ± 0.2 65.5°C×3 min: 42.1 ± 0.3 70°C×1 min: 40.6 ± 0.3 67°C×4.5 min: 40.7 ± 0.3	Batch 1: n.d. Batch 2: 61.0 ± 0.3 Batch 3: 60.6 ± 0.3 65.5°C×3 min: 59.4 ± 0.4 70°C×1 min: 60.0 ± 0.4 67°C×4.5 min: 59.9 ± 0.4	–	–	–	–	[75]

TABLE 7.4 *(Continued)*

Author/Year/County	Food Matrix	Process Condition	a*	b*	L*	TCD	BI	Chroma	Hue	References
Wang/2021/Japan	Liquid egg whites, yolks, and whole eggs	20 kHz and 50 V/cm	Approach 1: 0.51 ± 0.42 Approach 2: 0.31 ± 0.65 Whole Approach 1: 0.82 ± 0.18 Approach 2: 1.28 ± 0.42 Yolk Approach 1: 1.46 ± 0.08 Approach 2: 1.18 ± 0.20	White Approach 1: 0.00 ± 0.00 Approach 2: 0.00 ± 0.00 Whole Approach 1: 1.33 ± 0.28 Approach 2: 0.61 ± 0.29 Yolk Approach 1: 0.63 ± 2.42 Approach 2: 0.94 ± 1.53	White Approach 1: 1.56 ± 1.85 Approach 2: 2.01 ± 0.78 Whole Approach 1: 0.00 ± 1.01 Approach 2: 0.48 ± 0.33 Yolk Approach 1: 0.66 ± 2.28 Approach 2: 0.74 ± 1.33	–	–	White Approach 1: 0.31 ± 0.42 Approach 2: 0.44 ± 0.57 Whole Approach 1: 1.57 ± 0.14 Approach 2: 1.45 ± 0.25 Yolk Approach 1: 2.30 ± 0.61 Approach 2: 1.49 ± 0.02	–	[76]
Darvishi/2012/Iran	Liquid egg whites, yolks, and whole eggs	30 V/cm, 60 Hz	White: 0.80 ± 0.04 Whole: 6.78 ± 1.42 Yolk: 16.68 ± 1.17	White: 11.36 ± 1.34 Whole: 24.09 ± 2.60 Yolk: 40.21 ± 0.72	White: 80.16 ± 1.59 Whole: 77.93 ± 1.79 Yolk: 70.86 ± 0.93	–	–	–	–	[77]
Llave/2018/Japan	Egg yolks and whites	30 V/cm, 20 kHz	–	–	–	–	–	–	–	[78]
Gratz/2021/Austria	Carrot-based Purees	115°C 121°C 125°C 130°C	–	–	–	–	–	–	–	[19]
Hardinasinta/2021/Indonesia	50% and 100% Mulberry (*Morus nigra*) Puree	30–90°C	–	–	–	–	–	–	–	[95]

7.5 COLOR PRESERVATION OF FLUID FOOD IN OH

Ohmic heating works efficiently and effectively for the heating of pumpable foods. It is gaining tremendous attention for the processing of liquid foods such as milk, soymilk, whey, soups, syrup, stews, fruit, and vegetable juices, purees, pulps, and liquid egg for pasteurization, sterilization, and cooking [96]. It has the ability to provide continuous inline heating for the cooking and sterilization of liquid and semi-liquid food products. The rate of heating in OH processing is directly related to EC, and the factors affecting ohmic heating are temperature, applied frequency, and particle size. EC increases with an increase in temperature, an increase in frequency, and a decrease in particle diameter [17]. In comparison to conventional heating methods, ohmic heating technology is a promising efficient and cost-effective method for the pasteurization of liquid food [42, 97]. The properties of food, flow properties in the pumping system, and other design criteria govern the process control in the ohmic heating system. During processing on an industrial scale, the applied frequency, voltage gradient, electrode type used, EC differences, temperature range, flow properties, concentration, and type of food are all considered critical design parameters in general [22]. Good process control results in good product quality, which includes the preservation or enhancement of the color profile as well. Ohmic heating can decrease the lag period of fermentative bacteria in the food medium. Therefore, OH has huge potential to benefit wine or beer-like fermented products by reducing the processing time. However, more research and investigations are required to fully understand the efficacy of ohmic heating for the inactivation of pathogenic bacteria. In such scenarios, ohmic heating can also be complemented by other processing technologies to achieve better outcomes. For example, ohmic-assisted hydrodistillation is used to distill ethanol and essential oils, which is based on volumetric heating and has been proven to be more effective than distillation using conventional heating [98]. Similarly, other similar technologies can be combined with ohmic heating to obtain optimal color quality in fluid food. Some researchers have reported that the color difference (ΔE) is similar for carrot juices treated with thermal [99] and novel technologies [100]. On the other hand, some researchers claim that other fruits show minimal color differences (ΔE), the highest red color, and the highest total phenolics when compared to conventional thermal heating. Ohmic heating resulted in a lesser total color (E) change in watermelon juice compared to traditional heating at similar temperatures [4].

Changes in the color of food are somewhat related to its heat-treatment status. During thermal processing of fruits and vegetables, food color can be influenced by reactions such as non-enzymatic browning and pigment destruction. Color change indicates the extent of quality deterioration due to heat processing while evaluating food quality [17]. Therefore, all aspects of processing liquid food, especially the effects on color, should be extensively studied for every kind of beverage, syrup, soup, puree, and fruit juice, as color is the first component to degrade during heat processing. To efficiently and effectively process different kinds of food, specific ohmic heaters need to be designed with a specific EC range. Scaling up OH is a key point to consider for the advancement and imminent use of this method, as uniform heating is possible with just uniform electricity generation [101]. In the future, the development of adequate quality and safety protocols for obtaining FDA approval for all possible food materials can also help in the commercialization of ohmic heating technology.

7.6 SUMMARY

No doubt, ohmic heating has many advantages and is highly recommended for processing liquid or semi-liquid food. Since ohmic heating depends on the electrical conductivity (EC) of the food, different types of food with different EC values may require different treatment methods. The ideal method should be able to maintain all the quality attributes of the food, including its color.

When it comes to processing liquid food, ohmic heating treatment is better at preserving the natural food color compared to conventional heating methods. After ohmic heating, the color pigment lycopene content increases, carotene remains stable, and chlorophyll derivatives are retained better than with conventional heating methods. However, the content of anthocyanin (ACN) may vary depending on the product. Further research is needed to understand the changes in color profile that occur during ohmic heating processing of fluid food, in order to design ohmic heaters for a wider variety of foods in the future. Similarly, there is a lack of sufficient research comparing ohmic heating to other thermal processing methods under the same processing conditions. In the near future, advancements in ohmic heating for preserving the color component of liquid food are projected to become an important food processing technology.

KEYWORDS

- **carotene**
- **chlorophyll**
- **color compounds**
- **color difference**
- **color preservation**
- **food processing**
- **ohmic heating**

REFERENCES

1. Indiarto, R., & Rezaharsamto, B., (2020). A review on ohmic heating and its use in food. *Int. J. Sci. Technol. Res, 9*(2), 485–490.
2. Patel, A., & Singh, M., (2018). Ohmic heating for food products: A review. *Curr. J. Appl. Sci. Technol, 27*, 1–7.
3. Niakousari, M., Hedayati, S., Gahruie, H. H., Greiner, R., & Roohinejad, S., (2019). Impact of ohmic processing on food quality and composition. In: *Effect of Emerging Processing Methods on the Food Quality* (pp. 1–26). Springer.
4. Makrooa, H. A., & Srivastava, B., (2017). *Effects of Ohmic Heating on Different Liquid Food Materials.*
5. Gavahian, M., Sastry, S., Farhoosh, R., & Farahnaky, A., (2020). Ohmic heating as a promising technique for extraction of herbal essential oils: Understanding mechanisms, recent findings, and associated challenges. *Adv. Food Nutr. Res., 91*, 227–273.
6. Caruggi, N., Lucisano, M., Feyissa, A. H., Rahimi, Y. S., & Mohammadifar, M. A., (2019). Effect of ohmic heating on the formation and texture of acid milk gels. *Food Biophysics, 14*(3), 249–259.
7. Cappato, L. P., et al., (2017). Ohmic heating in dairy processing: Relevant aspects for safety and quality. *Trends in Food Science & Technology, 62*, 104–112.
8. Aghajanzadeh, S., Ziaiifar, A. M., & Verkerk, R., (2022). Effect of thermal and non-thermal treatments on the color of citrus juice: A review. *Food Reviews International*, 1–23.

9. Sandoval, J. R. M., Rosas, M. E. M., Sandoval, E. M., Velasco, M. M. M., & De Ávila, H. C., (2018). *Color Analysis and Image Processing Applied in Agriculture,* 32. Intech Rijeka.
10. Spence, C., & Piqueras-Fiszman, B., (2016). Food color and its impact on taste/flavor perception. In: *Multisensory Flavor Perception* (pp. 107–132). Elsevier.
11. Tomasević, I., et al., (2019). Comparison of a computer vision system vs. traditional colorimeter for color evaluation of meat products with various physical properties. *Meat Sci, 148*, 5–12.
12. Cabas, B. M., & Icier, F., (2021). Ohmic heating–assisted extraction of natural color matters from red beetroot. *Food and Bioprocess Technology, 14*(11), 2062–2077.
13. Chen, J., Du, J., Li, M., & Li, C., (2020). Degradation kinetics and pathways of red raspberry anthocyanins in model and juice systems and their correlation with color and antioxidant changes during storage. *LWT, 128*, 109448.
14. Inyang, U. E., Oboh, I. O., & Etuk, B. R., (2018). Kinetic models for drying techniques—Food materials. *Advances in Chemical Engineering and Science, 8*(2), 27.
15. Dash, K. K., Chakraborty, S., Kumar, R., & Ghosh, T., (2021). Rate kinetics in thermal food-processing operations. In: *Food Processing* (pp. 1–23). CRC Press.
16. Banti, M., (2020). *Review on Electrical Conductivity in Food, the Case in Fruits, and Vegetables, 4*, 80–89.
17. Kaur, N., & Singh, A. K., (2016). Ohmic heating: Concept and applications—A review. *Crit. Rev. Food Sci. Nutr., 56*(14), 2338–2351.
18. Rodríguez, L. M. N., et al., (2021). Comparison of the quality attributes of carrot juice pasteurized by ohmic heating and conventional heat treatment. *LWT, 145*, 111255.
19. Gratz, M., et al., (2021). Gentle sterilization of carrot-based purees by high-pressure thermal sterilization and ohmic heating and influence on food processing contaminants and quality attributes. *Front. Nutr., 8*, 643837.
20. Müller, W. A., Marczak, L. D. F., & Sarkis, J. R., (2020). Microbial inactivation by ohmic heating: Literature review and influence of different process variables. *Trends in Food Science & Technology, 99*, 650–659.
21. Alkanan, Z. T., Altemimi, A. B., Al-Hilphy, A. R. S., Watson, D. G., & Pratap-Singh, A., (2021). Ohmic heating in the food industry: Developments in concepts and applications during 2013–2020. *Applied Sciences, 11*(6), 2507.
22. Jaeger, H., et al., (2016). Opinion on the use of ohmic heating for the treatment of foods. *Trends in Food Science & Technology, 55*, 84–97.
23. Kubo, M. T. K., et al., (2020). Non-thermal effects of microwave and ohmic processing on microbial and enzyme inactivation: A critical review. *Current Opinion in Food Science, 35*, 36–48.
24. Jan, B., Shams, R., Rizvi, Q. E. H., & Manzoor, A., (2021). Ohmic heating technology for food processing: A review of recent developments. *J. Postharvest Technol., 9*, 20–34.
25. Achir, N., Dhuique-Mayer, C., Hadjal, T., Madani, K., Pain, J. P., & Dornier, M., (2016). Pasteurization of citrus juices with ohmic heating to preserve the carotenoid profile. *Innovative Food Science & Emerging Technologies, 33*, 397–404.

26. Aurina, K., & Sari, A., (2022). Ohmic heating: A review and application in food industry. In: *2nd International Conference on Smart and Innovative Agriculture (ICoSIA 2021)* (pp. 107–113).
27. Salari, S., & Jafari, S. M., (2020). The influence of ohmic heating on degradation of food bioactive ingredients. *Food Engineering Reviews, 12*(2), 191–208.
28. Abedi, A., Hemmati, F., Abedini, A. H., Mohammadi, A., & Moslemi, M., (2021). Application of thermal ultrasound-assisted liquid–liquid micro-extraction coupled with HPLC-UV for rapid determination of synthetic phenolic antioxidants in edible oils. *J. Am. Oil Chem. Soc., 98*(10), 969–978.
29. Zhang, Y., Dong, L., Zhang, J., Shi, J., Wang, Y., & Wang, S., (2021). Adverse effects of thermal food processing on the structural, nutritional, and biological properties of proteins. *Annual Review of Food Science and Technology, 12*, 259–286.
30. Yoo, S. R., Lee, S. W., & Jeon, H. M., (2020). The role of customer experience, food healthiness, and value for revisit intention in grocerant. *Sustainability, 12*(6), 2359.
31. Doan, K. N., Lai, D. Q., Kim Le, P. T., & Le, T. N., (2021). Inactivation of pectin methylesterase and lactobacillus plantarum by ohmic heating in pomelo juice. *International Journal of Food Science & Technology, 56*(4), 1987–1995.
32. Saberian, H., Hamidi-Esfahani, Z., Gavlighi, H. A., & Barzegar, M., (2017). Optimization of pectin extraction from orange juice waste assisted by ohmic heating. *Chemical Engineering and Processing: Process Intensification, 117*, 154–161.
33. Pereira, R. N., et al., (2018). Ohmic heating for the dairy industry: A potential technology to develop probiotic dairy foods in association with modifications of whey protein structure. *Current Opinion in Food Science, 22*, 95–101.
34. Sabanci, S., & Icier, F., (2022). Evaluation of an ohmic assisted vacuum evaporation process for orange juice pulp. *Food and Bioproducts Processing, 131*, 156–163.
35. Kumar, T., (2018). A review on ohmic heating technology: Principle, applications, and scope. *International Journal of Agriculture, Environment and Biotechnology, 11*(4), 679–687.
36. Bhat, S., Saini, C. S., & Sharma, H. K., (2017). Changes in total phenolic content and color of bottle gourd (*Lagenaria siceraria*) juice upon conventional and ohmic blanching. *Food Science and Biotechnology, 26*(1), 29–36.
37. Lasekan, O., Ng, S., Azeez, S., Shittu, R., Teoh, L., & Gholivand, S., (2017). Effect of pulsed electric field processing on flavor and color of liquid foods. *Journal of Food Processing and Preservation, 41*(3), e12940.
38. Mannozzi, C., et al., (2019). Influence of pulsed electric field and ohmic heating pretreatments on enzyme and antioxidant activity of fruit and vegetable juices. *Foods, 8*(7), 247.
39. Sagita, D., et al., (2022). Experimental investigation of heating pattern, energy requirement and electrical conductivity in a batch ohmic heating system for coffee fermentation. *Innovative Food Science & Emerging Technologies, 76*, 102946.
40. Makroo, H. A., Rastogi, N. K., & Srivastava, B., (2020). Ohmic heating assisted inactivation of enzymes and microorganisms in foods: A review. *Trends in Food Science & Technology, 97*, 451–465.
41. Tian, X., Yu, Q., Wu, W., & Dai, R., (2018). Inactivation of microorganisms in foods by ohmic heating: A review. *J. Food Prot., 81*(7), 1093–1107.

42. Misra, S., Mandliya, S., & Panigrahi, C., (2022). Ohmic heating: Principles and applications. *Thermal Food Engineering Operations*, 261–299.
43. Silva, R., et al., (2020). Advantages of using ohmic heating in Dulce de Leche manufacturing. *Innovative Food Science & Emerging Technologies, 65*, 102475.
44. Cheng, N., Barbano, D. M., & Drake, M. A., (2018). Hunter versus CIE color measurement systems for analysis of milk-based beverages. *J. Dairy Sci., 101*(6), 4891–4905.
45. Suebsiri, N., Kokilakanistha, P., Laojaruwat, T., Tumpanuvat, T., & Jittanit, W., (2019). The pasteurization of milk applying ohmic heating in comparison with conventional method and the quality attributes of lactose-free milk. *Research Journal Phranakhon Rajabhat: Science and Technology, 14*(1), 25–35.
46. Parmar, P., Singh, A. K., Meena, G. S., Borad, S., & Raju, P. N., (2018a). Application of ohmic heating for concentration of milk. *J. Food Sci. Technol, 55*(12), 4956–4963.
47. Parmar, P., Singh, A. K., Gupta, V. K., Borad, S., & Raju, P. N., (2018b). Effect of packaging material on shelf-life of basundi manufactured using ohmic heating. *Indian Journal. Dairy Sci., 71*(3), 215–222.
48. Chudy, S., Bilska, A., Kowalski, R., & Teichert, J., (2020). Color of milk and milk products in CIE L* a* b* space. *Medycyna Weterynaryjna, 76*(2), 77–81.
49. Buchweitz, M., (2016). Natural solutions for blue colors in food. In: *Handbook on Natural Pigments in Food and Beverages* (pp. 355–384). Elsevier.
50. Luzardo-Ocampo, I., Ramírez-Jiménez, A. K., Yañez, J., Mojica, L., & Luna-Vital, D. A., (2021). Technological applications of natural colorants in food systems: A review. *Foods, 10*, 634. [Note: MDPI stays neutral with regard to jurisdictional claims in published].
51. Dey, S., & Nagababu, B. H., (2022). Applications of food color and bio-preservatives in the food and its effect on the human health. *Food Chemistry Advances*, 100019.
52. Lu, W., et al., (2021). Antioxidant activity and healthy benefits of natural pigments in fruits: A Review. *International Journal of Molecular Sciences, 22*(9), 4945.
53. Ruxton, C. H. S., & Myers, M., (2021). Fruit juices: Are they helpful or harmful? An evidence review. *Nutrients, 13*(6), 1815.
54. Abedini, A., Mahdavi, A., Mirza, A. A., Hejazi, E., & Hosseini, H., (2022). A review on dietary additive, food supplement and exercise effects on the prevention of COVID-19. *Nutrition and Food Sciences Research, 9*(1), 1–14.
55. Dreher, M. L., (2018). Whole fruits and fruit fiber emerging health effects. *Nutrients, 10*(12), 1833.
56. Alkanan, Z. T., Altemimi, A. B., Al-Hilphy, A. R. S., Cacciola, F., & Ibrahim, S. A., (2021). Application and effects of ohmic-vacuum combination heating on the quality factors of tomato paste. *Foods, 10*(12), 2920.
57. Umair, M., et al., (2022). Thermal treatment alternatives for enzymes inactivation in fruit juices: Recent breakthroughs and advancements. *Ultrasonic's Sonochemistry*, 105999.
58. Ishita, C., & Athmaselvi, K. A., (2017). Changes in pH and color of watermelon juice during ohmic heating. *International Food Research Journal, 24*(2), 741.
59. Makroo, H. A., Saxena, J., Rastogi, N. K., & Srivastava, B., (2017). Ohmic heating assisted polyphenol oxidase inactivation of watermelon juice: Effects of the treatment

on pH, lycopene, total phenolic content, and color of the juice. *Journal of Food Processing and Preservation, 41*(6), e13271.
60. Li, X., Xu, X., Wang, L., & Regenstein, J. M., (2019). Effect of ohmic heating on physicochemical properties and the key enzymes of water chestnut juice. *Journal of Food Processing and Preservation, 43*(4), e13919.
61. Brochier, B., Mercali, G. D., & Marczak, L. D. F., (2019). Effect of moderate electric field on peroxidase activity, phenolic compounds, and color during ohmic heating of sugarcane juice. *Journal of Food Processing and Preservation, 43*(12), e14254.
62. Makroo, H. A., Srivastava, B., & Jabeen, A., (2022). Influence of mild electric field (MEF) on polyphenol oxidase and quality attributes of pineapple juice during ohmic heating. *LWT, 156*, 113021.
63. Torshizi, M. V., Azadbakht, M., & Kashaninejad, M., (2020). Application of response surface method to energy and exergy analyses of the ohmic heating dryer for sour orange juice. *Fuel, 278*, 118261.
64. Norouzi, S., Fadavi, A., & Darvishi, H., (2021). The ohmic and conventional heating methods in concentration of sour cherry juice: Quality and engineering factors. *Journal of Food Engineering, 291*, 110242.
65. Funcia, E. S., Gut, J. A. W., & Sastry, S. K., (2020). Effect of electric field on pectin esterase inactivation during orange juice pasteurization by ohmic heating. *Food and Bioprocess Technology, 13*(7), 1206–1214.
66. Müller, W. A., Pasin, M. V. A., Sarkis, J. R., & Marczak, L. D. F., (2021). Effect of pasteurization on *Aspergillus fumigatus* in apple juice: Analysis of the thermal and electric effects. *International Journal of Food Microbiology, 338*, 108993.
67. Fadavi, A., Yousefi, S., Darvishi, H., & Mirsaeedghazi, H., (2018). Comparative study of ohmic vacuum, ohmic, and conventional-vacuum heating methods on the quality of tomato concentrate. *Innovative Food Science & Emerging Technologies, 47*, 225–230.
68. Doan, N. K., Lai, Q. D., Le, T. K. P., & Le, N. T., (2021). Influences of AC frequency and electric field strength on changes in bioactive compounds in ohmic heating of pomelo juice. *Innovative Food Science & Emerging Technologies, 72*, 102754.
69. Tunç, M. T., Akdoğan, A., Baltacı, C., Kaya, Z., & Odabaş, H. İ., (2022). Production of grape pekmez by ohmic heating-assisted vacuum evaporation. *Food Science and Technology International, 28*(1), 72–84.
70. Rascón, L., Cruz, M., Rodríguez-Jasso, R. M., Neira-Vielma, A. A., Ramírez-Barrón, S. N., & Belmares, R., (2020). Effect of ohmic heating on sensory, physicochemical, and microbiological properties of 'aguamiel' of *Agave salmiana*. *Foods, 9*(12), 1834.
71. Attia, Y. A., Al-Harthi, M. A., Korish, M. A., & Shiboob, M. H., (2020). Protein and amino acid content in four brands of commercial table eggs in retail markets in relation to human requirements. *Animals, 10*(3), 406.
72. Kuang, H., Yang, F., Zhang, Y., Wang, T., & Chen, G., (2018). The impact of egg nutrient composition and its consumption on cholesterol homeostasis. *Cholesterol, 2018*.
73. Li, J., et al., (2021). Hen egg yolk in food industry-A review of emerging functional modifications and applications. *Trends in Food Science & Technology, 115*, 12–21.

74. Xiao, N., et al., (2021). A review on recent advances of egg byproducts: Preparation, functional properties, biological activities, and food applications. *Food Research International, 147*, 110563.
75. Alamprese, C., Cigarini, M., & Brutti, A., (2019). Effects of ohmic heating on technological properties of whole egg. *Innovative Food Science & Emerging Technologies, 58*, 102244.
76. Wang, C., Llave, Y., Sakai, N., & Fukuoka, M., (2021). Analysis of thermal processing of liquid eggs using a high frequency ohmic heating: Experimental and computer simulation approaches. *Innovative Food Science & Emerging Technologies, 73*, 102792.
77. Darvishi, H., Khoshtaghaza, M. H., Zarein, M., & Azadbakht, M., (2012). Ohmic processing of liquid whole egg, white egg, and yolk. *Agricultural Engineering International: CIGR Journal, 14*(4), 224–230.
78. Llave, Y., Fukuda, S., Fukuoka, M., Shibata-Ishiwatari, N., & Sakai, N., (2018). Analysis of color changes in chicken egg yolks and whites based on degree of thermal protein denaturation during ohmic heating and water bath treatment. *J. Food Eng., 222*, 151–161.
79. Koletzko, B., et al., (2019). Complementary foods in baby food pouches: Position statement from the nutrition commission of the German society for pediatrics and adolescent medicine (DGKJ, eV). *Mol. Cell Pediatr., 6*(1), 1–5.
80. Rapley, G., (2016). Are purėed foods justified for infants of 6 months? What does the evidence tell us? *Journal of Health Visiting, 4*(6), 289–295.
81. Rather, I. A., Koh, W. Y., Paek, W. K., & Lim, J., (2017). The sources of chemical contaminants in food and their health implications. *Front Pharmacol, 8*, 830.
82. Moon, K. M., Kwon, E. B., Lee, B., & Kim, C. Y., (2020). Recent trends in controlling the enzymatic browning of fruit and vegetable products. *Molecules, 25*(12), 2754.
83. Makroo, H. A., Prabhakar, P. K., Rastogi, N. K., & Srivastava, B., (2019). Characterization of mango puree based on total soluble solids and acid content: Effect on physico-chemical, rheological, thermal and ohmic heating behavior. *LWT, 103*, 316–324.
84. Rinaldi, M., Langialonga, P., Dhenge, R., Aldini, A., & Chiavaro, E., (2021). Quality traits of apple puree treated with conventional, ohmic heating and high-pressure processing. *European Food Research and Technology, 247*(7), 1679–1688.
85. Makroo, H. A., Rastogi, N. K., & Srivastava, B., (2017). Enzyme inactivation of tomato juice by ohmic heating and its effects on physico-chemical characteristics of concentrated tomato paste. *Journal of Food Process Engineering, 40*(3), e12464.
86. Saxena, J., Ahmad, M. H., & Srivastava, B., (2017). Effect of ohmic heating on polyphenol oxidase (PPO) inactivation and color change in sugarcane juice. *Journal of Food Process Engineering, 40*(3), e12485.
87. Kim, S. S., Choi, W., & Kang, D. H., (2017). Application of low frequency pulsed ohmic heating for inactivation of foodborne pathogens and MS-2 phage in buffered peptone water and tomato juice. *Food Microbiol., 63*, 22–27.
88. Kim, N. H., Ryang, J. H., Lee, B. S., Kim, C. T., & Rhee, M. S., (2017). Continuous ohmic heating of commercially processed apple juice using five sequential electric fields results in rapid inactivation of *Alicyclobacillus acidoterrestris* spores. *Int. J. Food Microbiol., 246*, 80–84.

89. Rodrigues, N. P., Brochier, B., De Medeiros, J. K., Marczak, L. D. F., & Mercali, G. D., (2021). Phenolic profile of sugarcane juice: Effects of harvest season and processing by ohmic heating and ultrasound. *Food Chemistry, 347*, 129058.
90. Darvishi, H., Salami, P., Fadavi, A., & Saba, M. K., (2020). Processing kinetics, quality and thermodynamic evaluation of mulberry juice concentration process using ohmic heating. *Food and Bioproducts Processing, 123*, 102–110.
91. Abedelmaksoud, T. G., Mohsen, S. M., Duedahl-Olesen, L., Elnikeety, M. M., & Feyissa, A. H., (2018). Optimization of ohmic heating parameters for polyphenoloxidase inactivation in not-from-concentrate elstar apple juice using RSM. *Journal of Food Science and Technology, 55*(7), 2420–2428.
92. Darvishi, H., Koushesh, S. M., Behroozi-Khazaei, N., & Nourbakhsh, H., (2020). Improving quality and quantity attributes of grape juice concentrate (molasses) using ohmic heating. *J. Food Sci. Technol., 57*(4), 1362–1370.
93. Park, I. K., Ha, J. W., & Kang, D. H., (2017). Investigation of optimum ohmic heating conditions for inactivation of *Escherichia coli*O157: H7, *Salmonella enterica* serovar *Typhimurium*, and *Listeria monocytogenes* in apple juice. *BMC Microbiology, 17*(1), 1–8.
94. Brochier, B., Mercali, G. D., & Marczak, L. D. F., (2016). Influence of moderate electric field on inactivation kinetics of peroxidase and polyphenol oxidase and on phenolic compounds of sugarcane juice treated by ohmic heating. *LWT, 74*, 396–403.
95. Hardinasinta, G., Mursalim, M., Muhidong, J., & Salengke, S., (2021). *Effect of Ohmic Heating on the Rheological Characteristics and Electrical Conductivity of Mulberry (Morus Nigra) Puree* (p. 289). By Institute of Animal Reproduction and Food Research of the Polish Academy of Sciences, Olsztyn.
96. Balthazar, C. F., et al., (2022). Conventional and ohmic heating pasteurization of fresh and thawed sheep milk: Energy consumption and assessment of bacterial microbiota during refrigerated storage. *Innovative Food Science & Emerging Technologies, 76*, 102947.
97. Priyadarshini, A., Rayaguru, K., & Nayak, P. K., (2020). Influence of ohmic heating on fruits and vegetables: A review. *Journal of Critical Reviews, 7*(19), 1952–1959.
98. Mesías, M., Wagner, M., George, S., & Morales, F. J., (2016). Impact of conventional sterilization and ohmic heating on the amino acid profile in vegetable baby foods. *Innovative Food Science & Emerging Technologies, 34*, 24–28.
99. Gouma, M., Álvarez, I., Condón, S., & Gayán, E., (2020). Pasteurization of carrot juice by combining UV-C and mild heat: Impact on shelf-life and quality compared to conventional thermal treatment. *Innovative Food Science & Emerging Technologies, 64*, 102362.
100. Szczepańska, J., Barba, F. J., Skąpska, S., & Marszałek, K., (2020). High pressure processing of carrot juice: Effect of static and multi-pulsed pressure on the polyphenolic profile, oxidoreductases activity and color. *Food Chemistry, 307*, 125549.
101. Singh, A. P., Yen, P. P. L., Ramaswamy, H. S., & Singh, A., (2018). Recent advances in agitation thermal processing. *Current Opinion in Food Science, 23*, 90–96.

CHAPTER 8

IMPACT OF UV IRRADIATION ON THE COLOR PROFILE OF FOOD

BHARATHIPRIYA RAJASEKARAN,[1] GOKULPRASANTH MURUGAN,[2] and KARTHIKEYAN VENKATACHALAM[3]

[1]*International Center of Excellence in Seafood Science and Innovation, Faculty of Agro-Industry, Prince of Songkla University, Hat Yai, Songkhla, Thailand*

[2]*Department of Fish Processing Technology, Dr. M.G.R. Fisheries College and Research Institute, Tamil Nadu Dr. J. Jayalalithaa Fisheries University, Ponneri, Tamil Nadu, India*

[3]*Faculty of Innovative Agriculture and Fishery Establishment Project, Prince of Songkla University, Surat Thani Campus, Makham Tia, Muang, Surat Thani, Thailand*

ABSTRACT

Ultraviolet (UV) irradiation is an effective nonthermal food processing technique, and its efficiency in the destruction of microorganisms is well known. However, it is equally important to consider the influence of UV irradiation on nutritional and sensory properties. The color of the product, in particular, plays a pivotal role in consumer preference. UV irradiation has been shown to have minimal influence on the physical, chemical, nutritional, and sensory properties of foods, while maintaining their quality. It has been effective in preventing enzyme-induced food discoloration. However, the impact of UV irradiation can vary widely depending

Nonthermal Food Processing Technologies: Impact on Color Profile.
Kaavya Rathnakumar and R. Pandiselvam (Eds.)

on the food product, such as fruits, vegetables, juices, milk, pork, beef, fish, turkey, and chicken. Additionally, radicals generated by UV radiation can contribute to undesirable effects on lipid-enriched muscle foods, such as off-flavor and color changes due to lipid oxidation. This chapter summarizes the influence of UV irradiation on the color profile of foods. Furthermore, it addresses the opportunities, limitations, and remedies associated with the utilization of UV for maintaining the sensory properties of food products without compromising microbial safety.

8.1 INTRODUCTION

In the modern world, food quality is a great concern among consumers. The prime target of processing methods in processing industries is to lengthen the product's shelf life through the destruction of microorganisms. Meanwhile, desirable sensory properties need to be maintained [39]. Thermal processing is a widely used technique to prevent spoilage by destroying pathogenic microorganisms and inactivating spoilage-causing enzymes [1]. Despite its benefits, the thermal process causes severe constraints such as changes in texture, aroma, color, and nutrient loss due to high-temperature application [2]. To overcome these limitations, nonthermal techniques can be adopted as potent alternatives to preserve food products. Nonthermal techniques include pulsed electric field (PEF), ultrasonication, high-pressure processing, cold plasma (CP), irradiation, etc. [3]. Among all, irradiation by ultraviolet (UV) is an emerging nonthermal technique, becoming popular since it has several beneficial aspects such as microbial decontamination, enzyme inactivation, shelf-life extension, maintenance of nutritional and sensorial quality, physicochemical, and functional properties [2]. This technique is economical, profitable, versatile, environmentally friendly, and also suitable for large-scale industrial applications. UV irradiation is primarily used in the food processing industry as a surface treatment for food products. Surface treatments can help prevent spoilage caused by microbes [4]. Traditional surface treatments for food products involve the use of disinfectants such as chlorine, chlorine dioxide, iodine, and trisodium phosphate. However, due to increasing consumer concerns about the safety of disinfectants and their residual levels in food, there is a growing demand for alternative techniques for surface treatment of fresh foods [2]. UV irradiation has been found to be effective against pathogenic microorganisms like *Staphylococcus, Salmonella,* and *E. coli*. However, exposure to high doses of UV irradiation can lead to undesirable changes in food products, particularly meat products. Excessive UV exposure can generate reactive oxygen species (ROS) that may cause lipid oxidation, resulting in changes in the color of the food product [5]. Consumers often consider the color of meat as an indicator of quality and freshness when making purchases. During UV irradiation, the color of the food product can be altered through a complex series of chemical reactions and interactions between UV-induced reactive species and the components of the food [6]. Therefore, this chapter aims to provide a comprehensive overview of UV irradiation technology and discuss recent updates and developments in various food processing industries. The impact of UV on the color aspects of food will be revisited. Additionally, the prerequisites and challenges of this technique will be addressed, and possible remedies to overcome the limitations associated with UV irradiation will be discussed.

8.2 FUNDAMENTALS OF UV IRRADIATION

Electronic radiation known as UV emits light with a wavelength longer than X-rays but shorter than visible light. While many sources of UV light are available, the sun is the only one that naturally emits 3% UV radiation [10].

UV light can be produced by a continuous beam in a monochromatic or polychromatic mode using mercury lamps operating at moderate pressure. UV radiation can cause nonthermal reactions that have been demonstrated to inactivate microbes under controlled conditions [7]. Based on their wavelength, UV is grouped into three (Figure 8.1). Within the electromagnetic (EM) spectrum, UV light is divided into three zones:

- **UV-A:** Long-length UV light (320 nm to 400 nm);
- **UV-B:** Medium-length UV light (280 nm to 320 nm); and
- **UV-C:** Short-length UV light (100 nm to 280 nm).

UV irradiation is artificially generated by mercury, halogen, and fluorescent lamps for use in food products [8]. The UV-C range exhibits germicidal properties in UV light. UV-C inactivation of microorganisms follows a bell-shaped curve, with the peak ranging from 254 to 264 nm, providing maximum efficiency. Mercury lamps emit UV light primarily at 254 nm and are suitable for use in food processing [9]. When food products are exposed to radiation, UV-C light penetrates the outer layer of the cell membrane and damages DNA by forming thymine dimers. This can hinder microbial cell transcription and replication, ultimately leading to cell death (Table 8.1 and Figure 8.2).

TABLE 8.1 Dose Requirements to Inactivate Various Microorganisms

Microorganisms	Inactivation Dose (mJ/cm2)
Bacteria	1–10
Yeast	2–8
Fungus	20–200
Protozoa	100–150
Algae	300–400

Sources: Ref. [12].

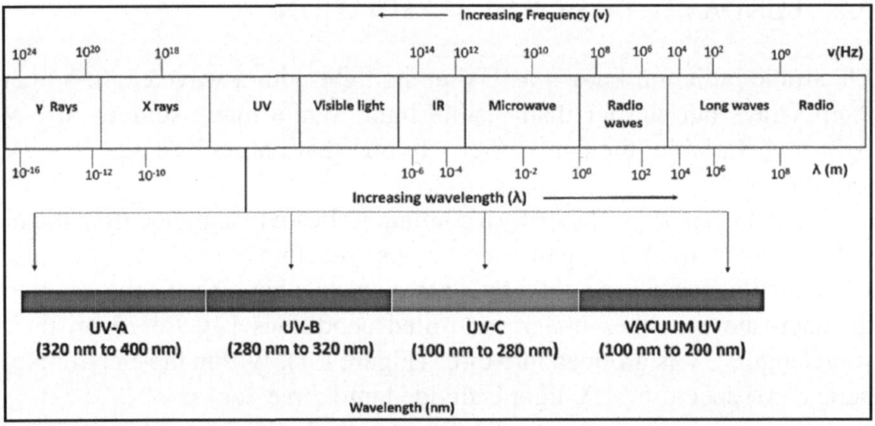

FIGURE 8.1 UV radiation and wavelength.

8.3 FACTORS INFLUENCING UV IRRADIATION ON FOODS

During irradiation, UV light travels within food products through absorption, reflection, refraction, and scattering effects [12]. UV light can only penetrate a few millimeters, which is determined by the optical properties of the product. Turbid foods such as milk and pulp juices cannot easily be penetrated by UV light and need to be exposed as a thin layer for effective microbial destruction [13]. In the case of water treatment, UV light is more effective due to its transparent nature. The color and turbidity of food products determine the absorption coefficient of UV radiation. As the absorption coefficient (15.9 cm^{-1}) increases, the inactivation efficiency of UV decreases by 10 times [14]. Generally, the level of UV light transmission is inversely proportional to the absorption coefficient of the system. Therefore, improving the penetration capacity of UV light is crucial for food systems with higher absorption coefficients. Additionally, the effectiveness of UV light is determined by various parameters including the sources of UV generation, process parameters, targeted microorganisms, and the type of foods to be disinfected. The resistance to UV light varies widely among different microorganisms [15]. For example, *Listeria monocytogenes* is one of the major pathogenic bacteria with high resistance to UV, while *Salmonella* spp. is most sensitive to UV irradiation. The variation among microorganisms is associated with factors such as cell thickness, size, and pigmentation [16]. Gram-positive bacteria, compared to gram-negative bacteria, have stronger and thicker peptidoglycan in their cell walls, which may hinder UV photons from penetrating into the cell membrane.

The majority of microorganisms, including yeast, molds, and bacteria, are all killed by UV light, which has a broad germicidal effect. Germs' nucleic acid absorbs UV radiation at short wavelengths (200-280 nm), particularly at 253.7 nm, where UV-C has the strongest germicidal impact [17]. UV-C photons are absorbed by nitrogenous bases in microbial DNA, causing a physical shifting of electrons, breaking certain bonds, and forming covalent cross-linking products such as pyrimidine dimers. These dimers are links between adjacent pairs of thymine and cytosine pyrimidines on the same strand (Figure 8.2) [18]. Furthermore, mutagenic damage caused by UV-C prevents DNA transcription, leading to mutagenesis and cell death. The microbial inactivation by UV depends on both the wavelength and the inactivation process, but it also varies depending on the species [19]. An organism's sensitivity to UV radiation is significantly influenced by the thickness and composition of the cell wall, as well as the presence of UV-absorbing proteins. Physiological, biochemical, and surface features, including thickness, viscosity, stiffness, and optical characteristics, may also impact the process efficiency [20]. UV radiation can only penetrate certain materials. Transparent liquids like water can be effectively disinfected by UV radiation, but opaque fluids like milk are less impacted due to weak light penetration depth, and germs cannot be directly affected [21].

Food-related processing factors:

1. **Safety Regulations:** As required by food safety legislation, the primary benefit of food processing is the destruction of microbes.
2. **Other Food Aspects:** Prolonging the keeping quality of products and inactivating toxin producing enzymes.
3. **Nutritional Value:** Enhanced digestion, prevention of nutrient losses after post-harvest.
4. **Sensory Quality:** Color, texture, taste, and flavor.
5. **Quality of Product:** The product needs to be improved because less time is being spent on food supply and preparation.

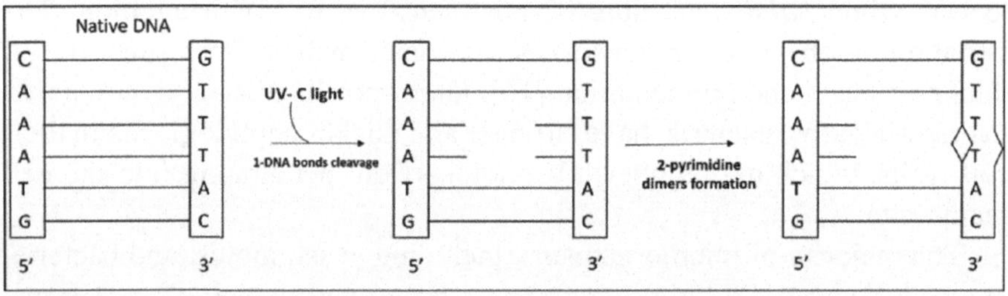

FIGURE 8.2 Effect of UV radiation in DNA of microorganisms.

8.4 EFFECT OF UV IRRADIATION ON DIFFERENT FOODS

8.4.1 EFFECT OF UV IRRADIATION ON FRUITS AND VEGETABLES

A balanced and nutritious diet should include both fruits and vegetables. Regular consumption of fruits can help prevent conditions such as cardiovascular disease, diabetes, and obesity [21]. However, fruits are prone to spoilage due to their high moisture content, which promotes enzymatic breakdown and microbial spoilage. Irradiation is an effective method for controlling the biological changes associated with the ripening, maturity, sprouting, and aging of fresh fruits and vegetables [22]. Enzymes present in fruits play a significant role in the production of juices and purees. Polyphenol oxidase (PPO) is an enzyme that contributes to enzymatic browning in fruits and fruit products. Therefore, it is important to deactivate this enzyme to prevent discoloration caused by the formation of melanin. UV irradiation has the potential to reduce the enzymatic activity of PPO, ATPase, and acid phosphate molecules by denaturing proteins [23]. Additionally, enzymes such as pectin methyl esterase, endo-polygalacturonase, and exo-polygalacturonase, which belong to the pectic group, can cause undesirable color changes and sedimentation of solid fractions in fruit juices, resulting in a cloudy appearance [24].

Tran & Farid [62] stated that orange juice treated with UV had a lower enzymatic activity of PPO, less than 25%, and 50% of other enzymes. In another study, apple juice irradiated with UV resulted in complete inactivation of peroxidase (POD) enzyme and PPO after treatment of 15 and 100 minutes, respectively, using a mercury vapor lamp (400 W) [28]. Furthermore, post-harvest loss occurs during the cold chain and distribution, affecting the overall acceptability of the product due to color discoloration [25]. One of the major post-harvest losses that occur in tomatoes during storage is color development and ripening. Tomatoes exposed to UV (4.2 kJ/m^2) prevented the discoloration while stored at 18°C for 35 days [26]. UV-C treatment effectively delays color development due to lessened chlorophyll and lycopene content degradation in UV-C exposed tomatoes [27]. This is related to a higher lycopene to β-carotene ratio in UV-C exposed food than untreated. UV radiation greatly slows down the rate at which green bananas ripen, stops potatoes and onions from sprouting, and stops endive and potatoes from turning green. The maximum limit for UV irradiation of fruits and vegetables is 1-3 kGy. Gomez et al. [27] studied the influence of UV-C light dose (5.6 ± 0.3; 8.4 ± 0.5 and 14.1 ± 0.9 kJ/m^2) on cut-apple's sensory quality (*Malus pumila*), the outer portion of the apple turned brown after 25 minutes of UV-C irradiation [54]. A slight color change was noticed in UV-C irradiated blanched apples after 20 minutes in comparison with the untreated sample.

Maharaj et al. [45] investigated the impact of UV on the aging of tomatoes (*Lycopersicon esculentum*) at various dose levels, such as 3.7 kJ/m^2 as the optimal dose and 24.4 kJ/m^2 as the hyper dose. The untreated sample was used as the control. The results showed that discoloration was delayed longer for samples treated with the optimal dosage of UV radiation compared to higher dosages and untreated samples after 21 days of storage [59]. Samples treated with a higher dosage of UV radiation exhibited enhanced surface browning [30]. Similarly, Frylinck et al. [24] reported the same findings on γ-irradiated mango. The surface browning effect at a higher dose of UV is due to a stress response. Cherry tomatoes (*S. lycopersicum*) treated with UV-C (4.2 kJ/m^2) turned a pink-reddish color compared to the normal orange-reddish color of tomatoes [32]. Fonseca & Rushing [23] demonstrated that UV-C irradiated fresh-cut watermelon at 1.40–13.70 kJ/m^2 significantly reduced bacteria populations, with a reduction of 4.1 kJ/m^2 without affecting color and overall acceptability. Manzocco et al. [46] demonstrated that UV-irradiated apple slices stored at 4°C had delayed color development of brown pigments.

8.4.2 EFFECT OF UV IRRADIATION ON MUSCLE AND MUSCLE PRODUCTS

The color of meat is attributed to the presence of hemoglobin and myoglobin, which are pigments consisting of globin and heme groups [36]. Unlike iron, which can exist in various oxidative forms, and porphyrin ring, which can be intact, polymerized, or oxidized, globin generally occurs in both native and denatured forms [36]. Fresh beef and pork may develop unfavorable brown, green, and bright red colors as a result of UV exposure. Among muscle foods, beef is reported to be highly susceptible to color changes caused by irradiation [37]. UV irradiation followed by packing in a modified atmosphere resulted in higher color retention of meat products [38]. Aerobically packed pork irradiated with UV had lower a* (redness) as well as higher gray and brown colors [40]. In another study, vacuum-packed irradiated beef resulted in a brown color due to decreased a* and increased b* (yellowness) values [41]. Meat naturally has reducing ability due to endogenous reductants, including electron transport intermediates, NADH, different enzyme systems, and substrates that delay the synthesis of metmyoglobin [29]. However, during storage, inactivation of the reducing system and consumption of reductants lead to the accumulation of metabolic by-products, which significantly affect the color of meat [42].

Metmyoglobin is the oxidized form of iron (present in myoglobin) that causes meat to turn brown. The discoloration into a brownish-red color is considered spoiled and unappealing to consumers [31]. Research by Nam and Ahn et al. [50] showed that irradiation of beef meat resulted in higher a* values and lower L* values during storage. Additionally, ionizing radiation reduces the brown color of cooked beef meat (globin myohemichromogen) into an unstable red color pigment (globin myohemochromogen) [51]. However, when exposed to air, the unstable red pigment gradually changes into ferric pigment (brown color) within 15 minutes [43]. This could be attributed to the denaturation of the enzyme system (reductants) during cooking, which separates the globin moiety from the myoglobin [33]. As a result, heme iron remains unstable in cooked meat and can easily lose electrons, converting into a ferric state when exposed to air [44]. The concentration of heme pigment in poultry products is generally low compared to other meats. Due to this low concentration, the oxyderivatives of myoglobin in poultry meat appear pale yellow and brown, but not prominently. UV irradiation has different effects on poultry meats. Nam et al. [52] reported that irradiated turkey meat appeared fresh and had higher a* values compared to untreated samples.

Kim et al. [37] reported a similar result in irradiated turkey meat. The mechanisms underlying the color change processes of irradiated meat vary widely between light and dark meat. The color changes of poultry meat are due to oxidation-reduction potential and carbon monoxide-myoglobin, which play a major role, whereas the color induced by irradiation of beef and pork meats is not clearly determined [46]. By applying UV-C radiation, the microbial load in fresh meat is reduced by two or three log cycles [60]. Furthermore, microbial reduction is enhanced with a higher dose of UV and duration of exposure. Stermer et al. [61] stated that UV-C resulted in no changes in color and overall acceptance of fresh meat. Wallner-Pendleton et al. [65] demonstrated that UV at 825–864 W/m^2 on chicken carcasses showed a 61% reduction of *Salmonella typhimurium* while not changing the sensory properties of the meat. According to Park et al. [55], the surface of dried fish fillets (*Stephanolepis cirrhifer*) treated with UV-C at 600 mW/cm^2 acts as an antifungal agent. However, *P. citrinum, C. cladosporioides,* and *A. Niger* were successfully inactivated up to 90% by UV radiation without any impact on the color of dried filefish. Yndestad et al. [67] reported that UV-C radiation (10,000 $\mu W/cm^2$) extended the shelf-life of poultry meat by decreasing the microbial load on the surface as well as maintaining the sensory properties (Figure 8.3).

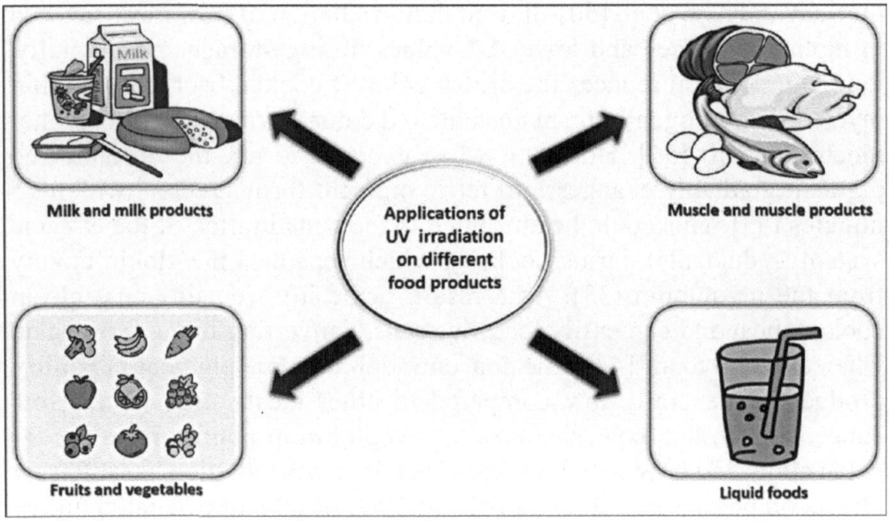

FIGURE 8.3 Applications of UV irradiation on different food products.

8.4.3 EFFECT OF UV IRRADIATION ON LIQUID FOODS

Dairy foods promote the growth of a wide range of microbes due to the presence of rich nutrients such as carbohydrates (particularly lactose), fatty acids, amino acids, proteins, vitamins, minerals, and beneficial enzymes [34]. As a result, manufacturing healthy dairy products is more difficult than making many other foods [50]. UV irradiation of milk was initially utilized for vitamin D enrichment in the mid-1900s [51]. Irradiated milk smells oxidized due to the oxidation of unsaturated fatty acid residues in milk. UV irradiation generates free radicals and catalyzes subsequent phases of oxidation [35]. Free radicals enhance the formation of free radicals, peroxides, H_2O_2, and superoxide radicals [52]. Superoxide radicals cause carbohydrate cross-linking, unsaturated fatty acid peroxidation, cross-linking, and fragmentation of proteins, and retard the functions of membrane fluidity [15]. Choudhary & Bandla [15] reported that UV denatures nutritious components of milk including proteins, enzymes, and amino acids, resulting in textural changes. Short-wavelength UV radiation produces vitamin D, but it has no effect on the concentrations of carotene, thiamine, vitamin A, or riboflavin. Moreover, it generates undesirable flavors and eliminates some nutrients [11]. In recent years, the effectiveness of UV irradiation has been investigated, and several research studies have been performed to understand in detail the prospective uses of UV radiation as an alternative to the conventional processing of milk [1].

UV irradiation has been developed to improve the sensory quality of the end product while maintaining microbiological safety [53]. Mold formation on the surface is a significant concern in cheese technology. Mold formation was effectively prevented by exposing it to UV before packaging. Lacivita et al. [42] stated that applying UV radiation ranging from 200 to 280 nm at various dosage levels (0.1, 0.6, 1.2, 6 kJ/m^2) at 8°C on the outer layer of cheese decreased the count of *Pseudomonas* spp. and Enterobacteriaceae (1–2 log cycle) without affecting the color, texture, and overall acceptability. In another study, Can et al. [12] investigated the impact of UV radiation on color change and lipid oxidation of cheese after being exposed to mild (5s at 13 cm), moderate (30s at 8 cm), and severe (40s at 5 cm) UV treatments [48]. The results demonstrated that mild treatments had no change in color or physicochemical properties of the cheeses, while the quality of the UV-irradiated cheese was increased compared to the untreated sample [56].

Juices are sensitive to thermal treatment since they lose their color, aroma, nutrients, etc., that are present in the product [57]. UV irradiation is an alternative treatment since it retains the aroma and color. When compared to typical pasteurization methods, UV-C did not affect the sensory properties of juice quality [58]. The effectiveness of enzyme inactivation and retention of nutrients primarily depends on the source and process parameters of UV irradiation [12]. Hakguder et al. [30] reported that UV had no effect on the color of orange juice (1.32 mW/cm^2). Zhang et al. [68] investigated the quality of watermelon juice subjected to different treatments including thermal, high-pressure, and UV-C. Juice treated with UV-C (4.8 kJ/l) decreased pectin methyl esterase activity by up to 75% within 5 minutes compared to those thermally treated at 60°C for 20 minutes. However, the L* value slightly changes in UV-irradiated juices compared to the untreated sample.

The efficacy of UV-C for the destruction of microbes (bacteria and yeast) in watermelon juice was investigated using a helix Teflon coil [40]. Watermelon juice samples were treated with UV-C at 2.7 J/mL and 37.5 J/mL and kept refrigerated for 37 days. The L* value of the treated and untreated juice was the same for the first two days of storage. After that, the L* value of the untreated juice increased slightly, while the b* value of the treated juices was considerably lower than that of the untreated samples, especially at 37.5 J/mL [41]. The value of a* remained unchanged in the treated juice. Similarly, Feng et al. [22] stated that the changes in color in the treated juice were not substantially different from the control values, indicating that UV-C improves juice quality and does not affect the juice's color during storage [61]. Donahue et al. [17] investigated UV radiation at 8.77-35.11 mJ/cm^2 on the color of apple cider, resulting in a higher L* value than the control during 21 days of storage. Muller et al. [49] reported that UV-C (100.47 kJ/l) on the color development of apple and grape juices at 4°C, in which the L* value was found to be lower than the untreated one after storage (12 days). Similar results were observed in the grape juice samples (Figure 8.4).

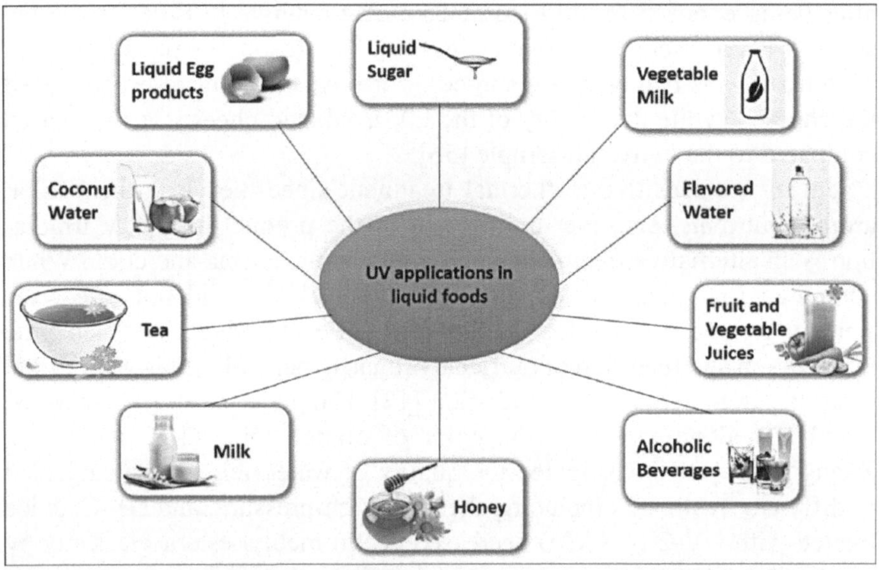

FIGURE 8.4 UV application in various liquid foods.

8.5 HURDLE TECHNOLOGY IN CONJUNCTION WITH UV IRRADIATION

UV irradiation, when combined with other available techniques, can maximize the benefits of germ reduction and increase product quality. This combination also helps to overcome the limitations associated with UV irradiation [64]. Further research is needed to explore the application of hurdle technology to UV-C in juice processing. UV can be combined with thermal processing, antioxidants, chemical preservatives, and other nonthermal processing methods such as ultrasonication and membrane filtration [65]. Hurdle technology has been tested on various liquid products. For example, a combined treatment of membrane filtration and UV irradiation resulted in a 5-log reduction in *E. coli.* UV combined with mild heat treatment (HT) increased the shelf-life of juice by inactivating enzymes [63]. UV-C and low heat combination successfully denaturized enzyme pectin methyl esterase, while maintaining similar bromelain and phenolic content compared to untreated samples [66]. However, in another study, the overall acceptability of juice decreased when UV was combined with ultrasonication due to oxidation induced by ultrasonication, which generated free radicals and resulted in undesirable flavor [67]. Therefore, the process parameters of hurdle techniques must be tailored to achieve desirable outcomes.

8.6 ADVANTAGES AND DISADVANTAGES OF UV IRRADIATION

➢ **Advantages:**

- UV irradiation is environmentally beneficial due to its non-usage of chemicals.
- Reduces the cost of water and electricity.
- Low repair, installation, and servicing expenses.
- UV irradiation is a reliable means of destroying microorganisms.
- It saves money on storage since it eliminates the need for specialized storage equipment.
- Maintaining the food's physicochemical properties and nutritional content without any undesirable changes in sensory quality.
- It prolongs the products' life without heat treatment.
- Combination with various nonthermal production technologies is possible.
- Operation in a continuous manner.
- Commercially suitable treatment in food processing industry.

➢ **Disadvantages:**

- UV irradiation requires that light reaches its target. As a result, when an external object interrupts its passage, it is rendered ineffective.
- Prolonged contact may cause injury to humans (eyes, burns, and skin cancer).
- UV transmission is limited by opaque substances.
- UV penetration of solid food content is low.
- It is challenging to quantify the rate of UV disinfection.

8.7 FUTURE PROSPECTS

A unique, nonthermal UV irradiation is a potential food processing method for microbial decontamination without compromising the sensory properties of the products (Figure 8.5). Generally, the majority of UV irradiation research has focused mainly on microbial inactivation. However, there are only a few reports regarding the quality of UV-treated foods. Several studies have documented that exposure to UV irradiation impacts the quality attributes of food products both positively and negatively during treatment and storage. Therefore, this scientific gap offers a great opportunity to explore the alteration in physicochemical characteristics of UV-treated foods at the molecular level. Furthermore, optimization of process parameters is still required to prevent or reduce the negative impact of UV treatment, including lipid oxidation, changes in color, and textural attributes. To overcome this drawback, the combination of UV treatment with other available techniques such as ultrasound and antioxidants has become a promising strategy. These hurdle technologies need to be explored more in terms of the quality of the treated foods. UV has a major advantage in combining with other novel nonthermal treatments, and this combination leads to better food preservation that effectively eliminates microbes under specific conditions and makes the food preferable and acceptable to consumers. The nonthermal nature of UV treatment enables the processing of heat-sensitive products and consumes low energy compared to conventional heat preservation techniques, thereby minimizing production costs.

Furthermore, further investigation is needed to gain detailed knowledge of the mechanisms of action, the interaction of UV and food components, and the process control over the quality attributes in order to optimize its potential on a commercial scale. Additionally, while the application of UV irradiation has been implemented at the laboratory scale, more research is still required on scale-up strategies, regulatory approval, and validation of UV-treated products, especially on an industrial scale. UV has been globally recognized as a novel technology category in food preservation, and there is a need for regulatory laws in each nation to support production at lower costs without compromising food safety levels. Specific legislation should be established regarding UV permissible levels, dosage, and duration of treatment, providing useful information for manufacturers and end users. The US FDA has defined UV treatment in the food additive category and set limits for monochromatic low-pressure mercury lamps at 254 nm to eliminate toxicological effects. In contrast, UV-LED lights are significantly smaller than standard lamps and outperform conventional lamps in terms of being environmentally friendly and easily incorporated into a wide range of equipment designs [64].

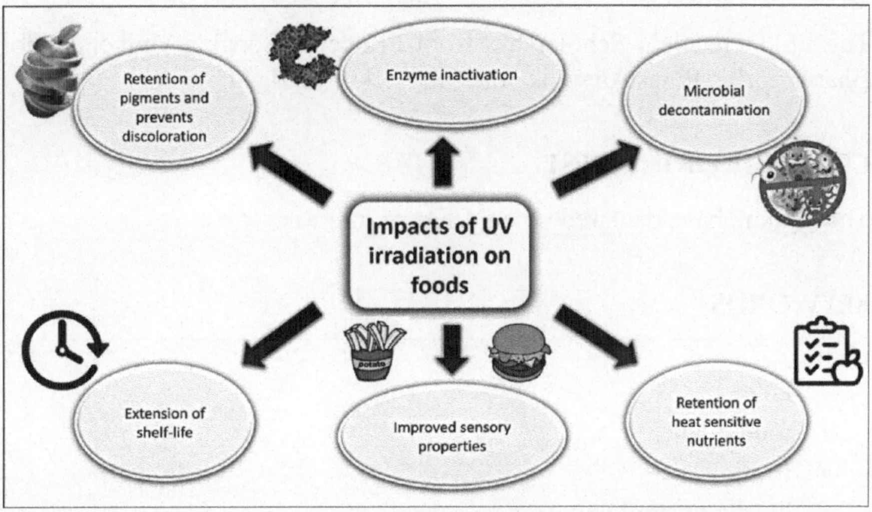

FIGURE 8.5 Impacts of UV irradiation on foods.

8.8 SUMMARY

Nowadays, the demand for nutritious, wholesome, and high-quality food products is expanding among consumers. To fulfill the growing needs, the keeping quality of foods needs to be extended by the application of several nonthermal processing techniques. Among all nonthermal techniques, UV irradiation in food processing has recently gained recognition worldwide for its excellent capacity in inhibiting a wide range of microbes without compromising the product's quality. UV irradiation has recently advanced, becoming a useful commercial tool in the food processing sector. UV radiation has the potential to be used for the pasteurization of milk and fruit juices. UV treatment enhances the product quality of solid and liquid foods without loss in the nutritional composition of processed foods. The food color of the product plays a crucial role in determining whether processed or unprocessed food is preferred by consumers. The color profile of foods has been maintained by the application of UV radiation as a surface treatment, as well as delaying the browning reaction caused by enzymes. Overall, UV irradiation appears to be a potential method for lengthening the time that processed foods can be stored after harvest.

ACKNOWLEDGMENT

The PSU President Scholarship from Prince of Songkla University for Bharathipriya Rajasekaran is gratefully acknowledged.

CONFLICT OF INTEREST

The authors have declared no conflicts of interest.

KEYWORDS

- **color**
- **food quality**
- **fruits**
- **hurdle technology**
- **juices**
- **meat**
- **milk**
- **nonthermal processing**
- **UV irradiation**

REFERENCES

1. Abdul, K. S. N. N., Shamsudin, R., Abdul, R. R., & Adzahan, N. M., (2016). Fruit juice production using ultraviolet pasteurization: A review. *Beverages, 2*(3), 22.
2. Acevedo, B. A., Sgroppo, S. C., & Dellacassa, E., (2018). *Effects of Ultraviolet Radiation on the Microbiological, Physicochemical, and Sensory Properties of Rangpur Lime Juice.*
3. Ahn, D. U., Olson, D. G., Jo, C., Chen, X., Wu, C., & Lee, J. I., (1998). Effect of muscle type, packaging, and irradiation on lipid oxidation, volatile production, and color in raw pork patties. *Meat Science, 49*(1), 27–39.
4. Akgün, M. P., & Ünlütürk, S., (2017). Effects of ultraviolet light emitting diodes (LEDs) on microbial and enzyme inactivation of apple juice. *International Journal of Food Microbiology*, *260*, 65–74.
5. Ansari, J. A., Ismail, M., & Farid, M., (2019). Investigate the efficacy of UV pretreatment on thermal inactivation of *Bacillus subtilis* spores in different types of milk. *Innovative Food Science & Emerging Technologies*, *52*, 387–393.
6. Bartosz, 2., Kołakowska, A., & Sikorski, Z. E., (2010). Lipid oxidation in food systems. *Chemical, Biological, and Functional Aspects of Food Lipids* (2ndedn., pp. 163–184). New York – USA: CRC Press.
7. Baysal, A. H., Molva, C., & Unluturk, S., (2013). UV-C light inactivation and modeling kinetics of *Alicyclobacillus acidoterrestris* spores in white grape and apple juices. *International Journal of Food Microbiology*, *166*(3), 494–498.
8. Beuchat, L. R., Nail, B. V., Adler, B. B., & Clavero, M. R. S., (1998). Efficacy of spray application of chlorinated water in killing pathogenic bacteria on raw apples, tomatoes, and lettuce. *Journal of Food Protection*, *61*(10), 1305–1311.
9. Brewer, S., (2004). Irradiation effects on meat color: A review. *Meat Science*, *68*(1), 1–17.
10. Bu, J., Yu, Y., Aisikaer, G., & Ying, T., (2013). Postharvest UV-C irradiation inhibits the production of ethylene and the activity of cell wall-degrading enzymes during softening of tomato (*Lycopersicon esculentum L.*) fruit. *Postharvest Biology and Technology, 86*, 337–345.
11. Caminiti, I. M., Noci, F., Morgan, D. J., Cronin, D. A., & Lyng, J. G., (2012). The effect of pulsed electric fields, ultraviolet light or high intensity light pulses in combination with manothermosonication on selected physico-chemical and sensory attributes of an orange and carrot juice blend. *Food and Bioproducts Processing*, *90*(3), 442–448.
12. Can, F. O., Demirci, A., Puri, V. M., & Gourama, H., (2014). Decontamination of hard cheeses by pulsed UV light. *Journal of Food Protection, 77*(10), 1723–1731.
13. Chawla, A., Lobacz, A., Tarapata, J., & Zulewska, J., (2021). UV light application as a mean for disinfection applied in the dairy industry. *Applied Sciences*, *11*(16), 7285.
14. Choi, L. H., & Nielsen, S. S., (2005). The effects of thermal and nonthermal processing methods on apple cider quality and consumer acceptability. *Journal of Food Quality, 28*(1), 13–29.
15. Choudhary, R., & Bandla, S., (2012). Ultraviolet pasteurization for food industry. *International Journal of Food Science and Nutrition Engineering*, *2*(1), 12–15.

16. Delorme, M. M., Guimarães, J. T., Coutinho, N. M., Balthazar, C. F., Rocha, R. S., Silva, R., & Cruz, A. G., (2020). Ultraviolet radiation: An interesting technology to preserve quality and safety of milk and dairy foods. *Trends in Food Science & Technology*, *102*, 146–154.
17. Donahue, D. W., Canitez, N., & Bushway, A. A., (2004). UV inactivation of *E. coli* O157: H7 in apple cider: Quality, sensory and shelf-life analysis. *Journal of Food Processing and Preservation, 28*(5), 368–387.
18. Dos, S. R. C., Magnani, M., Ramos, G. L. D. P. A., Bezerril, F. F., De Freitas, M. Q., Cruz, A. G., & Pimentel, T. C., (2022). Emerging technologies in food processing: Impacts on sensory characteristics and consumer perception. *Current Opinion in Food Science*, 100892.
19. Duran, A., & Kahve, H. Ġ., (2016). Ultraviolet radiation (UV) applications in milk industry. *International Conference on Engineering and Natural Sciences (ICENS)*.
20. Falguera, V., Pagán, J., & Ibarz, A., (2011). Effect of UV irradiation on enzymatic activities and physicochemical properties of apple juices from different varieties. *LWT-Food Science and Technology*, *44*(1), 115–119.
21. Falguera, V., Pagán, J., Garza, S., Garvín, A., & Ibarz, A., (2011). Ultraviolet processing of liquid food: A review: Part 2: Effects on microorganisms and on food components and properties. *Food Research International, 44*(6), 1580–1588.
22. Feng, M., Ghafoor, K., Seo, B., Yang, K., & Park, J., (2013). Effects of ultraviolet-C treatment in Teflon®-coil on microbial populations and physico-chemical characteristics of watermelon juice. *Innovative Food Science & Emerging Technologies, 19*, 133–139.
23. Fonseca, J. M., & Rushing, J. W., (2006). Effect of ultraviolet-C light on quality and microbial population of fresh-cut watermelon. *Postharvest Biology and Technology, 40*(3), 256–261.
24. Frylinck, L., Dubery, I. A., & Schabort, J. C., (1987). Biochemical changes involved in stress response and ripening behavior of γ-irradiated mango fruit. *Phytochemistry, 26*(3), 684–686.
25. Gayán, E., Condón, S., & Álvarez, I., (2014). Biological aspects in food preservation by ultraviolet light: A review. *Food and Bioprocess Technology, 7*(1), 1–20.
26. Gómez, P. L., Alzamora, S. M., Castro, M. A., & Salvatori, D. M., (2010). Effect of ultraviolet-C light dose on quality of cut-apple: Microorganism, color, and compression behavior. *Journal of Food Engineering, 98*(1), 60–70.
27. Gómez, P. L., García-Loredo, A., Salvatori, D. M., Guerrero, S., & Alzamora, S. M., (2011). Viscoelasticity, texture, and ultrastructure of cut apple as affected by sequential anti-browning and ultraviolet-C light treatments. *Journal of Food Engineering, 107*(2), 214–225.
28. Guerrero-Beltrn, J. A., & Barbosa-Cnovas, G. V., (2004). Advantages and limitations on processing foods by UV light. *Food Science and Technology International*, *10*(3), 137–147.
29. Guevara, M., Tapia, M. S., & Gómez-López, V. M., (2012). Microbial inactivation and quality of guava and passion fruit nectars treated by UV-C light. *Food and Bioprocess Technology*, *5*(2), 803–807.

30. Hakguder, T. B., Unluturk, S., Buzrul, S., & Alpas, H., (2015). The impact of UV-C irradiation on spoilage microorganisms and color of orange juice. *Journal of Food Science and Technology, 52*(2), 1000–1007.
31. Harry, S., (1928). Wisconsin alumni research foundation. *Antirachitic Product and Process*. U.S. Patent 1,680,818.
32. Ibarz, A., Garvín, A., & Falguera, V., & Suvendu, B., (2014). Ultraviolet in food preservation and processing; Chapter 17. *Conventional and Advanced Food Processing Technologies* (1stedn., pp. 411–436). India: Wiley Online Library.
33. Jagadeesh, S. L., Charles, M. T., Gariepy, Y., Goyette, B., Raghavan, G. S. V., & Vigneault, C., (2011). Influence of postharvest UV-C hormesis on the bioactive components of tomato during post-treatment handling. *Food and Bioprocess Technology, 4*(8), 1463–1472.
34. Kamarei, A. R., Karel, M., & Wierbicki, E., (1979). Spectral studies on the role of ionizing radiation in color changes of radappertized beef. *Journal of Food Science, 44*(1), 25–32.
35. Kasahara, I., Carrasco, V., & Aguilar, L., (2015). Inactivation of *Escherichia coli* in goat milk using pulsed ultraviolet light. *Journal of Food Engineering, 152*, 43–49.
36. Keyser, M., Műller, I. A., Cilliers, F. P., Nel, W., & Gouws, P. A., (2008). Ultraviolet radiation as a non-thermal treatment for the inactivation of microorganisms in fruit juice. *Innovative Food Science & Emerging Technologies, 9*(3), 348–354.
37. Kim, S. J., Kim, D. K., & Kang, D. H., (2016). Using UVC light-emitting diodes at wavelengths of 266 to 279 nanometers to inactivate foodborne pathogens and pasteurize sliced cheese. *Applied and Environmental Microbiology, 82*(1), 11–17.
38. Kim, T., Silva, J. L., & Chen, T. C., (2002). Effects of UV irradiation on selected pathogens in peptone water and on stainless steel and chicken meat. *Journal of Food Protection, 65*(7), 1142–1145.
39. Koca, N., Urgu, M., & Saatli, T. E., (2018). Ultraviolet light applications in dairy processing; Chapter. 1. *Technological Approaches for Novel Applications in Dairy Processing* (1stedn., pp. 3–22). London – United Kingdom: Intechopen.
40. Koutchma, T., Moraru, C. I., Forney, L. J., & Da-Wen, S., (2009). Ultraviolet light in food technology: Principles and applications. *Contemporary Food Engineering* series (1stedn., pp. 1–296). New York – USA: CRC Press.
41. Koutchma, T., (2009). Advances in ultraviolet light technology for non-thermal processing of liquid foods. *Food and Bioprocess Technology, 2*(2), 138–155.
42. Lacivita, V., Conte, A., Manzocco, L., Plazzotta, S., Zambrini, V. A., Del Nobile, M. A., & Nicoli, M. C., (2016). Surface UV-C light treatments to prolong the shelf-life of Fiordilatte cheese. *Innovative Food Science & Emerging Technologies, 36*, 150–155.
43. Liu, Y., Chen, Y. R., & Ozaki, Y., (2000). Characterization of visible spectral intensity variations of wholesome and unwholesome chicken meats with two-dimensional correlation spectroscopy. *Applied Spectroscopy, 54*(4), 587–594.
44. Luchsinger, S. E., Kropf, D. H., Garcia, Z. C. M., Hunt, M. C., Stroda, S. L., Marsden, J. L., & Kastner, C. L., (1997). Color and oxidative properties of irradiated whole muscle beef 1. *Journal of Muscle Foods, 8*(4), 427–443.
45. Maharaj, R., Arul, J., & Nadeau, P., (2010). UV-C irradiation of tomato and its effects on color and pigments. *Advances in Environmental Biology, 4*(2), 308–316.

46. Manzocco, L., Quarta, B., & Dri, A., (2009). Polyphenol oxidase inactivation by light exposure in model systems and apple derivatives. *Innovative Food Science & Emerging Technologies, 10*(4), 506–511.
47. Mditshwa, A., Magwaza, L. S., Tesfay, S. Z., & Mbili, N. C., (2017). Effect of ultraviolet irradiation on postharvest quality and composition of tomatoes: A review. *Journal of Food Science and Technology, 54*(10), 3025–3035.
48. Millar, S. J., Moss, B. W., MacDougall, D. B., & Stevenson, M. H., (1995). The effect of ionizing radiation on the CIELAB color co-ordinates of chicken breast meat as measured by different instruments. *International Journal of Food Science & Technology, 30*(5), 663–674.
49. Müller, A., Noack, L., Greiner, R., Stahl, M. R., & Posten, C., (2014). Effect of UV-C and UV-B treatment on polyphenol oxidase activity and shelf life of apple and grape juices. *Innovative Food Science & Emerging Technologies, 26*, 498–504.
50. Nam, K. C., & Ahn, D. U., (2002). Carbon monoxide-heme pigment is responsible for the pink color in irradiated raw turkey breast meat. *Meat Science, 60*(1), 25–33.
51. Nam, K. C., & Ahn, D. U., (2002). Carbon monoxide-heme pigment complexes are responsible for the pink color in irradiated raw turkey breast meat. *Meat Science, 60*(1), 49.
52. Nam, K. C., Min, B. R., Park, K. S., Lee, S. C., & Ahn, D. U., (2003). Effects of ascorbic acid and antioxidants on the lipid oxidation and volatiles of irradiated ground beef. *Journal of Food Science, 68*(5), 1680–1685.
53. Nanke, K. E., Sebranek, J. G., & Olson, D. G., (1998). Color characteristics of irradiated vacuum-packaged pork, beef, and turkey. *Journal of Food Science, 63*(6), 1001–1006.
54. O'grady, M. N., Monahan, F. J., Burke, R. M., & Allen, P., (2000). The effect of oxygen level and exogenous α-tocopherol on the oxidative stability of minced beef in modified atmosphere packs. *Meat Science, 55*(1), 39–45.
55. Park, S. Y., Lee, N. Y., Kim, S. H., Cho, J. I., Lee, H. J., & Ha, S. D., (2014). Effect of ultraviolet radiation on the reduction of major food spoilage molds and sensory quality of the surface of dried filefish (*Stephanolepis cirrhifer*) fillets. *Food Research International, 62*, 1108–1112.
56. Rosenfeldt, E. J., Linden, K. G., Canonica, S., & Von, G. U., (2006). Comparison of the efficiency of OH radical formation during ozonation and the advanced oxidation processes O_3/H^2O^2 and UV/H_2O_2. *Water Research, 40*(20), 3695–3704.
57. Santhirasegaram, V., Razali, Z., George, D. S., & Somasundram, C., (2015). Effects of thermal and non-thermal processing on phenolic compounds, antioxidant activity and sensory attributes of chokanan mango (*Mangifera indica* L.) juice. *Food and Bioprocess Technology, 8*(11), 2256–2267.
58. Seiji, M., & Iwashita, S., (1965). Enzyme inactivation by ultraviolet light and protective effect of melanin. The *Journal of Biochemistry, 57*(3), 457–459.
59. Singh, H., Bhardwaj, S. K., Khatri, M., Kim, K. H., & Bhardwaj, N., (2021). UVC radiation for food safety: An emerging technology for the microbial disinfection of food products. *Chemical Engineering Journal, 417*, 128084.
60. Song, X., Bredahl, L., Navarro, M. D., Pendenza, P., Stojacic, I., Mincione, S., & Giacalone, D., (2022). Factors affecting consumer choice of novel non-thermally

processed fruit and vegetables products: Evidence from a 4-country study in Europe. *Food Research International, 153*, 110975.
61. Stermer, R. A., Lasater-Smith, M., & Brasington, C. F., (1987). Ultraviolet radiation—An effective bactericide for fresh meat. *Journal of Food Protection, 50*(2), 108–111.
62. Tran, M. T. T., & Farid, M., (2004). Ultraviolet treatment of orange juice. *Innovative Food Science & Emerging Technologies, 5*(4), 495–502.
63. Vasuja, S., & Kumar, V., (2018). Ultra violet irradiation and its applications in food processing industries: A review. *Int. J. Trend Res. Dev., 5*(1), 343–346.
64. Vunnam, R., Hussain, A., Nair, G., Bandla, R., Gariepy, Y., Donnelly, D. J., Kubow, S., & Raghavan, G. S. V., (2014). Physico-chemical changes in tomato with modified atmosphere storage and UV treatment. *Journal of Food Science and Technology, 51*(9), 2106–2112.
65. Wallner-Pendleton, E. A., Sumner, S. S., Froning, G. W., & Stetson, L. E., (1994). The use of ultraviolet radiation to reduce *Salmonella* and psychrotrophic bacterial contamination on poultry carcasses. *Poultry Science, 73*(8), 1327–1333.
66. World Health Organization, (1988). *Food Irradiation: A Technique for Preserving and Improving the Safety of Food.*
67. Yndestad, M., Underdal, B., & Nordal, J., (1972). The effects of ultra-violet radiation on the bacterial counts and shelf life of chicken carcasses. *Acta Agriculturae Scandinavica, 22*(3), 169–172.
68. Zhang, C., Trierweiler, B., Li, W., Butz, P., Xu, Y., Rüfer, C. E., Ma, Y., & Zhao, X., (2011). Comparison of thermal, ultraviolet-c, and high-pressure treatments on quality parameters of watermelon juice. *Food Chemistry, 126*(1), 254–260.

CHAPTER 9

PULSED ELECTRIC FIELDS TECHNOLOGY AND ITS APPLICATIONS IN FOODS: EFFECTS ON COLOR PARAMETERS

DUARTE-CORREA YUDY,[1] OSORIO-ARIAS JUAN,[2] KARTIK SHAH,[3] and C. ASHOKKUMAR[4]

[1]*BIOALI Research Group, Food Department, Faculty of Pharmaceutical and Food Sciences, Universidad de Antioquia, Medellín, Colombia*

[2]*Corporación Universitaria Minuto de Dios – Uniminuto, Research Direction, Bogotá DC, Colombia*

[3]*Sargento Foods, Technology Principal, Elkhart Lake, Wisconsin, USA*

[4]*Department of Food Technology, College of Food and Dairy Technology, TANUVAS, Chennai, Tamil Nadu, India*

ABSTRACT

Pulsed electric field (PEF) processing technology is a rapidly emerging field in the food production industry, offering numerous advantages over traditional thermal food processing methods. By converting high electric supply into high oscillating pulse energy, the high electric volt power in the processing chamber produces biologically safe liquid food products and enhances their shelf life, providing longer-lasting freshness. Unlike thermally produced food products, many commercially available food

Nonthermal Food Processing Technologies: Impact on Color Profile.
Kaavya Rathnakumar and R. Pandiselvam (Eds.)

products lack certain unique physical characteristics. However, treating food with pulses of electric current through PEF overcomes these limitations and improves the visual and quality parameters of colorful foods. The impact of PEF treatment on colored food varies depending on the nature of the food matrix's color characteristics, resulting in either enhancement or reduction. As a result, PEF treatment increases the yield of bioactive compounds and preserves the natural color, antioxidant capacity, and vitamin content. When combined with other nonthermal food processing operations, PEF processing further enhances the quality characteristics of various liquids and foods. This chapter provides an in-depth discussion of the mechanism behind PEF food treatment and quantifies its impact on changes in color characteristics both instrumentally and visually.

9.1 INTRODUCTION

The use of innovative technologies that can inactivate foodborne bacteria has been crucial for food preservation. These technologies are essential and have served as a solid foundation for the industrialization and widespread commercialization of food. Pathogenic bacteria are often the primary target of many industrial food procedures due to the risks posed by foodborne illnesses. These bacteria may also possess the ability to withstand challenging conditions that arise during food production and storage, and adapt to them. Historically, the conventional methods used by the food industry to preserve food include heating (sterilization, thermization), cooling (freezing), reducing water activity (drying, adding salt and sugar), and adding preservatives, among others. While these technologies can effectively inactivate or prevent the growth of harmful microbes to ensure food safety, they often alter the nutritional, chemical/biochemical, and sensory aspects of foods, which can decrease customer satisfaction with the products [1]. Therefore, the new nonthermal and ecologically friendly method known as pulsed electric field (PEF) is highly appealing to food processors. PEF can process food matrices with less nutritional loss compared to conventional heat processes. As a result, PEF can provide consumers with microbiologically safe, minimally processed, and fresh-like foods. PEF can be categorized as electrical-based processing. In contrast to other electrically-based approaches, such as ohmic heating, the application of brief electric signals at increased voltages enables a minimal increase in the treated product temperature [2, 3]. Despite being introduced to the food industry in the late 1900s, PEF technology is still regarded as an evolving technology owing to recent advancements in its industrialized applications [1, 4]. Recent studies have investigated PEF and its contribution to the food preservation sector [1, 4–6]. Under the influence of PEF, which involves a lower intensity of the electrical field, the cellular membrane is electrically penetrated and temporarily or permanently loses its semi-permeability, enabling the selective recovery of value-added substances from diverse substrates. Additionally, PEF-assisted processing has recently gained considerable attention in the food engineering field for improving diffusion extraction, modification of enzymatic activity, disinfecting liquid waste, osmotic treatment, pressing extraction, drying, and freezing [7–9].

Using PEF technology, food samples are subjected to strong electric pulses (ranging from 20 to 80 kV/cm) across two electrodes. PEF systems require three main components: (i) a high voltage impulse generation system; (ii) a treatment chamber capable of converting pulsed voltage into PEFs; and (iii) a process parameter monitoring control system. PEF is only suitable for foodstuffs that can withstand strong electric fields, have low conductivity, and do not contain or produce bubbles. Therefore, it would be ideal if liquid food could have a longer shelf life while still maintaining its organoleptic qualities. Additionally, this technology is not effective against spores and enzymes. As a result, discoloration and off-flavors have been reported as negative impacts due to the presence of residual enzymes [10]. Several applications or operations have been reported using PEF systems, such as extraction, pressing, microorganism/enzyme inactivation, dehydration, freezing, thermal treatment, or mechanical disruption of cell elements prior to high-value compound recovery from agro-industrial products and byproducts, pasteurization of heat-sensitive liquids, wastewater treatment, among others. Additionally, several authors have also reported the use of PEF technology as a tool to alter the textural properties of foods during processing, resulting from modifications in the tissue structure [11, 12].

This chapter includes a brief understanding of the mechanism and components of PEF technology. It also includes the use of PEF food applications and its impact on the color of the food.

9.2 MECHANISM

The principal mechanism involved in the PEF system of food application is the electroporation phenomenon, which punctures the cell membrane with pores sized between 50 and 100 nm. A short high voltage pulse causes the disintegration of the cell membrane and excretes all intracellular components from the microorganisms, inactivating their survival. Short pulses of electric voltage (10 to 60 kV) pass through the food matrix placed between two electrodes with very short contact time, achieving the biological safety of the food products without difficulty. High voltage pulses are applied at a repetition rate of up to 1,000 per second to properly treat the food matrix. Based on the characteristics of the cell, such as plant, microbial, or animal, there is a field limit value between 1 and 10 kV/cm that, when exceeded, results in the creation of pores as a result of localized dielectric rupture of the outer cell membrane. These pores can act as conductive channels. The strong pulses of the electric field produce pores in the cellular membrane by expanding or forming new pores. Depending on the treatment conditions, these pores may be temporary or permanent, and they enhance membrane permeability [13]. In other words, when the cellular wall is exposed to an externally applied electric field (Ee), a potential difference is created. The significant transmembrane potential, also known as the critical electric field (Ec), represents the maximum tolerable value that the membrane can withstand. When the electric field exceeds this threshold, detachment occurs on the cellular membrane [14]. The disintegration of cell membrane components hinders several biochemical reactions and enzyme productions. The versatile application of PEF treatment provides better opportunities in the processing of liquid, semi-solid, and solid food products. PEF's ability to disrupt cells relies on various factors, including the duration of the treatment, the potential of the electrical field, microorganisms, temperature of the element, and the characteristics of the medium [7, 15].

9.3 SYSTEM COMPONENTS

Currently, no industrial system based on PEF technology is accessible. Researchers attribute this to a lack of treatment uniformity, comparability, and repeatability of PEF study outcomes [16]. However, some institutions and research centers around the world are working on its application. In that sense, two of the main pilot-scale applications are liquid food preservation and cell disintegration systems, which may be found at a university in Ohio, a commercial food processor with an international footprint, The Swedish Institute for Food and Biotechnology (SIK), Berlin University of Technology, and Forschungszentrum Karlsruhe; located in the United States, Netherlands, Sweden, and Germany, respectively [17]. PEF is a technique with benefits over other traditional thermal processes that can be quickly integrated into existing processing lines (continuous operability and short processing times). PEF requires three primary elements: a high voltage impulse generation system, a treatment chamber, and a processing parameter tracking system. Figure 9.1 depicts a simple PEF treatment unit.

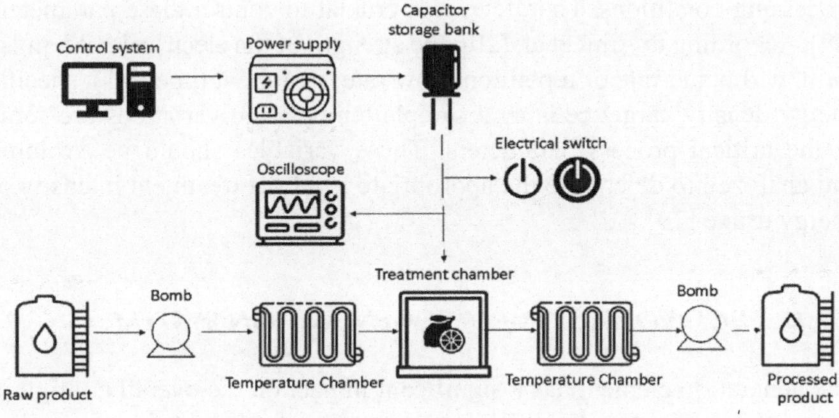

FIGURE 9.1 A diagram illustrating the operation of a pulsed electric field.

Figure 9.1 illustrates the components of the PEF system, which includes an oscilloscope, a button for energy release, a capacitor bank for energy storage, a power supply (providing high voltage), and a treatment chamber. The power supply converts the voltage from a utility line (110 V) into a pulsed, highly intense electric field. The power-source energy is gradually charged and stored in the capacitor bank before being rapidly released into the treatment vessel through the electrical switch. The switch allows for instant energy discharge in pulses lasting a millionth of a second, and it determines the system's maximum repetition rate, as well as the maximum current and voltage rating [18, 19]. The oscilloscope is used to measure the current and system voltage. The treatment vessel, with electrodes separated by an insulating material and distinct geometric configurations, is where the food products are processed in the previously created pulsed electric field [17]. Additionally, for the PEF process to be continuous, other components are required, such as a pump (used for food transportation) and heat exchangers (used to heat up the medium and dissipate electrical energy before and after the PEF treatment, respectively). It is also necessary to have specific control points for temperature, voltage, and current throughout the entire process.

Finally, it is necessary to implement aseptic packaging to prevent future microbiological contamination [17]. In order to fully utilize the technology, create a commercial device in the future, and most importantly, produce high-quality foods, each component of the PEF system shown in Figure 9.1 has several associated parameters that can affect the processing conditions. Therefore, it is crucial to control these parameters [19]. According to Amit et al. [20], the strength of the electrical field, pulse form, width and rate of repetition, flow rate, treatment time (TT), specific energy density, target cells, and sample temperature variations are some of the critical process parameters. These variables should be explored and analyzed to determine the appropriate levels of treatment intensity or energy usage [19].

9.3.1 HIGH VOLTAGE IMPULSE GENERATION SYSTEM

The generation element has a significant impact on the overall configuration of this method, as the PEF system must be designed to accommodate the repetition rate, adjustable operating voltage, duration, and pulse energy. This is necessary because the high voltage used in PEF processing requires strict safety measures. The generator system consists of a power supply, a fully sealed interlocking shield cabinet, a replaceable shunt resistor assembly, a series-limiting resistor, temperature sensors on the high-power resistor, and interactive computer control [21]. According to Góngora-Nieto et al., an electric field is generated when a higher intensity voltage is applied transversely to the electrodes of the processing vessel. Since the field amplitude is inversely proportional to the electrode gap and directly proportional to the voltage, it is recommended to use short gaps and higher intensity voltage to achieve high levels of mortality [22]. In this regard, two of the main parameters that need to be controlled and calculated are the charging voltage required to create pulses with an intense electric field [17] and the pulse repetition rate required to ensure a homogeneous process [21], which can be calculated using Equations (1) and (2) respectively.

$$E = \frac{U}{d} \tag{1}$$

$$f = \frac{nF}{v} \tag{2}$$

where 'E' represents the charging voltage, 'U' refers to the power supply in volts (kV), 'd' (m) refers to the length across electrodes, 'f' (Hz) is the repetition rate of the pulse, 'n' is the sum of pulses, 'v' refers to the treatment vessel volume, and 'F' is the food's flow rate in volume. An electric field ranging from 20 to 80 kV/cm is produced when a voltage from 20 to 40 kV and spacing across the electrodes ranging from 0.1 to 1 cm is applied [22]. The power supply switch plays a significant role in PEF processing. Several types of switches have been reported, with some of the main relevant ones being spark gap, thyratron switches, gate turn-off, silicon-controlled rectifier, insulated gate bipolar transistor, and high-power switches assembled with transistors or semiconductors. The type of switching power supply mentioned above has advantages, disadvantages, and limitations, and its selection for a PEF system must be based on the function of the food environment and the desired quality of the foodstuff.

9.3.2 TREATMENT VESSEL

The PEF treatment vessel contains electrodes and applies a powerful voltage to the food item. Typically, it consists of a pair of electrodes held together by an insulating substance to create an enclosure for food [23]. Therefore, the electrodes of the PEF treatment vessel should be designed to minimize electrolysis while maintaining a continuous electric field that causes only a slight temperature increase in the food items [24]. PEF treatment can be performed in either a static or continuous manner. In the treatment chamber, food is subjected to specified bumps and a consistent field strength is applied. In continuous processing, a pump continuously transports treated food into and out of the PEF treatment system [23]. Various schematic drawings of the PEF treatment chamber (both static and continuous) have been reported. Huang & Wang [23] have published several designs of PEF treatment chambers, providing detailed descriptions and examining the designs of steady-state and continuous-flow treatment vessels. They also discuss the impact of essential design elements and demonstrate the effectiveness of the ideal treatment system based on existing vessels. Schematic drawings of PEF treatment chambers, such as the continuous treatment vessel, U-shaped static treatment chamber, disk-shaped static treatment chamber, and increased electric field treatment vessel (continuous treatment vessel), can be found. Proper construction of the processing chamber for PEF will result in a lethal effect on various bacteria, achieved through homogeneous electric-field potency, an effective electrode area, and minimal temperature rise.

9.4 APPLICATIONS IN THE FOOD INDUSTRY AND THEIR EFFECT ON COLOR

PEF is distinguished for its minimal environmental impact, continuous operation with shortened processing times, and process with no waste. These characteristics enable the creation of novel, cost-effective, and energy-efficient application methods in the beverage and food industry, the biotech sector, and the pharma business [7, 25, 26]. Many food companies have already invested in PEF food processing machinery [27], which can process up to 10,000 L/h using various treatment chamber designs [28]. One of the biggest hurdles to the industrialization of PEF methodology is the rate of equipment [28]. PEF application can have an impact on product quality due to electro-induced pore development and growth, thus modifying the product appearance, including color. It is one of the major factors affecting food product acceptance, which is an important quality to be preserved in foods. The main applications reported in the literature are focused on fruits and vegetables. In this way, some color variations reveal details about tissue destruction and consumer perception [29]. However, its effect on PEF is still unclear. Since the L^* value is a color measurement along the light-dark axis, studies have generally found an increase in parameter L^*; hence, this increment suggests that PEF-processed items demonstrate an increase in brightness [30]. Since this parameter is closely associated with fruit enzymatic browning, researchers hypothesize that the tendency towards browning may be due to the deactivation of oxidative enzymes that cause browning [31]. However, according to other authors, color changes are a result of cellular membrane breakdown, which impairs functional cell compartmentalization, enhances enzyme-substrate interactions, and ultimately promotes tissue browning [32], leading to a decrease in the L* value. Some of these contradictions are as follows: Aguiló-Aguayo, Soliva-Fortuny, & Martín-Belloso [30] analyzed the color (Macbeth Color-Eye 3,000 colorimeter) of watermelon juice processed by PEF processing with increased intensity at 35 kV/cm for 1,727 μs and subjected to pulses of 4-μs in bipolar mode. There were small but significant variations in L* levels between the treatments, resulting in an increase in the L* value. Thus, high-intensity PEF (HIPEF) processed watermelon juice showed an increase in luminosity over time in storage. Additionally, the increase in h° values of samples processed with HIPEF was associated with an increase in red chromaticity intensity, which may be visible to the naked eye. Since color changes in fruit juices can be greatly influenced by oxidative enzymes such as peroxidase (POD) and polyphenol oxidase (PPO), the behavior of L* can be explained by the effective deactivation of POD in watermelon juice. López-Gámez et al. [29] studied the content of phenolic substances in carrots treated with PEF during the post-processing time, evaluating various quality attributes, including color. Carrots were exposed to different intensities of electric field ranging from 0.80 to 3.5 kV/cm and pulse frequencies ranging from 5 to 30. Both the administration of PEF and the time period after treatment resulted in significant color changes. The intensity of the electric field was the primary factor responsible for this outcome ($\eta p^2 = 0.848$; $p < 0.001$). Greater electric fields were associated with a significant reduction in carrot lightness compared to untreated carrots. They attribute the variations in L* value to two reasons: first, an initial reduction in lightness due to decompartmentalization and cellular covering rupture, which facilitates the interaction of an enzyme with phenolic compounds (previously found in the vacuoles); and secondly, the most recent modifications caused by metabolic changes resulting from structural cell injury.

Regarding the cumulative color variance (ΔE), which indicates the degree of color variation between untreated and PEF processed samples, it has been considered that a ΔE* value of two [17] would result in a noticeable visual distinction. In most of the articles found, this value was higher after PEF processing, suggesting that consumers could perceive such color modifications at first glance [33]. These variations are primarily associated with the changes in the L* value mentioned earlier. However, Zhu et al. [34], when evaluating blueberry juice for certain physicochemical characteristics using microchip PEF or MPEF (microchip pulsed electric field), observed no visual changes in color parameters immediately after the treatment. This was due to a total color variation of $\Delta E = 0.95 \pm 0.16$, which indicates no significant perceptible color variation. They employed an MPEF setup where a small gap between the two electrodes allows for the generation of a strong electric field with low voltage. Strawberry purée provides another example of how PEF can preserve the original color [35]. Aadil et al. [36] investigated the impact of sonication and PEF, both individually and in combination, on various quality parameters of grapefruit juice. The study was conducted using an electric field strength of 20 kV/m for a duration of 600 seconds. In comparison to the control group, the treatments did not have a considerable impact on color metrics. The study reported that the PEF process did not affect the color of grapefruit juice, as there were no remarkable differences in color parameters (L, a, b values) between the PEF processed samples and the control samples ($L^* = 3.89 \pm 0.06$; $a^* = 1.95 \pm 0.04$; $b^* = 2.20 \pm 0.05$). The variation in a* and b* values did not follow the same pattern, indicating that the behavior is product specific. In some cases, the value of a* shifted towards more yellow and less red color in PEF treated samples, while in other cases, products changed to redder and less yellow. Some parameters maintained the same levels without significant differences. These variations are likely attributable to the inherent features of each composition and the variances in processing parameters that each composition was subjected to. For example, Cortés et al. [33] studied the changes in color and carotenoid contents with various strengths of electric supply ranging from 25–40 kV/cm and different processing periods of up to 340 μs on orange juice. Results showed that b* values progressively shifted in a more positive direction in all the samples tested, and a* values reduced as the length of the treatment increased in all the studied samples. These variations presented statistically significant differences ($p < 0.05$), indicating that the orange juice processed with PEF, in comparison to unprocessed orange juice samples, showed a larger inclination toward yellow and a decreased tendency toward red. On the other hand, L* values, which indicate the juice's brightness, remained basically unchanged. It is worth mentioning that the authors found relationships between color and carotenoids. They also noticed that the occurrence and quantity of diverse carotenoids in juices influence the color characteristics. The authors found that the accumulation of total carotenoids (TC) decreases with increased duration of PEF field treatment compared to unprocessed orange fresh juice. Cortés et al. [37] analyzed the color of orange juice processed by High Intensity PEF in storage under chilled conditions, comparing it with pasteurized juice (heat treatment (HT) at 90°C, 20 s). The treatment lasted for 100 S, with the adjusted electric current of 30 kV/cm. When treating orange juice, the treatment chamber temperature does not rise above 49°C. The results of the color parameters showed a slight increase in parameter L* for the juice treated by PEF, with no statistically significant differences compared to 14.48 ± 0.58 and 14.73 ± 0.11 for the unprocessed and processed orange juice, respectively. The mean CIE a* value was significantly higher ($p < 0.05$) in the unprocessed juice (4.56 ± 0.40) than in the PEF processed juice (2.99 ± 0.08; $p < 0.05$), indicating significant differences among the samples. As for parameter b*, it was higher in the PEF processed orange juice samples (53.62 ± 0.57) than in the raw orange juice ones (50.73 ± 0.67). This color change suggests that the treated orange juices are shifting towards a more yellow color and away from a red color. The coloration of orange juice is mainly due to carotenoid pigments, and the primary obstacles preventing their release are cell walls, membranes, and chromoplasts. After PEF application, the diffusion of solutes is enhanced due to pore formations and structural rearrangements [38], resulting in a more yellow color. ΔE values were greater than two in all conditions, although the pasteurized orange juice resulted in higher values than those obtained for the juice processed by PEF. Regarding the browning, the authors discovered correlations indicating that when the juices turn brown, their color becomes greener and yellower, but less red compared to fresh orange juice. Timmermans et al. [39] compared various mild pasteurization processes of orange juice, including PEF, and their impact on overall quality attributes. The application of PEF with length pulses of 2 μs at 23 kV/cm electric field strength resulted in changes in orange juice's color during storage time, including a reduction in lightness (L* value), an increase in redness (a* value), and a decrease in yellowness (b* value).

Significant differences were found in the color parameters of orange juice between untreated orange juice (L*= 57.22 ± 0.08; a*= 1.95 ± 0.02; b*= 61.84 ± 0.11) and PEF treated orange juice (L*= 56.31 ± 0.02; a*= 2.07 ± 0.01; b*= 61.61 ± 0.05), for all color indication values of L*, a*, and b* shortly after processing.

Regarding other products, López-Gámez et al. [38] investigated the utilization of carrot with PEF processing using varying strengths of electric field (ranging from 0.8 to 3.5 kV/cm) and pulse count (ranging from 5 to 30). They found that the C* and h values were not changed by PEF processing in any of the assessed samples, and no significant difference in color value was found between the control and treated carrots. Some studies also reported the preservation of C* and h values in PEF processed juices made from processed grapes alone [40]. However, differences in the browning index (BI) were observed in carrots processed with different intensities. This was correlated to the carotenoid content, as carotenoids are mainly stored in the secondary phloem and can be affected by cell membrane disruption. Similarly, Wibowo et al. [31] investigated the influence of elevated pressure at 400 MPa and 600 MPa for 3 minutes, PEF under various conditions, and pasteurization at 72°C and 85°C for 15 and 30 seconds, respectively, on the color and other attributes of turbid apple juice.

As a result of the browning reaction caused by enzymes, either oxidase or POD or both on phenolic components, the results indicated that the untreated or control samples had the darkest color, whereas the treated samples had a higher level of lightness. The results indicated that the inactivation of oxidative enzymes by the applied treatments may have contributed to the retention of the color of apple juice. PEF-treated juices exhibited increased lightness and redness, whereas no considerable changes in yellowness were observed compared to the untreated control sample. Carbonell-Capella et al. [32] investigated the effect of PEF processing on the fruit juice quality of sweetened drinks with *Stevia rebaudiana*. The juices of papaya and mango were utilized; PEF treatment was performed with a TT ranging up to 360 μs and the intensity of electric supply ranged up to 40 kV/cm. They discovered that the use of PEF processing made a difference in the TC concentration of fruit juice, with an increase at lower and a decrease at higher electric fields, resulting in a decrease in yellowness. According to scientists, when PEF induces the permanent creation of pores in this line, more pronounced color changes may occur, but these can be reversed by resealing the pores.

Furthermore, it is determined by the content of colored pigments, which may be augmented or decreased as a result of the treatment intensity [41]. Therefore, PEF can act ambiguously.

In the winemaking process, PEF has also been used. El-Darra et al. [42] examined the outcome of three pretreatments, including PEF and enzyme treatment (ET), on the extraction of phenolic components and color properties in fermented wine characterized by the Cabernet Sauvignon model. The investigations were conducted using a peak electric current of 5 kV/cm, corresponding to 100 pulses for 10 μs, at a frequency of 0.5 Hz. The metrics a* and b*, which measure redness and yellowness hues, were found to be the highest for the PEF pretreatment (a* = 57.95; b* = 41.8), while the enzyme and control treatments had the lowest values. This is because, compared to the control, PEF treatments fix the ratios of anthocyanins (ACNs) in the freshly produced fermentation model of red wine. ACNs are red pigments primarily found in the exterior layers of the hypodermal tissue, mainly in the vacuoles, which enhance the color of the wine. PEF generates pores in cell membranes, a phenomenon known as electroporation, which increases membrane permeability and accelerates the release of intracellular chemicals. The overall color difference (E) in all pretreated freshly fermented model wines was greater than 3.0, indicating that a visual difference between samples could be clearly observed. In another study, Ying Leong et al. [40] examined the characteristic changes resulting from the application of PEF in relation to the composition of ACNs and the color of Merlot. The treatment was conducted at an increased intensity electric field. They employed a non-stop PEF system (500 kg/h) with energy inputs up to 49.4 kJ/L to achieve their goals. The color intensity of all musts treated with PEF was significantly higher than the color intensity of musts from untreated grapes (up to a 1.62-fold increase), proving that PEF treatment quickly increased the color intensity of musts from grapes. When grapes were processed at the upper range of energy inputs (>37.8 kJ/L), the most notable increase in color intensity between grape musts processed by PEF was seen. The amount of ACNs extracted from grape musts exceeded 100 mg/L, significantly intensifying the color. It is possible that this is due to the creation of irreversible holes at cell membranes when high-intensity electric fields are used. ACNs, color pigments, and phenolics could be readily removed from cell membranes and inside vacuoles after stable and irreversible holes have developed [40]. According to earlier studies, when Pinot Noir grapes receive adequate energy (> 35 kJ/kg at 35–40 kV/cm), irreversible holes widen, and the grape cell membrane and vacuoles are entirely open. Therefore,

PEF treatment seems to do such harm to grape cells, increasing the ACNs release, extracting them, and producing highly colored juice products with high functional value.

Regarding other beverages, Moussa-Ayoub et al. [43] examined how PEF processing affects the specific properties of cactus juice (*Opuntia dillenii*). The abundant purple to red color of the juice samples is due to their significantly increased concentration of betacyanin, primarily betanin, according to the color scales. PEF slightly increased the L* and a* values, from 21.72 ± 0.18 to 21.86 ± 0.20, and 7.47 ± 0.20 to 7.65 ± 0.25, respectively, while the b* value remained constant at 2.61 ± 0.08. Consequently, the authors noted that, despite enzymatic hydrolysis, the final *O. dillenii* juice exhibited a considerable impact on color [43]. Strawberry purée was pasteurized by Geveke et al. [35] using a pilot PEF unit with varied field strengths, temperatures, and a flow rate of 100 l/h. The authors observed that while the color of the PEF-treated samples remained the same as the untreated samples during the first three months of storage, the color of the beverages faded over time, which is a common phenomenon for these products. After 24 months of storage, the calculated color difference, E, was 1.4, indicating no discernible variation. Consequently, the beverage made from strawberry purée that underwent PEF treatment had a 24-month shelf life during which its color did not alter [35].

On the other hand, there have been reports of minor applications for meat-based products. For instance, Arroyo et al. evaluated the effect of changing PEF treatment parameters on certain characteristics of turkey breast meat in 2015 [44]. The author tested various voltages while keeping the pulse width at 20s. Under the experimental conditions, PEF treatments did not result in any difference in the color of fresh or frozen samples, as measured by an instrument (processed and unprocessed). A hedonic scale was also used to rank the hue, and there were no variations in scores between the PEF-treated and control samples, with averages of 5.11 and 5.29, respectively. Consequently, the judges rated the hue as satisfactory with no discernible differences [44].

Similarly, Arroyo et al. [44] evaluated the impact of pulsed electric field processing on the qualitative parameters of beef longissimus thoracis and lumborum (LTL) muscle at different ages, ranging from 2 to 26 days post-mortem [44]. The trial samples were treated with PEF (L*, a*, b*), and neither the aging process nor the treatment differences at each aging time resulted in a statistically significant difference. Therefore, the use of PEF had no effect on the color of LTL samples [44]. Chian et al. [45] confirmed that the treatment had an impact on the color values of bovine longissimus thoracic samples, as well as their effect on ultrastructure and in vitro protein digestibility, with similar outcomes. Low-intensity PEF treatment ranging from 1.00 to 1.25 kV/cm strength of electric current did not alter the color parameter ($p > 0.05$) of the meat, resulting in minimal changes between control and treated samples in terms of color characteristics. The untreated samples had a luminosity (L* value) ranging from 36.3 ± 0.3 to 36.5 ± 0.7, while the treated samples ranged from 36.04 ± 0.6 to 36.7 ± 0.5 [26]. Kantono et al. [46] assessed the physical, chemical, and sensorial characteristics of beef muscles after PEF application. They used two types of beef, Semitendinosus (ST) and bicep femoris (BF). The PEF treatment involved a 20 s pulse width, 50 hertz frequency, and a specific energy of 130 kJ/kg between 0.8 and 1.1 kV/cm, with varied electric field power. The L value of the ST muscle decreased immediately after the PEF processing, while the L value of the BF muscle increased soon after the treatment. Similarly, the PEF-treated samples showed different reactions after storage. On one hand, storage enhanced the color after 7 days since freshly prepared unprocessed ST beef muscles were darker. On the other hand, the PEF treatment on fresh BF beef muscles was reported to have decreased L values [46]. These results indicate that the type of muscle and the specifics of each study can influence the utilization of PEF in meat-based foodstuffs. Li et al. [47] examined the effect of PEF on frozen samples of salmon in terms of freezing and thawing behavior, comparing the freshness indicators of samples that underwent PEF treatment (1.0 kV/cm) to samples subjected to freezing and thawing without treatment [47]. The authors also reported a slight increase in the b* values, which they attributed to oxidation resulting from the use of PEF. The peroxidation value of the PEF treated group marginally increased compared to the untreated control group, and it has been documented in the literature that PEF can exacerbate sample oxidation, leading to color degradation [47].

Like any other food preservation technology, PEF technologies also have certain limitations. Nevertheless, despite these limitations, there are a number of benefits that have been compiled in Table 9.1.

TABLE 9.1 Advantages and Disadvantages of PEF Technologies

Advantages	Disadvantages	References
• Effective cell destruction. • Appropriate for heat-sensitive food matrices. • Compatible with both solid and liquid food products. • Shorter processing times. • Can be used in conjunction with other treatments.	• Increased capital cost. • Due to the heterogenicity of solid food, a higher intensity may be required. • Processing to be conducted without packaging, chances of post-processing contaminations. • Very limited information regarding the continuous process. • Not very effective against spored and enzymes.	[9, 10, 48–51]

9.5 SUMMARY

Pulsed electric field (PEF) is a cutting-edge, non-heating process that has potential uses in the food industry and the potential to lower the amount of energy needed for industrial processes while simultaneously increasing their energy efficiency. However, no commercial system based on PEF technology is currently available due to the lack of homogeneity, comparability, and repeatability of PEF treatment. Although examining the costs of investment and operation suggests that PEF may be able to replace or improve upon current processes, it is essential to understand how various matrices behave when processed by PEF. It is well acknowledged in the food industry that a product's color is a significant quality indicator since consumers notice it immediately and it might influence their purchasing decisions. However, very few studies have been done to date on how PEF processing affects the color characteristics of food items. The majority of the few studies that incorporate color are on beverages made from fruits and vegetables, with fewer studies on meat. PEF-treated samples were discovered to behave differently depending on the products to which they were applied. There were no notable changes in color characteristics in certain situations, whereas brightness rose or reduced depending on the material. Because natural pigments are known to engage in a variety of processes, they are typically considered unstable during processing or storage. As a result, PEF can inactivate the oxidative enzymes responsible for browning, resulting in less dark color; on the other hand, PEF can enhance enzyme-substrate interaction, resulting in more darkening due to cell membrane breakage. Additional research is required to understand the unique effect of the product and manage the reactions in order to maintain the overall quality attributes of color. It is also preferable to link instrumental color analysis with sensory assessment when possible to determine if these variations are perceptible to the senses, as this represents what would be experienced after consumption. This is important because the success of any new product or the use of any innovative technology relies on the long-term preservation of appealing sensory properties, including color.

KEYWORDS

- **color changes**
- **color variance**
- **electric field intensity**
- **food preservation**
- **food processing**
- **high voltage**
- **nonthermal food processing**
- **pulsed electric field**

REFERENCES

1. Barba, F. J., Koubaa, M., Do Prado-Silva, L., Orlien, V., & Sant'Ana, A. De. S., (2017). Mild processing applied to the inactivation of the main foodborne bacterial pathogens: A review. *Trends in Food Science & Technology*, *66*, 20–35. https://doi.org/10.1016/j.tifs.2017.05.011.
2. Gavahian, M., & Farahnaky, A., (2018). Ohmic-assisted hydrodistillation technology: A review. *Trends in Food Science & Technology*, *72*, 153–161. https://doi.org/10.1016/j.tifs.2017.12.014.
3. Gavahian, M., Farahnaky, A., Javidnia, K., & Majzoobi, M., (2012). Comparison of ohmic-assisted hydrodistillation with traditional hydrodistillation for the extraction of essential oils from *Thymus vulgaris* L. *Innovative Food Science & Emerging Technologies*, *14*, 85–91. https://doi.org/10.1016/j.ifset.2012.01.002.

4. Misra, N. N., Koubaa, M., Roohinejad, S., Juliano, P., Alpas, H., Inácio, R. S., Saraiva, J. A., & Barba, F. J., (2017). Landmarks in the historical development of twenty first century food processing technologies. *Food Research International*, *97*, 318–339. https://doi.org/10.1016/j.foodres.2017.05.001.
5. Soliva-Fortuny, R., Balasa, A., Knorr, D., & Martín-Belloso, O., (2009). Effects of pulsed electric fields on bioactive compounds in foods: A review. *Trends in Food Science & Technology*, *20*(11), 544–556. https://doi.org/10.1016/j.tifs.2009.07.003.
6. Gabrić, D., Barba, F., Roohinejad, S., Gharibzahedi, S. M. T., Radojčin, M., Putnik, P., & Bursać, K. D., (2018). Pulsed electric fields as an alternative to thermal processing for preservation of nutritive and physicochemical properties of beverages: A review. *Journal of Food Process Engineering*, *41*(1), e12638. https://doi.org/10.1111/jfpe.12638.
7. Barba, F. J., Parniakov, O., Pereira, S. A., Wiktor, A., Grimi, N., Boussetta, N., Saraiva, J. A., et al., (2015). Current applications and new opportunities for the use of pulsed electric fields in food science and industry. *Food Research International*, *77*, 773–798. https://doi.org/10.1016/j.foodres.2015.09.015.
8. Deng, Q., Zinoviadou, K. G., Galanakis, C. M., Orlien, V., Grimi, N., Vorobiev, E., Lebovka, N., & Barba, F. J., (2015). The effects of conventional and non-conventional processing on glucosinolates and its derived forms, isothiocyanates: Extraction, degradation, and applications. *Food Engineering Reviews*, *7*(3), 357–381. https://doi.org/10.1007/s12393-014-9104-9.
9. Barbosa-Cánovas, G. V., & Sepúlveda, D., (2005). Present status and the future of PEF technology. *Novel Food Processing Technologies*, 1–44.
10. Kumar, Y., Bashir, A. A., Indore, N., Vishwakarma, R. K., & Singh, R. K., (2021). Chapter 5—Pulsed electric field. In: Galanakis, C. M., (ed.), *Sustainable Food Processing and Engineering Challenges* (pp. 137–179). Academic Press. https://doi.org/10.1016/B978-0-12-822714-5.00005-X.
11. Fincan, M., & Dejmek, P., (2003). Effect of osmotic pretreatment and pulsed electric field on the viscoelastic properties of potato tissue. *Journal of Food Engineering*, *59*(2), 169–175. https://doi.org/10.1016/S0260-8774(02)00454-5.
12. Lebovka, N. I., Praporscic, I., & Vorobiev, E., (2004). Effect of moderate thermal and pulsed electric field treatments on textural properties of carrots, potatoes, and apples. *Innovative Food Science & Emerging Technologies*, *5*(1), 9–16. https://doi.org/10.1016/j.ifset.2003.12.001.
13. Chauhan, O. P., & Unni, L. E., (2015). 9—Pulsed electric field (PEF) processing of foods and its combination with electron beam processing. In: Pillai, S. D., & Shayanfar, S., (eds.), *Electron Beam Pasteurization and Complementary Food Processing Technologies* (pp. 157–184). Woodhead Publishing. https://doi.org/10.1533/9781782421085.2.157.
14. Qin, B. L., Barbosa-Canovas, G. V., Swanson, B. G., Pedrow, P. D., & Olsen, R. G., (1998). Inactivating microorganisms using a pulsed electric field continuous treatment system. *IEEE Transactions on Industry Applications*, *34*(1), 43–50. https://doi.org/10.1109/28.658715.
15. Morales-De La, P. M., Welti-Chanes, J., & Martín-Belloso, O., (2019). Novel technologies to improve food safety and quality. *Current Opinion in Food Science*, *30*, 1–7. https://doi.org/10.1016/j.cofs.2018.10.009.

16. Buchmann, L., Bloch, R., & Mathys, A., (2018). Comprehensive pulsed electric field (PEF) system analysis for microalgae processing. *Bioresource Technology*, *265*, 268–274. https://doi.org/10.1016/j.biortech.2018.06.010.
17. Toepfl, S., Siemer, C., Saldaña-Navarro, G., & Heinz, V., (2014). Chapter 6—Overview of pulsed electric fields processing for food. In: Sun, D. W., (ed.), *Emerging Technologies for Food Processing* (2^{nd}edn., pp. 93–114). Academic Press. https://doi.org/10.1016/B978-0-12-411479-1.00006-1.
18. Stefan, T., & Dietrich, K., (2008). Pulsed electric fields as a pretreatment technique in drying processes. *Stewart Postharvest Rev.*, *2*, 1–6. doi: 10.2212/spr.2006.4.3.
19. Arshad, R. N., Abdul-Malek, Z., Munir, A., Buntat, Z., Ahmad, M. H., Jusoh, Y. M. M., Bekhit, A. E. D., et al., (2020). Electrical systems for pulsed electric field applications in the food industry: An engineering perspective. *Trends in Food Science & Technology*, *104*, 1–13. https://doi.org/10.1016/j.tifs.2020.07.008.
20. Amit, S. K., Uddin, Md. M., Rahman, R., Islam, S. M. R., & Khan, M. S., (2017). A review on mechanisms and commercial aspects of food preservation and processing. *Agriculture & Food Security*, *6*(1), 51. https://doi.org/10.1186/s40066-017-0130-8.
21. Barbosa-Cánovas, G. V., Góngora-Nieto, M. M., Pothakamury, U. R., & Swanson, B. G., (1999). *Preservation of Foods with Pulsed Electric Fields*. Academic Press.
22. Góngora-Nieto, M. M., Younce, F., Hyde, G. M., Pedrow, P. D., Swanson, B. G., & Barbosa-Cánovas, G. V., (2002). Metrology system for pulsed electric fields processing. *Innovative Food Science & Emerging Technologies*, *3*(4), 337–348. https://doi.org/10.1016/S1466-8564(02)00068-1.
23. Huang, K., & Wang, J., (2009). Designs of pulsed electric fields treatment chambers for liquid foods pasteurization process: A review. *Journal of Food Engineering*, *95*(2), 227–239. https://doi.org/10.1016/j.jfoodeng.2009.06.013.
24. Butz, P., & Tauscher, B., (2002). Emerging technologies: Chemical aspects. *Food Research International*, *35*(2), 279–284. https://doi.org/10.1016/S0963-9969(01)00197-1.
25. Puértolas, E., Koubaa, M., & Barba, F. J., (2016). An overview of the impact of electro technologies for the recovery of oil and high-value compounds from vegetable oil industry: Energy and economic cost implications. *Food Research International*, *80*, 19–26. https://doi.org/10.1016/j.foodres.2015.12.009.
26. Puértolas, E., & Barba, F. J., (2016). Electro technologies applied to valorization of by-products from food industry: Main findings, energy, and economic cost of their industrialization. *Food and Bioproducts Processing*, *100*, 172–184. https://doi.org/10.1016/j.fbp.2016.06.020.
27. Jaeger, H., Balasa, A., & Knorr, D., (2008). Food industry applications for pulsed electric fields. In: *Electro technologies for Extraction from Food Plants and Biomaterials* (pp. 181–216). Springer. https://doi.org/10.1007/978-0-387-79374-0_7.
28. Toepfl, S., (2011). Pulsed electric field food treatment—Scale up from lab to industrial scale. *Procedia Food Science*, *1*, 776–779. https://doi.org/10.1016/j.profoo.2011.09.117.
29. López-Gámez, G., Elez-Martínez, P., Martín-Belloso, O., & Soliva-Fortuny, R., (2020). Enhancing phenolic content in carrots by pulsed electric fields during post-treatment time: Effects on cell viability and quality attributes. *Innovative Food Science & Emerging Technologies*, *59*, 102252. https://doi.org/10.1016/j.ifset.2019.102252.

30. Aguiló-Aguayo, I., Soliva-Fortuny, R., & Martín-Belloso, O., (2010). Color and viscosity of watermelon juice treated by high-intensity pulsed electric fields or heat. *Innovative Food Science & Emerging Technologies*, *11*(2), 299–305. https://doi.org/10.1016/j.ifset.2009.12.004.
31. Wibowo, S., Essel, E. A., De Man, S., Bernaert, N., Van, D. B., Grauwet, T., Van, L. A., & Hendrickx, M., (2019). Comparing the impact of high pressure, pulsed electric field and thermal pasteurization on quality attributes of cloudy apple juice using targeted and untargeted analyses. *Innovative Food Science & Emerging Technologies*, *54*, 64–77. https://doi.org/10.1016/j.ifset.2019.03.004.
32. Carbonell-Capella, J. M., Buniowska, M., Cortes, C., Zulueta, A., Frigola, A., & Esteve, M. J., (2017). Influence of pulsed electric field processing on the quality of fruit juice beverages sweetened with *Stevia rebaudiana*. *Food and Bioproducts Processing*, *101*, 214–222. https://doi.org/10.1016/j.fbp.2016.11.012.
33. Cortés, C., Esteve, M. J., Rodrigo, D., Torregrosa, F., & Frígola, A., (2006). Changes of color and carotenoids contents during high intensity pulsed electric field treatment in orange juices. *Food and Chemical Toxicology*, *44*(11), 1932–1939. https://doi.org/10.1016/j.fct.2006.06.026.
34. Zhu, N., Zhu, Y., Yu, N., Wei, Y., Zhang, J., Hou, Y., & Sun, A., (2019). Evaluation of microbial, physicochemical parameters and flavor of blueberry juice after microchip-pulsed electric field. *Food Chemistry*, *274*, 146–155. https://doi.org/10.1016/j.foodchem.2018.08.092.
35. Geveke, D. J., Aubuchon, I., Zhang, H. Q., Boyd, G., Sites, J. E., & Bigley, A. B. W., (2015). Validation of a pulsed electric field process to pasteurize strawberry purée. *Journal of Food Engineering*, *166*, 384–389. https://doi.org/10.1016/j.jfoodeng.2015.05.008.
36. Aadil, R. M., Zeng, X. A., Sun, D. W., Wang, M. S., Liu, Z. W., & Zhang, Z. H., (2015). Combined effects of sonication and pulsed electric field on selected quality parameters of grapefruit juice. *LWT – Food Science and Technology*, *62*(1, Part 2), 890–893. https://doi.org/10.1016/j.lwt.2014.10.025.
37. Cortés, C., Esteve, M. J., & Frígola, A., (2008). Color of orange juice treated by high intensity pulsed electric fields during refrigerated storage and comparison with pasteurized juice. *Food Control*, *19*(2), 151–158. https://doi.org/10.1016/j.foodcont.2007.03.001.
38. López-Gámez, G., Elez-Martínez, P., Martín-Belloso, O., & Soliva-Fortuny, R., (2021). Changes of carotenoid content in carrots after application of pulsed electric field treatments. *LWT*, *147*, 111408. https://doi.org/10.1016/j.lwt.2021.111408.
39. Timmermans, R. A. H., Mastwijk, H. C., Knol, J. J., Quataert, M. C. J., Vervoort, L., Der Plancken, I. V., Hendrickx, M. E., & Matser, A. M., (2011). Comparing equivalent thermal, high pressure and pulsed electric field processes for mild pasteurization of orange juice. Part I: Impact on overall quality attributes. *Innovative Food Science & Emerging Technologies*, *12*(3), 235–243. https://doi.org/10.1016/j.ifset.2011.05.001.
40. Leong, S. Y., Treadwell, M., Liu, T., Hochberg, M., Sack, M., Mueller, G., Sigler, J., Silcock, P., & Oey, I., (2020). Influence of pulsed electric fields processing at high-intensity electric field strength on the relationship between anthocyanins composition and color intensity of merlot (*Vitis vinifera* L.) musts during cold

maceration. *Innovative Food Science & Emerging Technologies*, *59*, 102243. https://doi.org/10.1016/j.ifset.2019.102243.

41. Odriozola-Serrano, I., Aguiló-Aguayo, I., Soliva-Fortuny, R., & Martín-Belloso, O., (2013). Pulsed electric fields processing effects on quality and health-related constituents of plant-based foods. *Trends in Food Science & Technology*, *29*(2), 98–107. https://doi.org/10.1016/j.tifs.2011.10.003.
42. El Darra, N., Turk, M. F., Ducasse, M. A., Grimi, N., Maroun, R. G., Louka, N., & Vorobiev, E., (2016). Changes in polyphenol profiles and color composition of freshly fermented model wine due to pulsed electric field, enzymes and thermovinification pretreatments. *Food Chemistry*, *194*, 944–950. https://doi.org/10.1016/j.foodchem.2015.08.059.
43. Moussa-Ayoub, T. E., Jäger, H., Knorr, D., El-Samahy, S. K., Kroh, L. W., & Rohn, S., (2017). Impact of pulsed electric fields, high hydrostatic pressure, and thermal pasteurization on selected characteristics of *Opuntia dillenii* cactus juice. *LWT – Food Science and Technology*, *79*, 534–542. https://doi.org/10.1016/j.lwt.2016.10.061.
44. Arroyo, C., Eslami, S., Brunton, N. P., Arimi, J. M., Noci, F., & Lyng, J. G., (2015). An assessment of the impact of pulsed electric fields processing factors on oxidation, color, texture, and sensory attributes of turkey breast meat. *Poultry Science*, *94*(5), 1088–1095. https://doi.org/10.3382/ps/pev097.
45. Chian, F. M., Kaur, L., Oey, I., Astruc, T., Hodgkinson, S., & Boland, M., (2019). Effect of pulsed electric fields (PEF) on the ultrastructure and in vitro protein digestibility of bovine longissimus thoracis. *LWT*, *103*, 253–259. https://doi.org/10.1016/j.lwt.2019.01.005.
46. Kantono, K., Hamid, N., Oey, I., Wang, S., Xu, Y., Ma, Q., Faridnia, F., & Farouk, M., (2019). Physicochemical and sensory properties of beef muscles after pulsed electric field processing. *Food Research International*, *121*, 1–11. https://doi.org/10.1016/j.foodres.2019.03.020.
47. Li, J., Shi, J., Huang, X., Zou, X., Li, Z., Zhang, D., Zhang, W., & Xu, Y., (2020). Effects of pulsed electric field on freeze-thaw quality of Atlantic salmon. *Innovative Food Science & Emerging Technologies*, *65*, 102454. https://doi.org/10.1016/j.ifset.2020.102454.
48. Gómez, B., Munekata, P. E. S., Gavahian, M., Barba, F. J., Martí-Quijal, F. J., Bolumar, T., Campagnol, P. C. B., et al., (2019). Application of pulsed electric fields in meat and fish processing industries: An overview. *Food Research International*, *123*, 95–105. https://doi.org/10.1016/j.foodres.2019.04.047.
49. Jan, A., Sood, M., Sofi, S., & Norzom, T., (2017). *Non-thermal Processing in Food Applications: A Review*.
50. Hirneisen, K. A., Black, E. P., Cascarino, J. L., Fino, V. R., Hoover, D. G., & Kniel, K. E., (2010). Viral inactivation in foods: A review of traditional and novel food-processing technologies. *Comprehensive Reviews in Food Science and Food Safety*, *9*(1), 3–20. https://doi.org/10.1111/j.1541-4337.2009.00092.x.
51. Barba, F. J., Grimi, N., & Vorobiev, E., (2015). New approaches for the use of non-conventional cell disruption technologies to extract potential food additives and nutraceuticals from microalgae. *Food Engineering Reviews*, *7*(1), 45–62. https://doi.org/10.1007/s12393-014-9095-6.

CHAPTER 10

EFFECT OF LIGHT EMITTING DIODE (LED) TREATMENT ON THE COLOR OF FOOD PRODUCTS

PRANAV VASHISHT,[1] SACHIN SHARMA,[2] LOVEPREET SINGH,[2] and VENKATESWARLU SUNKESULA[3]

[1]*Food Biosciences and Technology Program, Department of Agricultural and Environmental Sciences, Tennessee State University, Nashville, Tennessee, USA*

[2]*Department of Animal, Veterinary, and Food Sciences, University of Idaho, Moscow*

[3]*Department of Food and Dairy Science, Midwest Dairy Foods Research Center, South Dakota State University, Brookings, South Dakota, USA*

ABSTRACT

Light Emitting Diode (LED) treatment is a novel nonthermal technology that ensures the safety of food products. Researchers have investigated this technique to overcome the disadvantages of Ultra-Violet (UV) irradiation. LEDs can effectively inactivate foodborne pathogens, spoilage-causing microorganisms, and enzymes without significantly affecting their nutritional and sensory parameters. Numerous studies have been conducted to test the inactivation efficiency of LED treatment, regardless of its impact on the quality parameters of the product. In terms of food quality, color plays an important role in customer preference. This chapter reviews the

Nonthermal Food Processing Technologies: Impact on Color Profile.
Kaavya Rathnakumar and R. Pandiselvam (Eds.)

changes in color observed in LED-treated food products, including fresh produce (strawberries, peppers, tomatoes, pears, mandarins, oranges, and pineapples), dairy products (milk and cheese), juices (orange and apple juice), seafood (salmon), and poultry (chicken). Most literature studies have reported a non-significant effect of LED treatment using different wavelengths on the color of the treated food product. Overall, this efficient process is capable of providing food safety to a variety of food products (both solid and liquid) while preserving their natural properties.

10.1 INTRODUCTION

Presently, foodborne illnesses are of great concern, driving industries towards strict processing norms for the safety of food products [4]. However, consumer awareness of healthy and wholesome food is forcing industries to investigate and explore ways to produce minimally processed food with intact nutritional composition [32, 72]. Heat treatments (HTs) are commonly used in industrial processing technologies to ensure food safety. However, they are known to have a significant effect on the organoleptic as well as nutritional profile of the products [26]. In recent years, authors have investigated numerous alternate technologies to traditional HTs. Several common examples include HPP (high pressure processing), ultrasound, pulse electric field, and UV irradiation [11, 73, 98, 99, 102, 103, 110]. Among these light-based technologies, especially UV irradiation, has been explored with keen interest due to its effectiveness against a range of microbes from bacteria to viruses, no significant effect on the product's natural characteristics, lower energy consumption, and lower processing cost [74]. This technique has proven its potential for a range of products, but its limited mass transfer and penetration inside solid food and opaque fluids are the major challenges [72]. Another disadvantage is the use of mercury lamps, which increases the risk of mercury leakage in food products and can be hazardous [52]. LED technology has been investigated and demonstrated as an efficient and safe alternative. Figure 10.1 showcases numerous applications of LEDs in the food industry. LEDs are known for their longer lifespan, energy efficiency, and lower energy consumption [52]. Studies have shown that LED treatment can enhance the bioactive components in various crops, such as Chinese cabbage, watercress, lettuce, sprouted broccoli, and garden peas [7, 45]. Choi et al. [17] reported increased free sugar and ACN content in LED-treated strawberry fruit. Figure 10.1 also demonstrates the various applications of LEDs in the food industry. LEDs are semiconductor-based lamps that emit light energy in the wavelength range of 200 to 780 nm [77]. They are recognized for their high performance, good shelf stability, low energy consumption, durability, and cost-effectiveness [91]. Literature has reported that LEDs emitting light in the visible wavelength range have shown efficient inactivation of various foodborne microbes [48–50, 54]. Some authors have also explored the use of multi-wavelength LEDs as a synergistic microbial killing treatment [53].

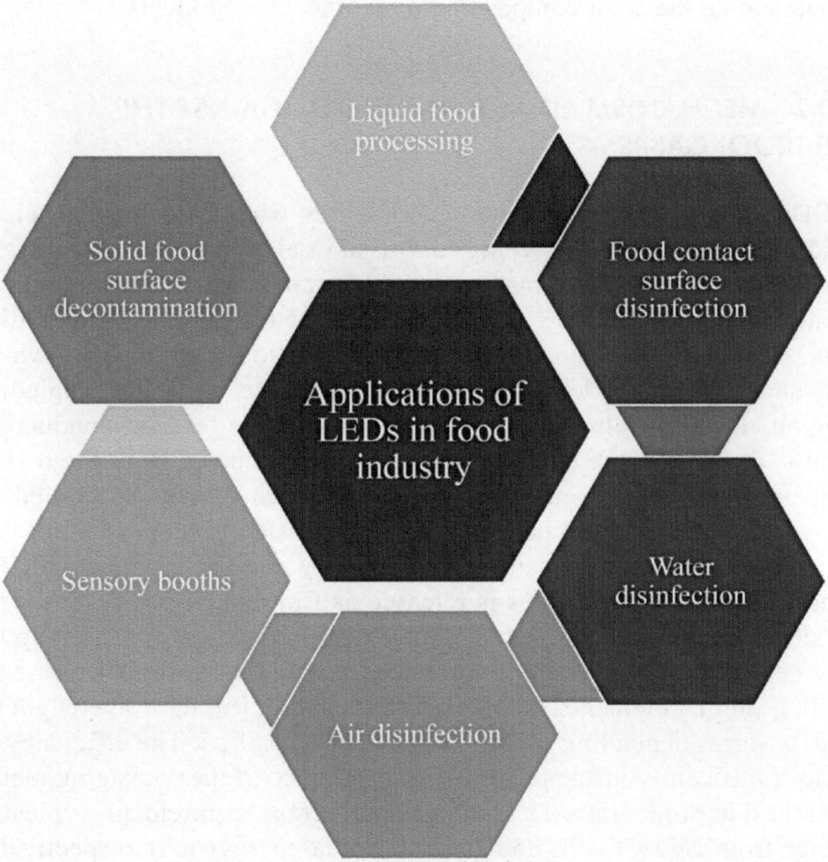

FIGURE 10.1 Various applications of LEDs in food industry.

Quality parameters (microbial, chemical, and sensory) are highly important for any food product. Consumers evaluate sensory properties of food products such as flavor, color, texture, appearance, and odor to assess the quality [63]. Additionally, the color of a product significantly affects consumer preference or choice [94]. Therefore, it is crucial to assess the impact of processing on the sensory parameters of products, including color. There is ample literature on the effectiveness of LEDs in inactivating foodborne microorganisms [34, 48–50]. However, most of the research in this field has been conducted without considering the potential alteration of the natural characteristics and structure of the food [77]. This chapter discusses the properties and efficiency of different wavelengths of LED lights in ensuring food safety and the impact of treatment on the color composition of various food products.

10.2 MECHANISM OF ACTION OF LED AGAINST THE MICROORGANISMS

LEDs are semiconductors that generate light when a magnetic or electric field is applied. According to Prasad et al. [77], when an electron encounters an electric or magnetic field, it transitions to a lower energy state by releasing energy in the form of electromagnetic (EM) radiation. The LED system consists of the p-side and n-side, also known as the anode and cathode, respectively [35] (Figure 10.2). The depletion region, located between the anode and cathode, is non-conductive and is formed by the combination of additional holes from the p-side and free electrons from the n-side [85]. When current is applied to the LED, independent electrons from the n-side combine with holes on the p-side, which are in a lower energy state. The energy difference between the two sides is released as EM energy or photons. The p-side of LEDs is made by combining group II elements with group III elements, while the n-side is formed by combining group IV elements with group III elements. According to Bao et al. [8], the wavelength of LEDs varies depending on the chemical composition. The efficiency is also significantly influenced by the absorbance of the packaging material used in production. The voltage and current requirements typically range from 1.5 to 3 volts and 10 to 30 milliamperes (mA), respectively, depending on the material used for diode production and the emitted wavelength [35].

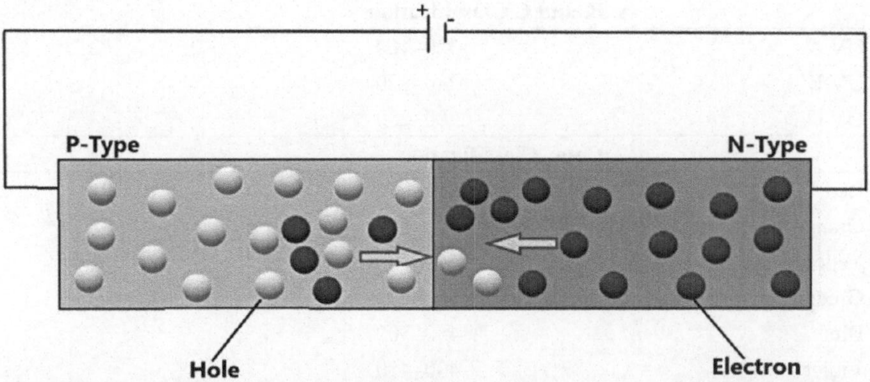

FIGURE 10.2 Working mechanism of LEDs.

10.3 TYPES OF LEDS

Safety concerns regarding the leakage of mercury lamps have prompted researchers to explore LED options [52]. LEDs have proven to be eco-friendly and have experienced significant growth in recent years, with numerous applications that heavily rely on the output and wavelength used. The classification of LEDs is described in Table 10.1. UV-A irradiation triggers the formation of hydroxyl and oxidative substances, which further damage microbial cell membranes and kill them [87]. These LEDs are beneficial for processing liquid food products, including opaque and turbid ones, due to their deeper penetration power. There is limited literature available on the bactericidal application of UV-B LEDs. Argyraki et al. [5] reported higher efficiency of UV-B LEDs compared to UV-C against *P. aeruginosa* biofilm. UV-C literature has reported the germicidal nature of these waves against a variety of microorganisms, from low-resistant bacteria to highly resistant spores [101]. Inactivation has been reported throughout the spectrum, but higher efficiency has been observed below 230 nm and between 260 and 270 nm, with some reports at 254 nm [13, 72].

TABLE 10.1 Summarizing the Classification of LEDs Based Upon the Wavelength Regions in the Light Spectrum

Classification Type	Wavelength Range (nm)	References
	A, B, and C Classification	
UV-A	320–400	[72]
UV-B	280–320	
UV-C	200–280	
	Color Classification	
Red	610–670	[22]
Orange	590–610	
Yellow	570–590	
Green	500–570	
Blue	450–500	
Violet	400–450	

Blue LEDs are known to work based on an indirect mechanism, where reactive oxidative species damage DNA and other essential components, leading to the inactivation of microbes. These LEDs are effective against numerous foodborne microorganisms in clear buffer and broth solutions [28, 48–51].

Literature also reports the symbiotic effect of multiwavelength LEDs against foodborne pathogens. Akgün & Ünlütürk [3] reported the combined effect of UV-A and C LEDs, which was more efficient than using UV-C alone against the polyphenol oxidase (PPO) enzyme present in clear apple juice.

10.4 EFFICIENCY OF LEDS AGAINST FOODBORNE MICROORGANISMS

Microbial log reduction during the LED processing depends on various factors. The wavelength of the LED source significantly affects the microbial reduction [53]. UV-C has been reported as an efficient inactivation source compared to UV-A and UV-B LEDs [53]. The optical properties of the fluid, such as absorption and scattering coefficient, also play a significant role [33]. According to Sizer & Balasubramaniam [89], a major portion of the UV-C waves gets absorbed at the surface of the liquid food products, leading to lower inactivation efficiency. However, streamlining the UV reactor geometry can assist in equivalent dose distribution, hence higher microbial inactivation at a lower dose [102, 103]. Researchers have demonstrated the potential of different wavelengths of LEDs in the inactivation of pathogenic as well as spoilage-causing microorganisms suspended in a clear solution (like phosphate buffer solution/PBS and water), opaque fluids (like fruit juice and milk), and solid surfaces (like fresh produce, dairy, and meat products) [29, 48–50, 69]. Inactivation studies in various food products conducted by various authors are highlighted in Table 10.2. About 405 nm was widely used in food applications. Srimagal, Ramesh, & Sahu [93] reported that 405 nm has a higher inactivation efficiency than 433 and 460 nm when the targeted *E. coli* bacteria were suspended in skim milk and kept at 5, 10, and 15°C. Sommers, Gunther IV, & Sheen [90] reported a 1.7–2.1 reduced population range of *Campylobacter* species on the chicken skin using the same wavelength LED. Kim, Bang, & Yuk [48] demonstrated its efficiency against *S. typhi*, *S. agona*, *S. saintpaul*, and *S. newport* on papaya pieces. Kim, Tang, Bang, & Yuk [50] reported population reduction of *E. coli, S. typhi*, and *L. monocytogenes* on mango pieces. Use of 460 nm LED against *S. typhi*, *S. montevideo*, *S. gaminara*, *S. newport*, and *S. saintpaul* inoculated in orange juice at a temperature of 4, 12, and 20°C was found [30]. Niu et al. [71] demonstrated its effectiveness against *S. cerevisiae* inoculated in orange juice where the UV-C LED was used for the treatment, and delivered dose was 1,420 mJ/cm^2. Ghate et al. [29] demonstrated the inactivation of *S.* Montevideo *S.* Newport and *S.* Saintpaul on the surface of pineapple pieces at a different temperature of 7 and 16°C. Josewin et al. [44] demonstrated the reduction of a cocktail of *L. monocytogenes* strains on smoked salmon.

TABLE 10.2 Summarizing the LED Treatment Studies on Various Wavelengths Against Different Microbial Strains in Different Food Products

Product	Microorganism	Wavelength Used (nm)	Log Reduction	Dose Delivered/ Treatment Time	References
Papaya pieces	*S.saintpaul* *S.agona* *S.newport* *S. typhimurium*	405 ± 5	0.3–1.3	1.3–1.7 kJ/cm^2	[48]
Pineapple pieces	*S. gaminara* *S. montevideo* *S. newport* *S. saintpaul* *S. typhimurium*	460	0.61–0.72	7,950 J/cm^2	[30]

TABLE 10.2 *(Continued)*

Product	Microorganism	Wavelength Used (nm)	Log Reduction	Dose Delivered/ Treatment Time	References
Pineapple pieces	*E. coli* *S. aureus*	462	0.46 of *E. coli* 1.17 of *S. aureus*	50 J/cm^2	[12]
Cooked chicken	*S. enteritidis*	405 ± 5	0.8–0.9	4.8 kJ/cm^2	[49]
Fresh salmon	*L. monocytogenes*	405	2–2.8	8 hours	[58]
Fresh salmon	*L. monocytogenes* *S. enteritidis* *S. newport* *S. typhimurium*	405	0.4 and 0.3 of *Listeria* 0.5 and 0.4 in *Salmonella*	8 hours	[57]
Smoked salmon	*L. monocytogenes*	460	0.7–1.2	2.4 kJ/cm^2	[44]
Sliced camembert cheese	*E. coli*O157:H7 *S. typhimurium* *L. monocytogenes*	266, 270, 275, and 279	4–5 of *E. coli* and *Salmonella* 3–4 of *Listeria*	3 mJ/cm^2	[51]
Sliced camembert cheese	*E. coli*O157:H7 *S. typhimurium* *L. monocytogenes*	266	4.88 of *E. coli* 4.72 of *Salmonella* 3.52 of *Listeria*	3 mJ/cm^2	[51]
Cabbage	*E. coli* DH 5α	365	3.23	90 minutes	[2]
Orange juice	*E. coli* DH 5α	365	0.35–1.58	126 J/cm^2	[60]
Orange juice	*S. cerevisiae*	–	4.44	1,420 mJ/cm^2	[71]
Apple juice	*E. coli* K12	280	2	–	[3]
Apple juice	*E. coli* K12	280/365	2	–	[3]
Apple juice	*E. coli* K12	280	4.4	40 min	[3]
Apple juice (clear)	*E. coli* K12	280	3.9	40 min	[3]
Chicken skin	*C. jejuni* *C. coli*	405	1.7 of *C. jejuni* 2.1 of *C. coli*	184–186 J/cm^2	[90]
Chicken skin	*Staphylococcus* *Salmonella* *Listeria* *E. coli*	405	0.40	180 J/cm^2	[90]
Chicken purge	*Staphylococcus* *Salmonella* *Listeria* *E. coli*	405	0.23–0.1.01	180 J/cm^2	[90]
Milk	*E. coli*	405–460	>5	–	[93]
Mixed beverage	*E. coli* K 12	280/365	>4	40 minutes	[9]

Some authors also tested the efficiency of multiwavelength LEDs against foodborne microbes. Lu, Yang, Zhang, & Lai [62] reported the symbiotic effect of 288 and 271 nm LEDs against *E. coli, S. epidermidis,* and *S. Typhimurium* inoculated on solid agar. Baykuş, Akgün, & Unluturk [9] used 280 and 365 nm LEDs against *E. coli* K12 in a mixed beverage. McSharry, Koolman, Whyte, & Bolton [65] found that among 255, 265, and 285 nm, 285 nm was the most efficient against *L. monocytogenes, S. typhimurium*, total viable, and Enterobacteriaceae count inoculated in meat broth, and further synergistic effect can be created by collaborating it with 255 or 265 nm. Overall, LEDs have proved to be an effective processing treatment against a range of foodborne microbes as well as enzymes [3].

10.5 IMPORTANCE OF COLOR IN FOOD PRODUCTS AND ITS DETERMINATION

Color refers to a memorable visual element, which is a crucial advertising and promotion tool [16]. It is an essential element of the sensory properties of the food product [94]. The food industry should consider the product color as the utmost part of marketing, but that is not the case. Despite adequate literature, the importance of food color has been abandoned because of its little understanding by the food promoters. The history of color can be traced back to ancient Egypt, where candies were colored while wines were known to be colored since 400 BC [21]. Previously, manufacturers relied on natural sources to provide color to the food products [94]. Some common examples are beetroot extract, paprika, flower petals, saffron, squid ink, and cochineal insects. But increased competition in the food market is deviating manufacturers from the use of natural coloring agents. The use of cheaper colors like copper arsenate, red lead, and vermillion has become a common practice. Presently, colors are mainly classified as natural, nature identical, and artificial. Natural can be claimed for the food that is not processed or mechanically processed. Processing can lead to alterations in the properties of the coloring elements [94]. Nature identical colors are formed using substances similar to ones found in nature [36]. Artificial, also known as manmade colors, are the ones whose molecular arrangement cannot be found in nature or originated from the natural source [70]. Various natural and artificial coloring components present in different food products are summarized in Table 10.3.

TABLE 10.3 Summarizing the Various Natural and Artificial Color Substances Present in the Different Food Products

Name	Color	Products	References
Natural			
Paprika	Dark red	Confectionary products.	[100]
Annatto	Bright yellow to orange red	Cheddar cheese, sausages, fish, margarine, snacks, dressings, sauces, and confections.	[46]
Caramel	Dark brown	Beverages, soda, dark breads, cider, desserts, meats, seasoning, and confectionery.	[105]
Carotene	Red to brown	Ice cream, mayonnaise, and chocolates.	[19]
Curcumin	Orange yellow	Mustard, pastries, dairy products, and canned fish.	[24]
Riboflavin	Whitish yellow	Cereal based foods.	[23]
Carmine	Bright red	Dairy based drinks, ripened cheese.	[15]
Artificial			
Sunset yellow	Orange or red	Swiss roll, beverages, candies, ice creams, and jam.	[82]
Tartrazine	Light orange	Bakery and dairy products, candies, and beverages.	[31]
Ponceau	Red	Soft drink and juices.	[27]
Allura red	Deep red	Soups and meat products.	[83]
Amaranth	Reddish brown	Caviar, sweets, and beverages.	[76]
Indigo carmine	Bluish brown	Beverages and food.	[6]
Quinoline yellow	Yellow to orange	Jellies, caramels, processed seafood.	[64]

Many food products contain natural coloring substances that can be used to color other food products. For example, grapes contain ACN, red beet contains betaine, tomatoes contain lycopene, and turmeric contains curcumin [94]. The water-soluble ACNs provide color to the food depending on its pH. At higher pH levels, the color can range from yellow-red to purple/brilliant blue [38]. Some authors have also reported using alfalfa parts, grass containing chlorophyll, marigold flowers, Bixa Orellana, β-carotene, and astaxanthin as coloring agents [10, 55, 88]. Other edible coloring materials mentioned in the literature include sweet potatoes, grapes, blueberries, spirulina, and safflower [94]. Controlled heating can result in the production of caramel color in food products [21]. The main component responsible for caramelization is the carbohydrate or sugar component of the product [96]. Caramel is classified as a separate category that does not fall into the artificial or natural category [25]. Insufficient combustion of nut shells, cellulose residue, and peat can produce vegetable carbon [41]. Some common artificial colors used in the food industry are iron oxide, calcium carbonate, and titanium oxide [70].

The color of food products is usually measured using a colorimeter or spectrophotometer [29]. L*, a*, and b* values are evaluated, where L* represents the lightness, a* represents the red or green color, and b* represents the yellow or blue color of the product [12]. The yellow index and color difference of the product are evaluated using the following equations:

$$\text{Yellow index} = 142.86\frac{b^*}{L^*} \tag{1}$$

$$\text{Color difference} = \sqrt{\left(L^* - L_c\right)^2 + \left(a^* - a_c\right)^2 + \left(b^* - b_c\right)^2} \tag{2}$$

where; L_c, a_c, and b_c are the values of controlled or untreated samples.

10.6 CHANGES IN THE COLOR OF FOOD PRODUCTS TREATED WITH LEDS

10.6.1 FRESH PRODUCE

According to Carle, Reinhold, and Schweiggert [15], fresh produce, such as fruits and vegetables, is often transported to distant places. This creates a need for treatments that can delay the ripening of these products. Literature has reported the use of various wavelengths of LEDs to improve the shelf life of fresh produce [20, 45, 47, 56]. ACNs, a group of flavonoids, contribute to the color of many fresh produce items, such as strawberries and red lettuce [95, 107]. Other components responsible for the coloration of fresh produce include chlorophyll, carotenoids, and lycopene [37, 40, 59]. It has been reported that blue LEDs at 470 nm are effective in enhancing the color intensity of strawberries compared to 525 nm and 630 nm [47]. This effect was observed after 3 to 4 days and was attributed to the increased ACN content in the treated product. Similarly, Kanazawa et al. [45] observed intense red color in strawberries exposed to green LED lights (500–600 nm). Stutte, Edney, and Skerritt [95] used 440 nm, 530 nm, 640 nm, and 740 nm LEDs for the treatment of red lettuce, and found that the lettuce treated with blue LEDs had approximately twice the ACN content compared to the lettuce treated with red LEDs. Jin et al. [42] treated broccoli florets with 630 nm wavelength red LEDs and observed an increase in chlorophyll content corresponding to an increased L value. Habanero peppers are a distinct category of peppers that exhibit colors ranging from light to dark green, orange yellow, and red, depending on their maturity level [14, 18]. Pérez-Ambrocio et al. [75] stated that neither UV-C nor blue LEDs were able to immediately change the color of habanero peppers. However, during storage, the untreated samples showed significant color changes, while the samples exposed to the highest UV dose showed no major color changes. The major component responsible for the color of tomatoes is lycopene [40]. In the studies mentioned in Ref. [20], blue and red LEDs with wavelengths of 440 nm and 450 nm, respectively, were used to treat green tomatoes. Samples treated with a 450 nm LED and stored for a period of seven days (under darker conditions) were found to have a higher lycopene content compared to those treated with a 440 nm LED. However, samples treated with both 440 nm and 450 nm LEDs and stored for 21 days were reported to have approximately similar lycopene content. Sun et al. [97] reported the improvement in the redness of pear fruit (Red Chinese sand) treated with 315 nm and 765 nm LEDs, respectively. Ghate et al. [29] found discoloration in pineapple pieces treated with a 460 nm LED. Papaya pieces treated under 405 ± 5 nm showed no major alteration in color [48]. Kim et al. [49] treated mango pieces using a similar LED system and observed the same. Literature has reported that carotenoids absorb light between 400 and 500 nm [66]. Ghate et al. [29] found that the absorption peak of β-carotene is between 450 and 478 nm, which aligns with the results of [12] studies. They observed a significant alteration in the color of pineapple pieces treated with a blue LED (462 nm), which was attributed to the degradation of β-carotene. Johkan et al. [43] found a high carotenoid content in lettuce plantlets processed with a blue LED. The same was observed in the studies of Satsuma mandarin and Valencia orange in Ref. [108].

10.6.2 DAIRY PRODUCTS

Milk is an important source of nutrition. It consists of a significant amount of fat, protein, minerals, vitamins, and sugars (carbohydrates) [68]. This nutrition benefits consumers but also poses a threat of microbial growth and contamination. However, heat pasteurization, sterilization, and UHT are potential treatments followed by the dairy industry to eliminate pathogens from milk. Literature reports the non-desirable effect of these treatments on natural parameters [33]. This calls for the application of nonthermal technology in the dairy industry. Srimagal, Ramesh, & Sahu [93] used LEDs ranging from 405–460 nm for the treatment of milk. It was found that as the wavelength decreased, there was an increased alteration in the color of milk; however, there was no significant distinction.

Cheese is a highly nutritious dairy product consisting of a potential amount of fat, protein, and essential amino acids [1, 104]. Cheese has a significant demand in the dairy market because of its ease of consumption, either alone or with other food products [67]. Kim, Kim, & Kang [51] used four wavelengths of LEDs, namely 279, 275, 270, and 266 nm, for the treatment of camembert cheese surface. They observed an insignificant change in the color of all the products, including the control (maximum dose reported was 3 mJ/cm^2).

10.6.3 JUICE

Another important segment of the food sector is the juice market. This market is consistently growing because consumers are increasingly aware of health and healthy products [78]. The largest producers and sellers of juice products include Germany, China, France, the UK, and the US. According to Ghate et al. [30], fruit juices consist of an abundant amount of carbohydrates and have high water activity, which makes them a potential source of microbial growth. The juice industry also follows the same food safety processes, namely pasteurization and sterilization, which significantly affect the sensory and nutritional composition, especially the vitamin C content [80, 84]. Ghate, Kumar, Zhou, & Yuk [30] used 460 nm for orange juice processing, where significant alteration was reported at 500 J/cm^2 of delivered dose, while the color didn't significantly get affected up to 3,500 and 2,500 J/cm^2 of dose at 147 mW/cm^2 irradiance and 20°C and 4°C temperature, respectively. Whereas at 254.7 mW/cm^2, the color held up well up to 1,000, 2,000, and 1,000 J/cm^2 of delivered dose at 20, 12, and 4°C. The alteration was attributed to the degradation of carotenoids due to a light reaction. The result was consistent with the literature statement regarding the absorption of light by carotenoids in the range of 400–500 nm [66]. Niu et al. [71] reported no change in the color of orange juice when treated with a delivered dose of 1,420 mJ/cm^2 using

UV-C LED. Baykuş, Akgün, & Ünlütürk [9] observed browning in apple juice treated with multi-wavelength LEDs of 280 and 365 nm for more than 40 minutes. Similarly, Akgün & Ünlütürk [3] reported a significant ($p < 0.05$) alteration in the color of apple juice treated with UV-A LED for 50 minutes. The minimum color alteration in cloudy apple juice was observed with the symbiotic treatment of 280/365 nm.

10.6.4 POULTRY

In the span of 50 years, the consumption of meat products has tripled, reaching 340 million [109]. Myoglobin, a protein, is responsible for the color of meat products [39]. The color of muscles varies from dark-purple red to light gray depending on the amount of myoglobin present. Breast muscles contain less myoglobin, resulting in a pale or white color, while leg muscles have a higher myoglobin content, giving them a dark purple color. Different authors have used various wavelengths of LEDs to deactivate foodborne pathogens in meat products. Some studies have assessed the quality of treated products, with color being an important parameter. Soro et al. [92] used LEDs at 255, 280, 300, and 365 nm for different durations. Samples treated with 280 nm (for 6 minutes) and 365 nm (for 10 minutes) showed similar color. However, compared to untreated samples, products treated at 365 nm for 6 and 10 minutes exhibited a darker color. Additionally, both untreated and 365 nm treated samples showed color deterioration after seven days of storage. Wang et al. [106] reported no significant change in the color of chicken breast when exposed to a UV-C dose of 4,000 mJ/cm^2.

10.6.5 SEA FOOD

Salmon fish is considered an important ingredient in sushi [79]. There is a growing demand for raw fish in sushi, which raises significant food safety concerns [61]. Li, Kim, & Yuk [57]; and Li, Kim, Bang, & Yuk [58] treated salmon pieces with a 405 nm LED, where no change in the a and L values of untreated and LED-treated salmon was reported. However, there was a difference in the b value, but the overall change in color was insignificant. Another popular variety of salmon is smoked salmon. Consumers like smoked salmon because of its different taste and special texture [81]. It has been reported as a significant medium for the growth of *L. monocytogenes* [81], which necessitates processing methods that can ensure the safety of this product. Josewin et al. [44] reported major alterations in the color of treated and untreated samples containing riboflavin content treated with an LED of wavelength 460 nm. A major difference was observed between the samples in which riboflavin was absent. This leads to the conclusion that LED treatment can significantly degrade the yellow-orange color of riboflavin.

10.7 SUMMARY

LEDs are an efficient processing technique that can kill a range of microbes in solid and liquid food products. This processing can significantly inactivate pathogenic and spoilage microorganisms, including bacteria, spores, fungi, yeast, molds, and viruses. The processing is not only effective against the microbes but also inactivates spoilage-causing enzymes like PPO. The majority of the literature reports no major changes in the color profile of products under treatment. An increase in several coloring contents like chlorophyll, ACN, and carotenoid was observed in strawberries, lettuce, and broccoli, which proves that these components can significantly absorb UV light. Overall, this processing technology has the potential to deliver microbially safe food products with overall intact quality. However, the lack of literature is impeding the scaling up approvals required in this area. More focus on the quality assessment of LED-treated food products is needed to gain the confidence of food processors to use this technology.

KEYWORDS

- **anthocyanin**
- **chlorophyll**
- **color**
- **food products**
- **foodborne microorganisms**
- **light emitting diode**
- **microbes**
- **nonthermal processing**
- **ultraviolet**

REFERENCES

1. Ah, J., & Tagalpallewar, G. P., (2017). Functional properties of Mozzarella cheese for its end use application. *Journal of Food Science and Technology*, *54*(12), 3766–3778.
2. Aihara, M., Lian, X., Shimohata, T., Uebanso, T., Mawatari, K., Harada, Y., & Takahashi, A., (2014). Vegetable surface sterilization system using UVA light-emitting diodes. *The Journal of Medical Investigation*, *61*(3, 4), 285–290.
3. Akgün, M. P., & Ünlütürk, S., (2017). Effects of ultraviolet light emitting diodes (LEDs) on microbial and enzyme inactivation of apple juice. *International Journal of Food Microbiology*, *260*, 65–74.
4. Antonio-Gutiérrez, O., López-Díaz, A., Palou, E., López-Malo, A., & Ramírez-Corona, N., (2019). Characterization and effectiveness of short-wave ultraviolet irradiation reactors operating in continuous recirculation mode to inactivate *Saccharomyces cerevisiae* in grape juice. *Journal of Food Engineering*, *241*, 88–96.
5. Argyraki, A., Markvart, M., Nielsen, A., Bjarnsholt, T., Bjørndal, L., & Petersen, P. M., (2016). Comparison of UVB and UVC irradiation disinfection efficacies on *Pseudomonas Aeruginosa* (*P. aeruginosa*) biofilm. In: *Biophotonics: Photonic Solutions for Better Health Care V* (Vol. 9887, pp. 303–309). SPIE.
6. Arvand, M., Saberi, M., Ardaki, M. S., & Mohammadi, A., (2017). Mediated electrochemical method for the determination of indigo carmine levels in food products. *Talanta*, *173*, 60–68.
7. Avercheva, O., Berkovich, Y. A., Smolyanina, S., Bassarskaya, E., Pogosyan, S., Ptushenko, V., & Zhigalova, T., (2014). Biochemical, photosynthetic, and productive parameters of Chinese cabbage grown under blue–red LED assembly designed for space agriculture. *Advances in Space Research*, *53*(11), 1574–1581.
8. Bao, X., Sun, P., Liu, S., Ye, C., Li, S., & Kang, J., (2015). Performance improvements for AlGaN-based deep ultraviolet light-emitting diodes with the p-type and thickened last quantum barrier. *IEEE Photonics Journal*, *7*(1), 1–10.
9. Baykuş, G., Akgün, M. P., & Unluturk, S., (2021). Effects of ultraviolet-light emitting diodes (UV-LEDs) on microbial inactivation and quality attributes of mixed beverage made from blend of carrot, carob, ginger, grape and lemon juice. *Innovative Food Science & Emerging Technologies*, *67*, 102572.
10. Ben-Amotz, A., & Avron, M., (1990). The biotechnology of cultivating the halotolerant alga *Dunaliella*. *Trends in Biotechnology*, *8*, 121–126.
11. Bhargava, N., Mor, R. S., Kumar, K., & Sharanagat, V. S., (2021). Advances in application of ultrasound in food processing: A review. *Ultrasonics Sonochemistry*, *70*, 105293.
12. Bhavya, M. L., Shewale, S. R., Rajoriya, D., & Hebbar, H. U., (2021). Impact of blue LED illumination and natural photosensitizer on bacterial pathogens, enzyme activity and quality attributes of fresh-cut pineapple slices. *Food and Bioprocess Technology*, *14*(2), 362–372.
13. Bolton, J. R., (2017). Action spectra: A review. *IUVA News*, *19*(2), 10–12.
14. Butcher, J. D., Crosby, K. M., Yoo, K. S., Patil, B. S., Ibrahim, A. M. H., Leskovar, D. I., & Jifon, J. L., (2012). Environmental and genotypic variation of capsaicinoid and

flavonoid concentrations in habanero (*Capsicum chinense*) peppers. *HortScience*, *47*(5), 574–579.

15. Carle, R., & Schweiggert, R. M., (2016). *Handbook on Natural Pigments in Food and Beverages: Industrial Applications for Improving Food Color*. Woodhead Publishing.
16. Cheskin, L., (1957). *How to Predict What People Will Buy*.
17. Choi, H. G., Kwon, J. K., Moon, B. Y., Kang, N. J., Park, K. S., Cho, M. W., & Kim, Y. C., (2013). Effect of different light emitting diode (LED) lights on the growth characteristics and the phytochemical production of strawberry fruits during cultivation. *Horticultural Science & Technology*, *31*(1), 56–64.
18. Crosby, K. M., Jifon, J. L., Villalon, B., & Leskovar, D. I., (2007). 'TAM Dulcito,' a new, multiple virus-resistant sweet jalapeño pepper. *HortScience*, *42*(6), 1488–1489.
19. De Carvalho, A. S., De Rezende, S. C., Caleja, C., Pereira, E., Barros, L., Fernandes, I., & Barreiro, M. F., (2021). β-Carotene coloring systems based on solid lipid particles produced by hot melt dispersion. *Food Control*, *129*, 108262.
20. Dhakal, R., & Baek, K. H., (2014). Short period irradiation of single blue wavelength light extends the storage period of mature green tomatoes. *Postharvest Biology and Technology*, *90*, 73–77.
21. Downham, A., & Collins, P., (2000). Coloring our foods in the last and next millennium. *International Journal of Food Science & Technology*, *35*(1), 5–22.
22. D'Souza, C., Yuk, H. G., Khoo, G. H., & Zhou, W., (2015). Application of light-emitting diodes in food production, postharvest preservation, and microbiological food safety. *Comprehensive Reviews in Food Science and Food Safety*, *14*(6), 719–740.
23. Dufossé, L., (2006). Microbial production of food grade pigments. *Food Technology and Biotechnology*, *44*(3), 313–323.
24. Esatbeyoglu, T., Huebbe, P., Ernst, I. M., Chin, D., Wagner, A. E., & Rimbach, G., (2012). Curcumin—From molecule to biological function. *Angewandte Chemie International Edition*, *51*(22), 5308–5332.
25. EUTECA, (2010). *European Technical Caramel Association. Classification of Caramel Colors for Labelling Purposes*. (Online) 2020-EUTECA-final-position-on-natural-20.11.2009.pdf (accessed on 23 January 2024).
26. Falguera, V., Pagán, J., Garza, S., Garvín, A., & Ibarz, A., (2011). Ultraviolet processing of liquid food: A review: Part 2: Effects on microorganisms and on food components and properties. *Food Research International*, *44*(6), 1580–1588.
27. Ghalkhani, M., Zare, N., Karimi, F., Karaman, C., Alizadeh, M., & Vasseghian, Y., (2022). Recent advances in ponceau dyes monitoring as food colorant substances by electrochemical sensors and developed procedures for their removal from real samples. *Food and Chemical Toxicology*, 112830.
28. Ghate, V. S., Ng, K. S., Zhou, W., Yang, H., Khoo, G. H., Yoon, W. B., & Yuk, H. G., (2013). Antibacterial effect of light emitting diodes of visible wavelengths on selected foodborne pathogens at different illumination temperatures. *International Journal of Food Microbiology*, *166*(3), 399–406.
29. Ghate, V., Kumar, A., Kim, M. J., Bang, W. S., Zhou, W., & Yuk, H. G., (2017). Effect of 460 nm light emitting diode illumination on survival of *Salmonella* spp.

on fresh-cut pineapples at different irradiances and temperatures. *Journal of Food Engineering*, *196*, 130–138.
30. Ghate, V., Kumar, A., Zhou, W., & Yuk, H. G., (2016). Irradiance and temperature influence the bactericidal effect of 460-nanometer light-emitting diodes on *Salmonella* in orange juice. *Journal of Food Protection*, *79*(4), 553–560.
31. Ghoreishi, S. M., Behpour, M., & Golestaneh, M., (2012). Simultaneous determination of sunset yellow and tartrazine in soft drinks using gold nanoparticles carbon paste electrode. *Food Chemistry*, *132*(1), 637–641.
32. Gopisetty, V. V. S., Patras, A., Kilonzo-Nthenge, A., Yannam, S., Bansode, R. R., Sasges, M., & Xiao, H., (2018). Impact of UV-C irradiation on the quality, safety, and cytotoxicity of cranberry-flavored water using a novel continuous flow UV system. *LWT*, *95*, 230–239.
33. Gunter-Ward, D. M., Patras, A., Bhullar, M. S., Kilonzo-Nthenge, A., Pokharel, B., & Sasges, M., (2018). Efficacy of ultraviolet (UV-C) light in reducing foodborne pathogens and model viruses in skim milk. *Journal of Food Processing and Preservation*, *42*(2), e13485.
34. Hamamoto, A., Mori, M., Takahashi, A., Nakano, M., Wakikawa, N., Akutagawa, M., & Kinouchi, Y., (2007). New water disinfection system using UVA light-emitting diodes. *Journal of Applied Microbiology*, *103*(6), 2291–2298.
35. Held, G., (2016). *Introduction to Light Emitting Diode Technology and Applications*. Auerbach publications.
36. Henry, B. S., (1996). Natural food colors. In: Hendry, G. A., & Houghton, J. D., (eds.), *Natural Food Colorants* (2ndedn.) Blackie A&P, Glasgow.
37. Hofmann, C. M., Cronin, T. W., & Omland, K. E., (2006). Using spectral data to reconstruct evolutionary changes in coloration: Carotenoid color evolution in new world orioles. *Evolution*, *60*(8), 1680–1691.
38. Horbowicz, M., Kosson, R., Grzesiuk, A., & Debski, H., (2008). Anthocyanins of fruits and vegetables-their occurrence, analysis, and role in human nutrition. *Vegetable Crops Research Bulletin*, *68*, 5.
39. Hui, Y. H., Nip, W. K., & Rogers, R., (2001). *Meat Science and Applications*. CRC press.
40. Ilahy, R., Siddiqui, M. W., Tlili, I., Montefusco, A., Piro, G., Hdider, C., & Lenucci, M. S., (2018). When color really matters: Horticultural performance and functional quality of high-lycopene tomatoes. *Critical Reviews in Plant Sciences*, *37*(1), 15–53.
41. JECFA, (1990). *Vegetable Carbon.* (Online) http://www.fao.org/ag/agn/jecfa-additives/specs/Monograph1/Additive-486.pdf (accessed on 23 January 2024).
42. Jin, P., Yao, D., Xu, F., Wang, H., & Zheng, Y., (2015). Effect of light on quality and bioactive compounds in postharvest broccoli florets. *Food Chemistry*, *172*, 705–709.
43. Johkan, M., Shoji, K., Goto, F., Hashida, S. N., & Yoshihara, T., (2010). Blue light-emitting diode light irradiation of seedlings improves seedling quality and growth after transplanting in red leaf lettuce. *Hort Science*, *45*(12), 1809–1814.
44. Josewin, S. W., Ghate, V., Kim, M. J., & Yuk, H. G., (2018). Antibacterial effect of 460 nm light-emitting diode in combination with riboflavin against *Listeria monocytogenes* on smoked salmon. *Food Control*, *84*, 354–361.
45. Kanazawa, K., Hashimoto, T., Yoshida, S., Sungwon, P., & Fukuda, S., (2012). Short photoirradiation induces flavonoid synthesis and increases its production

in postharvest vegetables. *Journal of Agricultural and Food Chemistry*, *60*(17), 4359–4368.

46. Kang, E. J., Campbell, R. E., Bastian, E., & Drake, M. A., (2010). Invited review: Annatto usage and bleaching in dairy foods. *Journal of Dairy Science*, *93*(9), 3891–3901.
47. Kim, B. S., Lee, H. O., Kim, J. Y., Kwon, K. H., Cha, H. S., & Kim, J. H., (2011). An effect of light emitting diode (LED) irradiation treatment on the amplification of functional components of immature strawberry. *Horticulture, Environment, and Biotechnology*, *52*(1), 35–39.
48. Kim, M. J., Bang, W. S., & Yuk, H. G., (2017a). 405±5 nm light emitting diode illumination causes photodynamic inactivation of *Salmonella* spp. on fresh-cut papaya without deterioration. *Food Microbiology*, *62*, 124–132.
49. Kim, M. J., Ng, B. X. A., Zwe, Y. H., & Yuk, H. G., (2017b). Photodynamic inactivation of *Salmonella enterica* enteritidis by 405±5-nm light-emitting diode and its application to control salmonellosis on cooked chicken. *Food Control*, *82*, 305–315.
50. Kim, M. J., Tang, C. H., Bang, W. S., & Yuk, H. G., (2017c). Antibacterial effect of 405±5 nm light emitting diode illumination against *Escherichia coli* O157: H7, *Listeria monocytogenes*, and *Salmonella* on the surface of fresh-cut mango and its influence on fruit quality. *International Journal of Food Microbiology*, *244*, 82–89.
51. Kim, S. J., Kim, D. K., & Kang, D. H., (2016). Using UVC light-emitting diodes at wavelengths of 266 to 279 nanometers to inactivate foodborne pathogens and pasteurize sliced cheese. *Applied and Environmental Microbiology*, *82*(1), 11–17.
52. Koutchma, T., (2009). Advances in ultraviolet light technology for non-thermal processing of liquid foods. *Food and Bioprocess Technology*, *2*(2), 138–155.
53. Koutchma, T., & Popović, V., (2019). UV light-emitting diodes (LEDs) and food safety. *Ultraviolet LED Technology for Food Applications*, 91–117.
54. Lacombe, A., Niemira, B. A., Sites, J., Boyd, G., Gurtler, J. B., Tyrell, B., & Fleck, M., (2016). Reduction of bacterial pathogens and potential surrogates on the surface of almonds using high-intensity 405-nanometer light. *Journal of Food Protection*, *79*(11), 1840–1845.
55. Leal, F., & De Clavijo, C. M., (2011). Annatto: Botany and horticulture. *Horticultural Reviews*, *39*, 389–419.
56. Lee, S. W., Seo, J. M., Lee, M. K., Chun, J. H., Antonisamy, P., Arasu, M. V., & Kim, S. J., (2014). Influence of different LED lamps on the production of phenolic compounds in common and Tartary buckwheat sprouts. *Industrial Crops and Products*, *54*, 320–326.
57. Li, X., Kim, M. J., & Yuk, H. G., (2018a). Influence of 405 nm light-emitting diode illumination on the inactivation of *Listeria monocytogenes* and *Salmonella* spp. on ready-to-eat fresh salmon surface at chilling storage for 8 h and their susceptibility to simulated gastric fluid. *Food Control*, *88*, 61–68.
58. Li, X., Kim, M. J., Bang, W. S., & Yuk, H. G., (2018b). Anti-biofilm effect of 405-nm LEDs against *Listeria monocytogenes* in simulated ready-to-eat fresh salmon storage conditions. *Food Control*, *84*, 513–521.

59. Li, Y., Sun, Y., Jiang, J., & Liu, J., (2019). Spectroscopic determination of leaf chlorophyll content and color for genetic selection on *Sassafras tzumu*. *Plant Methods*, *15*(1), 1–11.
60. Lian, X., Tetsutani, K., Katayama, M., Nakano, M., Mawatari, K., Harada, N., & Takahashi, A., (2010). A new colored beverage disinfection system using UV-A light-emitting diodes. *Biocontrol Science*, *15*(1), 33–37.
61. Liu, C., Mou, J., & Su, Y. C., (2016). Behavior of *Salmonella* and *Listeria monocytogenes* in raw yellowfin tuna during cold storage. *Foods*, *5*(1), 16.
62. Lu, Y., Yang, B., Zhang, H., & Lai, A. C. K., (2021). Inactivation of foodborne pathogenic and spoilage bacteria by single and dual wavelength UV-LEDs: Synergistic effect and pulsed operation. *Food Control*, *125*, 107999.
63. Lyon, B. G., & Lyon, C. E., (2000). Meat quality: Sensory and instrumental evaluations. In: *Poultry Meat Processing* (pp. 107–130). CRC Press.
64. Macioszek, V. K., & Kononowicz, A. K., (2004). The evaluation of the genotoxicity of two commonly used food colors: Quinoline yellow (E 104) and brilliant black BN (E 151). *Cellular and Molecular Biology Letters*, *9*(1), 107–122.
65. McSharry, S., Koolman, L., Whyte, P., & Bolton, D., (2022). Inactivation of *Listeria monocytogenes* and *Salmonella typhimurium* in beef broth and on diced beef using an ultraviolet light emitting diode (UV-LED) system. *LWT*, *158*, 113150.
66. Miller, E. S., (1934). Quantitative absorption spectra of the common carotenoids. *Plant Physiology*, *9*(3), 693.
67. Mohd, S. A., Alias, A. K., Murad, M., Yen, K. S., & Cheng, L. H., (2022). A review of natural cheese and imitation cheese. *Journal of Food Processing and Preservation*, *46*(1), e16112.
68. Munir, M., Nadeem, M., Qureshi, T. M., Leong, T. S., Gamlath, C. J., Martin, G. J., & Ashokkumar, M., (2019). Effects of high pressure, microwave and ultrasound processing on proteins and enzyme activity in dairy systems—A review. *Innovative Food Science & Emerging Technologies*, *57*, 102192.
69. Murdoch, L. E., McKenzie, K., Maclean, M., Macgregor, S. J., & Anderson, J. G., (2013). Lethal effects of high-intensity violet 405-nm light on *Saccharomyces cerevisiae*, *Candida albicans*, and on dormant and germinating spores of *Aspergillus Niger*. *Fungal Biology*, *117*(7, 8), 519–527.
70. NATCOL, (2013). *Position on the Term "Natural Color" and the Categorization of Food Colors.* (Online) Microsoft Word – NATCOL Position Paper on Natural color Final _2_.doc.
71. Niu, L., Wu, Z., Yang, L., Wang, Y., Xiang, Q., & Bai, Y., (2021). Antimicrobial effect of UVC light-emitting diodes against *Saccharomyces cerevisiae* and their application in orange juice decontamination. *Journal of Food Protection*, *84*(1), 139–146.
72. Patras, A., Bhullar, M. S., Pendyala, B., & Crapulli, F., (2021). *Ultraviolet Treatment of Opaque Liquid Foods: From Theory to Practice.*
73. Pendyala, B., Patras, A., Gopisetty, V. V. S., Sasges, M., & Balamurugan, S., (2019). Inactivation of *Bacillus* and *Clostridium* spores in coconut water by ultraviolet light. *Foodborne Pathogens and Disease*, *16*(10), 704–711.
74. Pendyala, B., Patras, A., Gopisetty, V. V. S., Vashisht, P., & Ravi, R., (2022). Inactivation of *B. cereus* spores in whole milk and almond milk by serpentine path

coiled tube UV-C system: Numerical simulation of flow field, lipid peroxidation and volatiles analysis. *Food Research International*, 111652.

75. Pérez-Ambrocio, A., Guerrero-Beltrán, J. A., Aparicio-Fernández, X., Ávila-Sosa, R., Hernández-Carranza, P., Cid-Pérez, S., & Ochoa-Velasco, C. E., (2018). Effect of blue and ultraviolet-C light irradiation on bioactive compounds and antioxidant capacity of habanero pepper (*Capsicum chinense*) during refrigeration storage. *Postharvest Biology and Technology*, *135*, 19–26.
76. Pourreza, N., & Elhami, S. (2009). Cloud point extraction and spectrophotometric determination of amaranth in food samples using nonionic surfactant Triton X-100 and tetrabutylammonium hydrogen sulfate. *Journal of the Iranian Chemical Society*, *6*, 784–788.
77. Prasad, A., Du, L., Zubair, M., Subedi, S., Ullah, A., & Roopesh, M. S., (2020). Applications of light-emitting diodes (LEDs) in food processing and water treatment. *Food Engineering Reviews*, *12*(3), 268–289.
78. Priyadarshini, A., & Priyadarshini, A., (2018). Market dimensions of the fruit juice industry. In: *Fruit Juices* (pp. 15–32). Academic Press.
79. Puah, S. M., Chua, K. H., & Tan, J. A. M. A., (2016). Virulence factors and antibiotic susceptibility of *Staphylococcus aureus* isolates in ready-to-eat foods: Detection of S. aureus contamination and a high prevalence of virulence genes. *International Journal of Environmental Research and Public Health*, *13*(2), 199.
80. Rawson, A., Patras, A., Tiwari, B. K., Noci, F., Koutchma, T., & Brunton, N., (2011). Effect of thermal and non-thermal processing technologies on the bioactive content of exotic fruits and their products: Review of recent advances. *Food Research International*, *44*(7), 1875–1887.
81. Rørvik, L. M., (2000). *Listeria monocytogenes* in the smoked salmon industry. *International Journal of Food Microbiology*, *62*(3), 183–190.
82. Rovina, K., Prabakaran, P. P., Siddiquee, S., & Shaarani, S. M., (2016a). Methods for the analysis of sunset yellow FCF (E110) in food and beverage products-a review. *TrAC Trends in Analytical Chemistry*, *85*, 47–56.
83. Rovina, K., Siddiquee, S., & Shaarani, S. M., (2016b). Extraction, analytical and advanced methods for detection of Allura red AC (E129) in food and beverages products. *Frontiers in Microbiology*, *7*, 798.
84. Santhirasegaram, V., Razali, Z., George, D. S., & Somasundram, C., (2015). Comparison of UV-C treatment and thermal pasteurization on quality of Chokanan mango (*Mangifera indica* L.) juice. *Food and Bioproducts Processing*, *94*, 313–321.
85. Shannon, M. A., Bohn, P. W., Elimelech, M., Georgiadis, J. G., Marinas, B. J., & Mayes, A. M., (2010). Science and technology for water purification in the coming decades. *Nanoscience and Technology: A Collection of Reviews from Nature Journals*, 337–346.
86. Shao, M., Liu, W., Zha, L., Zhou, C., Zhang, Y., & Li, B., (2020). Differential effects of high light duration on growth, nutritional quality, and oxidative stress of hydroponic lettuce under red and blue LED irradiation. *Scientia Horticulturae*, *268*, 109366.
87. Sinha, R. P., & Häder, D. P., (2002). UV-induced DNA damage and repair: A review. *Photochemical & Photobiological Sciences*, *1*(4), 225–236.

88. Šivel, M., Kráčmar, S., Fišera, M., Klejdus, B., & Kubáň, V., (2014). Lutein content in marigold flower (*Tagetes erecta* L.) concentrates used for production of food supplements. *Czech Journal of Food Sciences*.
89. Sizer, C. E., & Balasubramaniam, V. M., (1999). New intervention processes for minimally processed juices. *Food Technology (Chicago)*, *53*(10), 64–67.
90. Sommers, C., Gunther, IV. N. W., & Sheen, S., (2017). Inactivation of *Salmonella* spp., pathogenic *Escherichia coli*, Staphylococcus spp., or *Listeria monocytogenes* in chicken purge or skin using a 405-nm LED array. *Food Microbiology*, *64*, 135–138.
91. Song, K., Taghipour, F., & Mohseni, M., (2018). Microorganisms inactivation by continuous and pulsed irradiation of ultraviolet light-emitting diodes (UV-LEDs). *Chemical Engineering Journal*, *343*, 362–370.
92. Soro, A. B., Whyte, P., Bolton, D. J., & Tiwari, B. K., (2021). Application of a LED-UV based light technology for decontamination of chicken breast fillets: Impact on microbiota and quality attributes. *LWT*, *145*, 111297.
93. Srimagal, A., Ramesh, T., & Sahu, J. K., (2016). Effect of light emitting diode treatment on inactivation of *Escherichia coli* in milk. *LWT-Food Science and Technology*, *71*, 378–385.
94. Stich, E., (2016). Food color and coloring food: Quality, differentiation and regulatory requirements in the European Union and the United States. In: *Handbook on Natural Pigments in Food and Beverages* (pp. 3–27). Woodhead Publishing.
95. Stutte, G. W., Edney, S., & Skerritt, T., (2009). Photo regulation of bioprotectant content of red leaf lettuce with light-emitting diodes. *Hort Science*, *44*(1), 79–82.
96. Suárez-Pereira, E., Rubio, E. M., Pilard, S., Ortiz, M. C., & Garcia, F. J. M., (2010). Di-D-fructose dianhydride-enriched products by acid ion-exchange resin-promoted caramelization of D-fructose: Chemical analyses. *Journal of Agricultural and Food Chemistry*, *58*(3), 1777–1787.
97. Sun, Y., Qian, M., Wu, R., Niu, Q., Teng, Y., & Zhang, D., (2014). Postharvest pigmentation in red Chinese sand pears (*Pyrus pyrifolia* Nakai) in response to optimum light and temperature. *Postharvest Biology and Technology*, *91*, 64–71.
98. Thakur, B. R., & Nelson, P. E., (1998). High-pressure processing and preservation of food. *Food Reviews International*, *14*(4), 427–447.
99. Toepfl, S., Siemer, C., Saldaña-Navarro, G., & Heinz, V., (2014). Overview of pulsed electric fields processing for food. In: *Emerging Technologies for Food Processing* (pp. 93–114). Academic Press.
100. Topuz, A., Feng, H., & Kushad, M., (2009). The effect of drying method and storage on color characteristics of paprika. *LWT-Food Science and Technology*, *42*(10), 1667–1673.
101. Vashisht, P., (2021). *Performance Evaluation of a Pilot Scale Dean Flow UV System: Fundamentals and Applications*. Doctoral dissertation, Tennessee State University.
102. Vashisht, P., Pendyala, B., Gopisetty, V. V. S., & Patras, A., (2021). Modeling and validation of delivered fluence of a continuous Dean flow pilot scale UV system: Monitoring fluence by biodosimetry approach. *Food Research International*, *148*, 110625.
103. Vashisht, P., Pendyala, B., Patras, A., Gopisetty, V. V. S., & Ravi, R., (2022). Design and efficiency evaluation of a mid-size serpentine Dean flow UV-C system for the

processing of whole milk using computational fluid dynamics and biodosimetry. *Journal of Food Engineering*, 111168.

104. Vlahović, B., Popović-Vranješ, A., & Mugoša, I., (2014). International cheese market–current state and perspective. *Economic Insights-Trends & Challenges*, *66*(1).
105. Vollmuth, T. A., (2018). Caramel color safety – An update. In: *Food and Chemical Toxicology* (Vol. 111, pp. 578–596). Elsevier Ltd.
106. Wang, W., Zhao, D., Li, K., Xiang, Q., & Bai, Y., (2021). Effect of UVC light-emitting diodes on pathogenic bacteria and quality attributes of chicken breast. *Journal of Food Protection*, *84*(10), 1765–1771.
107. Xu, F., Cao, S., Shi, L., Chen, W., Su, X., & Yang, Z., (2014). Blue light irradiation affects anthocyanin content and enzyme activities involved in postharvest strawberry fruit. *Journal of Agricultural and Food Chemistry*, *62*(20), 4778–4783.
108. Zhang, L., Ma, G., Kato, M., Yamawaki, K., Takagi, T., Kiriiwa, Y., & Nesumi, H., (2012). Regulation of carotenoid accumulation and the expression of carotenoid metabolic genes in citrus juice sacs in vitro. *Journal of Experimental Botany*, *63*(2), 871–886.
109. Zhang, S., & Wang, H., (2019). Current progress in 3D bioprinting of tissue analogs. *SLAS Technology: Translating Life Sciences Innovation*, *24*(1), 70–78.
110. Zhang, Z. H., Wang, L. H., Zeng, X. A., Han, Z., & Brennan, C. S., (2019). Non-thermal technologies and its current and future application in the food industry: A review. *International Journal of Food Science & Technology*, *54*(1), 1–13.

CHAPTER 11

IMPACT OF VACUUM FRYING ON COLOR PROPERTIES OF FRIED SNACKS

GAYATHRI BALAKRISHNAN

Department of Food Science and Human Nutrition, University of Florida, Gainesville, Florida, USA

ABSTRACT

The food industry has utilized atmospheric deep fat frying for several years to transform perishable fruits and vegetables into shelf-stable chips/crisps with a unique flavor, taste, mouthfeel, and texture. Although the process results in a finished product that consumers widely appreciate for its sensory attributes, it has also been shown to be accompanied by a high intake of fat/oil, loss of nutrients, and the formation of acrylamide, which is deleterious to our bodies. Moreover, the high temperature involved in the frying process causes browning and color deterioration, negatively impacting product quality. With increased consumer awareness of adopting a healthy lifestyle, it has become imperative to develop alternative methods to produce fried products that are superior in terms of their sensory and nutritional value. Vacuum frying technology, along with pretreatments/hybrid technology combinations with microwave (MW), ultrasound, and pulsed electric field (PEF), are innovative approaches for manufacturing healthy low-fat fried products. The processing parameters of vacuum frying, such as frying time, vacuum pressure level, and frying temperature, significantly influence the color attributes of the finished

Nonthermal Food Processing Technologies: Impact on Color Profile.
Kaavya Rathnakumar and R. Pandiselvam (Eds.)

product and must be optimized based on the fruit or vegetable being fried. This chapter discusses the vacuum frying process and the hybrid methodologies that have been researched to improve overall frying efficiency and finished product characteristics. There are different classes of color pigments present in commonly fried fruit/vegetable chips; the vacuum frying conditions that enhance their color stability compared to atmospheric deep fat frying are also discussed in detail.

11.1 INTRODUCTION

11.1.1 DEEP-FRIED FRUIT/VEGETABLE SNACKS

Deep-fried snack products have been preferred by consumers due to their delicious taste, texture, and flavor. Frying is an age-old cooking method; the process has been used for several decades to enhance the palatability and extend the shelf life of perishable products. *Deep fat frying* is a processing method that involves immersing the food product in oil/fat at a high temperature (>100°C) that exceeds the boiling point of water. Exposure to such temperature causes the surface water to evaporate from the material being fried into the surrounding oil or fat. Since the oil temperature is raised above the boiling point, this process can also be considered a method of dehydration to reduce moisture from the product.

The immersed food product changes its cellular structure, leading to the formation of pores. The oil penetrates these pores, and the amount of oil uptake correlates to the moisture content of the food. In other words, more oil uptake occurs in the presence of high moisture content. Potato chips/crisps are the most notable product in the category of deep-fried foods, but there is also a wide variety of fried fresh produce, fruits, vegetables, and animal-based products such as fish sticks and chicken. Frying food products results in the creation of unique colors, flavors, and textures in the final product, enhancing its sensory characteristics. This category of food products is a dominant part of the Western diet, and it is very likely to remain so in the future.

11.1.2 ATMOSPHERIC IMMERSION FRYING

Atmospheric frying is the process of frying food products in hot oil between 150°C and 190°C in atmospheric conditions (OHACs). During the cooking process, several heat-induced chemical reactions occur, which impact the flavor, texture, and color of the fried material. Among the heat-driven reactions, Maillard browning and caramelization impact the product color. The fundamental phenomenon during atmospheric frying is dehydration/moisture loss, wherein the high temperature of the oil enables rapid heat and mass transfer and a short cooking period. There are four phases in the frying process:

- In the food product to be fried is immersed in oil during the initial (1st) heating phase, and heat transfer occurs through convection.
  - The surface of the food material reaches the boiling point of the water.
- In the second stage, rapid heat transfer occurs from the oil to the food both at the surface level (convective heat transfer coefficient) and within the food (conductive heat transfer coefficient).
  - Due to water evaporation from the surface and the high temperature of the process, crust formation is initiated during this phase.
- The third (3rd) phase, also known as the falling rate period, is the primary phase of the frying process.
  - During this phase, there is an increase in crust layer thickness, starch gelatinization, protein denaturation, and production of Maillard browning compounds.
  - The core temperature of the food raises steadily to match the boiling point of the water.
- In the final (4th) phase of frying (bubble endpoint stage), surface water vaporization stops completely. The bubble endpoint occurs due to the complete removal of water from the fried product or a decrease in heat transfer to the crust/core interface [20].

The oil absorption during atmospheric frying is an unavoidable phenomenon that occurs throughout the frying process and also during the cooling phase. During the frying phase, the cellular structure of the food undergoes a drastic change, accompanied by rapid water evaporation from the food surface leading to crust formation. Moisture within the food product is transformed into steam, and the steam escapes through open pores and cracks. As the frying progresses, oil adheres to the food and migrates into the spaces created by water due to evaporation. Once the frying is complete, and the food is taken out from the oil, the product undergoes a cooling phase. As the product cools down, water vapor condenses, causing a subsequent decline in the internal pressure of the product. Because of the pressure drop, the surface oil is pulled into the product ('vacuum effect'). These two mechanisms are the essential drivers of oil absorption during atmospheric frying. The amount of oil absorbed depends on the physical composition of the food product, such as size, shape, porosity, moisture content, oil quality, viscosity, and frying parameters (time and temperature). The amount of oil absorbed due to atmospheric frying varies from 15% to 40% of the final product weight. In addition to substantial oil uptake, the process involves intense heat transfer that causes surface darkening and scorching, which adversely affect the color of the finished product.

11.1.2.1 DISADVANTAGES OF DEEP-FAT ATMOSPHERIC FRYING

The main disadvantage of deep-fat fried foods is the excessive amount of oil absorbed during the process. Although this cooking method enhances the sensory properties of the food, consuming such foods results in a high intake of fat, which has numerous negative health implications. Fried foods typically contain 15–40% fat, and this excessive fat consumption is associated with the development of various diseases, including obesity, hypertension, diabetes, heart disease, and stroke. As a result, consumers have consciously made an effort to switch to low-fat products that still maintain desirable sensory characteristics. Vacuum frying offers an excellent alternative to deep-fat frying for creating snack products with reduced oil content, while still achieving similar flavor, color, and texture to the traditionally deep-fried products.

11.1.3 VACUUM FRYING

Vacuum frying is a deep-frying method performed below atmospheric pressure, often at 6.65 kPa or 50 Torr. The reduction of pressure leads to a substantial reduction in the boiling point of free water present in the food being fried. This, in turn, causes a decrease in the frying temperature. The process occurs within an enclosed system with minimal exposure to oxygen during frying. The low temperature and absence of oxygen offer several benefits to the organoleptic and nutritional quality of the finished product:

- Protection of nutritional compounds;
- Less oil uptake;
- Decrease in oil oxidation;
- Decrease in the formation of detrimental compounds such as acrylamide;
- Preservation of color, texture, and flavor.

Several fried snacks today, especially fruit and vegetable chips/crisps, are produced through vacuum frying to avoid excessive oil absorption and to enhance product color and flavor retention. Vacuum frying is an effective method for manufacturing low-fat chips with the desired level of dehydration, without causing excessive darkening of the color.

11.1.3.1 PROCESS SETUP FOR VACUUM FRYING

The frying operation in a traditional vacuum fryer (VF) occurs as a batch process (Figure 11.1). The processing setup consists of:

- An airtight chamber with a heating element;
- A frying basket;
- A condenser;
- A vacuum pump.

A frying basket is used to contain the food material within the airtight chamber. The chamber is closed after the oil reaches the target temperature, and then the entire system is depressurized to the required level. The basket is then lowered into the frying oil inside the chamber for a specific amount of time using a lift rod. Thc product is fried for the necessary duration; once finished, the frying basket is lifted, and the chamber is pressurized using the pressure valve. Low pressure is created during the process with a vacuum pump.

During frying, there is considerable moisture removal as steam; a condenser is attached to the system to collect the steam. The preferable pressure for vacuum frying is 7 kPa or less, which has shown to drastically reduce the frying temperature to as low as 90°C. While reduction of temperature is beneficial, it is also equally important not to have too low temperature as this would hinder the structural changes in the product and adversely impact the finished product texture (crispiness). The mechanism involved in oil absorption during vacuum frying has not been well understood. Most absorption has been proven to occur during the cooling phase, i.e., after removal of the food material from the fryer. Factors such as the percentage of water in the food, the rate of moisture loss during frying, influence the total oil uptake; both these factors are directly correlated to the quantity of oil that is absorbed by the final product.

The pressurization step during vacuum frying is also vital in determining the amount of oil that is absorbed. It influences the oil absorption rate based on the extent of surface oil present after frying. Hence, removing excess oil is crucial upon completion of the vacuum frying process. Therefore, an additional centrifugation step is followed in the commercial vacuum frying operation to accomplish the separation of excess oil.

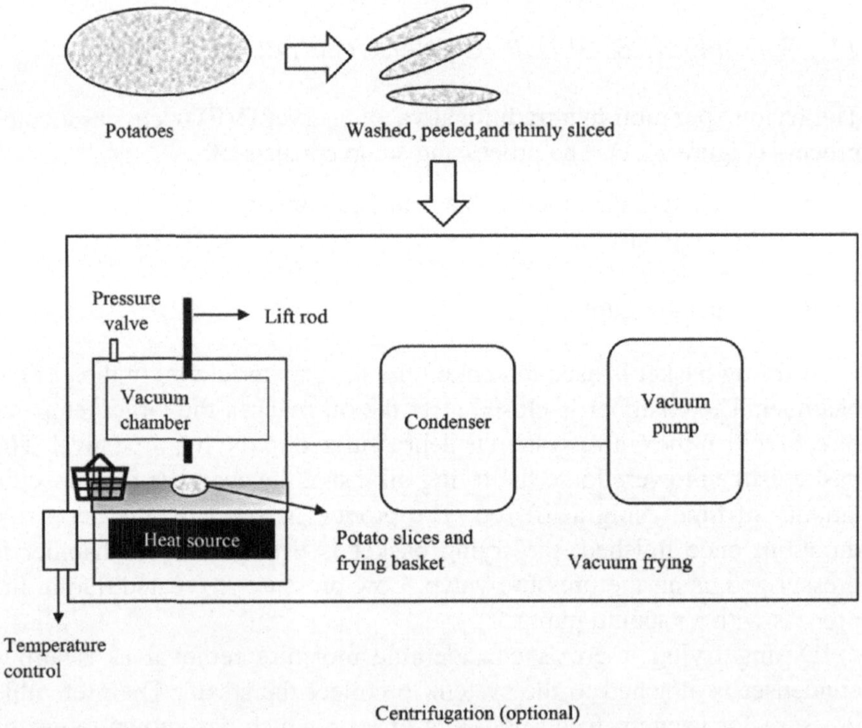

FIGURE 11.1 Schematic representation of the vacuum frying process for potato chips.

11.1.3.2 DISADVANTAGES OF VACUUM FRYING

Although the application of VFs presents an excellent option to traditional atmospheric fryers, the process still has some disadvantages. Table 11.1 lists the advantages and disadvantages of vacuum frying. Due to the lower frying temperature and longer duration, the process is often not time or energy efficient. The amount of surface oil is also an issue, especially in the absence of a centrifugation step. While the procedure has been successfully used to manufacture fried products with low oil, there are often challenges in meeting the oil requirement targets. The rapid pressure changes from vacuum to atmospheric during the pressurization step may promote oil penetration. Despite the low temperature, the longer frying duration could also impact sensory quality, particularly the product's color and texture. Several innovative modifications/pretreatments are continuously being explored to improve the efficiency of vacuum frying systems. The details of these techniques will be discussed later in this chapter.

TABLE 11.1 Advantages and Disadvantages of Vacuum Frying

Advantages	Disadvantages
• Process performed at less than typical frying temperature (<150°C). • Nutritional superiority (better retention of antioxidants and heat-labile nutrients). • Less color deterioration due to caramelization, Maillard, and enzymatic oxidation. • Minimal scorching or darkening. • Crispy texture. • Low oil uptake and better shelf stability.	• Time consuming due to low frying temperature. • May not be energy or cost efficient. • Longer frying time impacts sensory characteristics and oil uptake. • Pressurization step leads to increased oil uptake in the absence of centrifugation step.

11.2 COLOR QUALITY FOR THE SENSORY ACCEPTABILITY OF FRIED SNACKS

For the quality evaluation of fried products, color is a key measure that should meet the finished product specification. It is the first and foremost parameter evaluated by consumers to decide whether the product is acceptable for consumption. Several factors such as frying conditions (temperature, duration), oil type, and the composition of the product matrix, particularly the presence of reducing sugars, amino acid moieties, and other functional groups impact product color when frying. Therefore, accurate measurement of food color is critical when considering the overall quality evaluation of fried foods.

11.2.1 COLOR ASSESSMENT

Color assessment in the commercial setting is often performed through visual evaluation, which involves matching a test product to a reference color or standard color chart. However, this method is prone to operator bias and may not produce accurate results. Therefore, in order to objectively quantify color in the fried food industry, a color system is selected from various tristimulus formats: XYZ, RGB (red, yellow, green), and L*, a*, b*.

The L*, a*, and b* system, adopted by the Commission Internationale de l'Elcairage (CIE), is currently the industry standard for measuring color. The three dimensions refer to lightness (L*), green to red (a*), and blue to yellow (b*) values. In the LAB system, color distribution is uniform, and hence the values are close to what is usually perceived by the human eye. The color measurement using these units is accomplished with a colorimeter, which is one of the most popular tools for color analysis due to the wide range of applications, ease of use, and rapidness of the test method.

A severe limitation of the conventional analysis using a colorimeter is the assumption of homogeneity in the food product. Food matrices and surfaces are non-uniform and have different hues. The measurement of color distribution throughout the surface will provide accurate and more detailed information than an average color reading. Hence, using a computer vision system (CVS) that scans the surface color of products is a valuable tool for understanding color irregularities; they help analyze the entire food product and compare two samples or a sample to a reference standard. Their use is becoming progressively more popular in the food and agriculture industry despite their high equipment and installation cost. Some applications where they are being routinely used are to sort raw fruits and vegetables based on maturity, assess the quality of processed food products, including fried snacks, where browning or discolored product is a commonly observed defect [5].

11.2.2 CHEMICAL REACTIONS THAT IMPACT THE COLOR OF FRIED SNACKS

11.2.2.1 NON-ENZYMATIC BROWNING

Non-enzymatic browning refers to the occurrence of brown pigments in foods without the involvement of enzymes during or after processing. Two types of non-enzymatic browning observed during frying are Maillard reactions and caramelization. The Maillard reaction involves two substrates, namely amino acids and reducing sugar, while caramelization is primarily a sugar reaction triggered by high temperature and other physical parameters such as water activity, pH, and heating conditions.

Maillard reactions involve the condensation of amino groups from proteins with carbonyl groups from reducing sugars, resulting in the formation of intermediate Schiff bases and subsequent rearrangement to Heyns/Amadori products [16]. As the reaction progresses, several intermediary products such as aldehydes, furanones, and dicarbonyls are produced through aldol condensation. In later stages, these aldol products interact with amino acids, forming low molecular weight compounds that subsequently polymerize into high molecular weight melanoidins. The advanced stage encompasses a range of reactions, including cyclization, retroadolization, isomerization, rearrangement, dehydration, and additional condensations. Ultimately, the formation and quantity of high molecular weight compounds through the Maillard reaction determine the extent of browning and color deterioration in fried products.

11.2.2.2 ENZYMATIC BROWNING

Enzyme-induced browning refers to the oxidation of phenolic compounds in fruits and vegetables by the polyphenol oxidase enzyme. It is commonly observed in light-colored materials such as potatoes, apples, bananas, and pineapples before and during frying. Processing steps prior to frying, such as peeling and slicing the material, are sufficient to trigger enzymatic browning. The initial products of enzymatic oxidation are quinones that react through rapid condensation to form brown polymers called melanins. The oxidation rate often increases when cellular integrity is disrupted in the food material. The concentration of phenolic compounds and enzymes, pH, temperature, and oxygen availability are the main components that determine the browning rate. In the case of fried products, temperature and oxygen are critical as they have a tremendous influence on the formation of the undesirable brown pigments.

11.2.2.3 PRETREATMENTS TO IMPROVE FRIED SNACK COLOR DURING VACUUM FRYING

Before the frying process, there are options to pretreat the food material in order to improve the frying efficiency (Figure 11.2). These pretreatments play a crucial role in determining the oil uptake kinetics, regulating the process parameters including frying time and temperature, and influencing the final moisture content and sensory characteristics. Blanching, osmotic dehydration (OD), freezing, and coating with hydrocolloids are the essential pretreatment techniques that have been investigated in combination with vacuum frying.

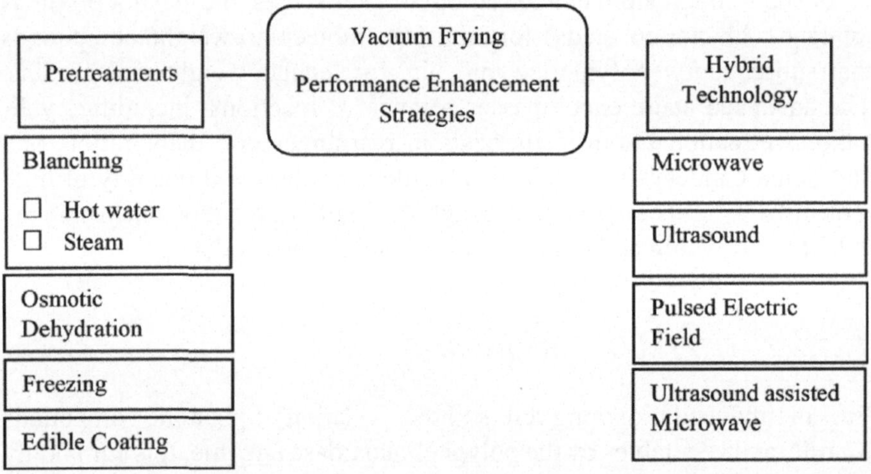

FIGURE 11.2 Pretreatment and hybrid technology to improve the efficiency of vacuum frying.

11.2.2.3.1 Blanching

This thermal technique refers to immersing the food material in boiling water for a few minutes to inactivate the enzymes responsible for quality deterioration during frying. Blanching involves two steps: (i) dipping the food in boiling water for a specified time (varies from a few seconds to less than 10 minutes); and (ii) rapidly cooling the blanched material to stop any chemical reactions. The time required for blanching often depends on the physical characteristics of the food material, the ratio of blanching water to the food product, the amount of sugars, as well as the presence of enzymes like polyphenol oxidase that cause browning and development of off-color in fried products. The use of hot water and steam are the traditional blanching methods. Commercially, hot water blanching is the most used because of its simplicity and ease of operation. The procedure, as mentioned earlier, involves product immersion in boiling water. Since the blanching water tends to saturate with leached nutrients from the food material, the water must be replaced after a few minutes depending on the total blanching time used. While hot water blanching has traditionally been used in food industries to stabilize the food material prior to frying, the disadvantage to the method is the loss of valuable water-soluble nutrients such as ascorbic acid (AA), B-vitamins, minerals, carbohydrates, and amino acids.

In steam blanching, superheated steam is used as the heating medium instead of boiling water. During the early phase of the process, the steam condenses on the product surface. A large amount of latent heat transfers to the food material due to the lower temperature of the food being blanched. There is a gradual increase in the temperature until it matches the critical temperature of enzymes, thereby leading to enzyme inactivation. Steam blanching is carried out by placing the product on a conveyor belt, where steam is blown onto the product surface as it moves through the steam chamber. The method effectively blanches small or cut fruits and vegetables and requires less time than hot water blanching. However, in the case of thicker food products, due to temperature differences between the surface and the central region, there could be uneven blanching leading to under-blanched and over-blanched areas. When comparing the two blanching techniques, the loss of nutrients is less in the latter method.

While blanching can successfully inactivate enzymes responsible for browning, it is likewise important to optimize blanching parameters as excessive blanching time may cause significant color degradation.

11.2.2.3.2 Osmotic Dehydration (OD)

An alternative to thermal blanching pretreatment is osmotic dehydration (OD). This method involves immersing or spraying an osmotic solution onto the fruit or vegetable to induce partial dehydration. The desired pieces of fruits or vegetables are soaked in a solution with high osmotic pressure, causing water to move from the food material to the osmotic solution. Meanwhile, the dehydrated product absorbs solutes through osmosis. Sucrose, at a high concentration, is the most commonly used hypertonic substance. However, other ingredients such as sodium chloride, maltodextrin, fruit juices and syrups, and sorbitol, which can generate high osmotic pressure, are also becoming more prevalent. One drawback of this process is the low rate of mass transfer, which can be enhanced by combining newer technologies like high-power ultrasound sonication. The retention of color and the preservation of colorful antioxidants after partial dehydration through osmosis, followed by vacuum frying, have been extensively studied for fruits and vegetables such as potato, shiitake mushroom, purple sweet potato, and pumpkin [12, 21, 23, 28].

11.2.2.3.3 Edible Coating

The application of edible coating is another nonthermal pretreatment used before vacuum frying. It refers to the method of surface coating by carbohydrate-based films that create a barrier for the diffusion of moisture, oxygen, and solutes into the food product [24]. The thickness of such coating is usually as thin as 0.3 mm or less. Food hydrocolloids belonging to the category of cellulose derivatives, gums, and other polysaccharides such as pectin, carrageenan, guar gum, starch, chitosan, carboxymethyl cellulose, and hydroxypropyl methylcellulose with good gelling properties are chosen as the coating material. This method creates a barrier on the food surface, which reduces moisture loss and leads to better color and flavor retention [32]. The use of edible coating before vacuum frying is less investigated than other pretreatments like blanching and OD. However, the positive impact of edible coating in improving the sensory quality of the finished product is well-founded for atmospheric frying.

11.2.2.3.4 Freezing

Freezing is a well-known method for reducing the free water content without significantly affecting the sensory quality of the product. However, freezing may not be suitable for products with high moisture content, so the feasibility of freezing should be evaluated before implementing this pretreatment strategy. This technique alters the microstructure and increases porosity, resulting in higher activation energies and less shrinkage during frying [32]. In a study conducted by Maity et al. [17], the pre-freezing effect on vacuum-fried jackfruit chips was examined, and it was found that the frozen product had a crispy texture and a lighter color compared to the unfrozen control.

11.2.2.4 HYBRID VACUUM FRYING TECHNOLOGIES FOR COLOR STABILITY

11.2.2.4.1 Microwave-Assisted Vacuum Frying (MVF)

The use of microwave (MW) technology involves the absorption of energy and the generation of heat from within the food material. This results in a pressure gradient that increases the rate of moisture loss. In the industry, MW technology is applied in two different ways: (i) the frying process takes place directly using MW as the heating source, and (ii) the product is precooked/pretreated using MW for a few minutes before the actual frying process.

In a standard vacuum frying (VF) process, the heating element typically consists of hot plates, which can be difficult to control and require a large surface area. As a result, the unit operation is slow, inefficient, and costly to install. Microwave-assisted vacuum frying (MVF) is a unique process that replaces the hot plates with microwave (MW) heating. MW heating has been proven to be an effective method for achieving faster dehydration in fried fruits and vegetables, resulting in a final product with improved sensory and nutritional characteristics compared to traditional VF [30]. The MVF process utilizes electromagnetic (EM) waves within the infrared and radio frequency (RF) range, operating at a parameter of 300 MHz–300 GHz. This significantly reduces the frying time in the VF. The mechanism of MW vacuum frying involves generating heat within the interior of the food product through molecular friction, which is then transferred to the exterior surface by the rapid spinning of polar water molecules with a rapidly changing electric field [1]. Consequently, the direction of heat transfer is opposite to conventional frying, as heat is generated from the core of the product to the exterior surface. This unique heating mechanism allows for the production of high-quality products in a shorter time and with less energy.

11.2.2.4.2 Ultrasound Assisted Vacuum Frying (UVF)

Ultrasonication is an innovative technology in the food processing industry with a broad range of applications, from retaining the quality of fresh produce to reducing allergenicity and ensuring microbial safety. Ultrasound refers to cavitation caused by sound waves at frequencies varying from 20 kHz to 10 MHz that bring about physical, chemical, and cellular changes to food products while they are subjected to vacuum-frying. Ultrasound causes alternating expansion and compression effects that enhance mass transfer through the formation of minuscule channels within the food structure [19]. The process provides advantages of reducing the cost of the operation, improving processing time, convenience, and preservation of the nutritional and sensory quality. In the frying process, combining ultrasound with vacuum frying preserves the organoleptic properties of the fried product in addition to lowering oil absorption.

11.2.2.4.3 Pulsed Electric Field (PEF)

The pulsed electric field (PEF) is a method of applying short voltage pulses (ranging from microseconds to milliseconds) with a high electric field strength to a product placed between two electrodes. The electric field causes permeabilization/modification of the cell membrane, known as electroporation. This process results in changes to the cell structure that increase the rate of heat and mass transfer during frying [3]. The creation of electroporation due to PEF technology offers better preservation of color and texture in the finished fried snacks, along with reduced oil intake. Additionally, there are production advantages due to the cost and energy efficiency of the PEF procedure.

11.3 FRUIT/VEGETABLE COLOR PIGMENTS AND THEIR STABILITY DUE TO VACUUM-FRYING

Phenolic compounds are responsible for the characteristic color of various fresh vegetables and fruits. Table 11.2 lists the commonly vacuum-fried products from this category, along with the color changes observed during frying, which are often seen as a 'quality defect.' The following section describes the polyphenols in fried vegetables and fruits and the application of vacuum frying/hybrid methodologies to stabilize these compounds.

TABLE 11.2 Color Changes Observed in Fried Fruits/Vegetables Chips

Fruits/Vegetables	Compounds Responsible for Color	Color Changes That Impact Consumer Acceptability
Potato	Colorless phenolic compounds/flavones.	Visible browning
Apple		Color darkening.
Zucchini		Loss of characteristic vegetable/fruit color.
Mushroom		
Tapioca		Faded/dull appearance.
Carrot	Carotenoids	
Mango		
Papaya		
Sweet potato		
Purple sweet potato	Anthocyanins	
Purple yam		
Strawberry		
Edamame	Chlorophyll	
Green peas		
Green beans		

11.3.1 LIGHTLY COLORED FRUITS AND VEGETABLES

11.3.1.1 POTATO

Fried chips/crisps made from potatoes are the most common and popular salty snacks among consumers of all ages. Potato chips are thin slices of raw potato deeply fried to an oil content of 35-40%. The color of the chips is a critical measure in the snack processing industry and is mainly related to consumer experience. Consumers associate color with more than just product appearance; it is an indicator of the sensory quality, safety, and nutritional value of the chips. The initial steps in the production of potato chips include cleaning, sorting, peeling, and thinly slicing the raw potatoes. After these steps, the slices undergo atmospheric frying in hot oil (temperature above 160°C); slices are removed from the oil once the moisture level reaches less than 2% of the finished product weight. Immersion frying under OHACs leads to rapid cooking, which has color browning effects along with the development of a unique flavor and texture. When frying under vacuum, the same process is performed at a lower pressure, with the temperature varying between 80°C and 150°C.

The color of fried potato chips is primarily driven by the Maillard reaction. The primary amino acid in potatoes, asparagine, is responsible for interacting with reducing sugar, resulting in the formation of browning compounds. Therefore, the final color of the chips depends on the quantity of these reactive substances, as well as the temperature and duration of frying. To control browning during frying, it is necessary to have a low content of reducing sugar.

Additionally, color development begins only after sufficient moisture has been removed from the potato slices. Therefore, besides the quantity of reactive substances, the rate of moisture removal and the heat transfer coefficient during different stages of frying also impact the color formation. One of the earlier studies compared the color parameters of potato chips from vacuum frying versus atmospheric frying. This study highlighted the criticality of having optimal pressure and temperature conditions to limit color defects [14]. The authors evaluated the effects of temperature and pressure using a 3x3 factorial design, with three temperature levels (118°C, 132°C, 144°C) and three pressure levels (3.12 kPa, 9.88 kPa, and 16.66 kPa). They observed that vacuum-fried chips processed at 144°C and 3.12 kPa had a higher L* value, and lower a* and b* values compared to chips fried using OHACs at 160°C. A higher L* value, which represents lightness, is often desirable for sensory acceptance of potato chips. Lower values of a* and b* indicate a reduced Maillard reaction in vacuum-fried chips compared to atmospheric frying. Another study by the same authors investigated color differences at different temperatures during vacuum frying, where the pressure was maintained below 1.33 kPa (10 Torr). The research group reported a clear increase in the b* value (blue-yellow hues) of the chips that were vacuum fried at 130°C and 140°C compared to those fried at 120°C. Therefore, the application of vacuum allows for much lower processing temperatures in potato chip production, resulting in a lighter, high-quality product without scorching or non-enzymatic Maillard browning.

While the use of vacuum frying has been shown to preserve color, the additional use of pretreatments in combination with vacuum frying could result in an even lighter and more appealing color, signifying the 'premium quality' of potato chips. Microwave (MW) frying using 16–20 W/g under vacuum conditions (OHVCs) achieved a much lower frying temperature of 100°C, enhancing the lightness of potato chips compared to those simply vacuum fried. The total color difference, represented by ΔE, indicates the color variation between unfried and fried chips in the L*, a*, b* system. This value was comparatively lower for MW pretreated samples [29].

11.3.1.2 APPLE

Apples are a nutritious and delicious fruit snack with worldwide popularity. Since they are a seasonal fruit, they are processed into juice, jams, jellies, and dehydrated chips to increase their shelf life. Apples are a rich source of polyphenols, catechin, epicatechin, quercetin, chlorogenic acid, and procyanidins. Similar to potatoes, they are prone to enzymatic browning. Blanching and immersion into chemical solutions such as AA, sodium chloride, or sugar are standard practices to prevent undesired browning reactions. Fried apple chips are prepared by peeling, washing, and cutting them into thin slices pretreated by one of the above techniques before directly frying or frying under a vacuum. The three coordinates of the color measurement have been shown to decrease over increasing frying time of 2–15 minutes in atmospheric frying, while these values remained consistent under vacuum frying conditions. The ΔE value was twice as high in atmospheric frying, indicating an overall darker sample [18]. Although vacuum frying is usually carried out at various temperatures (100–150°C) and time combinations (10–30 minutes), the optimal vacuum frying parameters for apple chips were determined to be 100–110°C for 20–25 minutes with a pretreatment of immersing the slices in a sugar solution [25].

An advanced vacuum frying technology that adds value by reducing frying processing time while enhancing the quality of the finished product is ultrasound-assisted microwave vacuum frying (UMVF). This procedure combines the power of ultrasound to initiate cell lysis with the superior heating ability of MW for vacuum frying. Faruq et al. [13] compared the color parameters of UMVF apple chips with MW vacuum fried (MVF) chips. UMVF apple chips had a higher L* value than MVF chips due to a more significant reduction in non-enzymatic browning. The authors noted that ultrasound energy alters the protein structure (secondary and tertiary) by disrupting hydrogen bonds, which may prevent the proteins from participating in Maillard reactions. For both UMVF and MVF, the frying temperature was 90°C, significantly lower than atmospheric frying.

11.3.1.3 MUSHROOM

Mushrooms are the fruiting bodies of fungal species that are safe for consumption. Among the edible mushrooms, button and shiitake are two popular varieties commonly consumed globally. Fresh mushrooms are a perishable commodity with a shelf life of 2–3 days. Processing them into fried snacks is a strategy followed to preserve the nutritional quality and improve the shelf stability of mushrooms. A recent study by Devi et al. [9] investigated two technologies: ultrasound MW power and MW power without ultrasound. MW power of 1,000 W was combined with temperatures of 80°C and 90°C, and a pressure of 12 kPa. UMVF retained the most color, providing the lightest color with significantly less red and yellow hue to the product. UMVF also reduced the frying time by 2 minutes compared to MVF and VF. 5-hydroxymethylfurfural is a brown pigment that has been shown to be present in vacuum fried mushrooms; the lighter color of UMVF could be attributed to the lesser degree of formation of 5-HMF than MVF and VF samples. Among the pretreatment techniques done before frying, the combination of OD, blanching, and coating with hydrocolloids was found to be the most suitable for obtaining the best quality of vacuum fried shiitake mushrooms [23].

Other lightly colored fruits and vegetables that have benefitted using pretreatments and hybrid technologies such as MVF, PEF, and UVF include zucchini, jackfruit, banana, and tapioca.

11.3.2 CAROTENOID-RICH VEGETABLES AND FRUITS

Carotenoids are a wide range of molecules synthesized by plants that contribute to orange, yellow, and red colors in fruits and vegetables. There are more than 600 types of carotenoids, and their molecular structure, such as the presence of conjugated double bonds and the type of functional groups, influences the final color. The important classes of carotenoids include carotene, lycopene, and xanthophyll. They cannot be synthesized by the human body and must be obtained through diet. Some rich dietary sources of carotenoids include carrots, mangoes, papayas, apricots, tomatoes, sweet potatoes, and pumpkins.

11.3.2.1 CARROT

Carrot is a leading source of dietary carotenoids with worldwide appeal. Carrots occur in a wide variety of colors ranging from white, yellow, orange, and purple. The traditional orange color is mainly due to the presence of two carotenoid compounds – α and β carotene. Aside from imparting the bright, aesthetically attractive orange color, carotene has various health benefits such as anticancer, anti-inflammatory, and antioxidant. Both α- and β-carotene (also called provitamin A) produce retinol (one or two molecules of vitamin A) in our body, and hence act as its precursor [31]. α and β carotene account for nearly 100% of the total carotenoids (TC) in carrots; they occur naturally in trans-form. Although carotenoid compounds are more stable than other phenolic compounds, the presence of conjugated double bonds in their structure makes them prone to thermal degradation. Several previous reports have shown that exposure of carrots to high temperatures (>100°C) results in geometrical isomerization of their carotene compounds to cis-isoforms that negatively impact the color of the finished product.

Preparing vegetable snacks, such as carrot chips, involves frying the product at temperatures greater than 140°C. However, these high temperatures often result in the trans-cis isomerization of carotene compounds, leading to color deterioration. In a comparative study on the quality of carrot chips, significant differences in color values were observed between atmospheric and vacuum fried chips. The atmospheric frying temperatures were 160°C and 180°C, while the vacuum frying temperatures were 98°C and 118°C. The fried chips at higher temperatures showed a higher isomerization of trans carotenes to cis forms, resulting in a significant decrease in the a* and b* color dimensions associated with orange and yellow pigments. The lightness of the chips, indicated by L*, also notably decreased during atmospheric frying, while it remained constant under vacuum frying [11]. Another study by the same authors found that vacuum fried chips had a significantly lower occurrence of brown pigments compared to the control [10]. These observed effects were attributed to the production of melanoidins, heterocyclic amines, and hydroxymethylfurfural (HMF) through the browning reaction.

11.3.2.2 PAPAYA

Papaya fruit is another rich source of carotenoids, including lycopene, beta-carotene, and beta-cryptoxanthin. It is widely grown and consumed in tropical and subtropical countries and serves as an essential dietary source for pro-vitamin A as well as other vitamins and minerals. A multitude of research has focused on optimizing the parameters of vacuum frying to preserve the papaya's heat-sensitive phytochemicals that contribute to its yellow-orange color. The temperature and duration of vacuum frying significantly impact the color of papaya chips; when the frying temperature and time were above 140°C and more than 14 minutes, there was a significant degradation of the product color [26]. The color degradation is both because of the loss of carotenoids and the activation of caramelization reactions at high heat. Since papaya is naturally low in amino acids, there is less chance for the occurrence of Maillard reactions during heating. However, the fruit is rich in reducing sugars such as glucose and fructose, which are prone to caramelization when heated at temperatures above 120°C. Nonetheless, in the same study, maintaining lower frying temperatures such as 100°C and 120°C reduced browning, with the best color results seen (without fading or browning) at 100°C.

11.3.2.3 MANGO

Another tropical fruit that is becoming popular is mango, which gets its bright yellow color from carotenoids. The product color depends on the ripening stage, fruit maturity, overall moisture level, sugar content, and TC. Both unripe and fully ripe mangoes are processed into mango chips. Vacuum-fried mango chips are lighter with higher red (a*) and yellow values (b*) than atmospheric frying due to the reduction of excessive darkening observed in atmospheric frying. Vacuum-fried mango chips can be produced at low temperatures of 100°C–130°C with a holding time of less than 200s; however, in atmospheric frying, higher temperature (160°C–180°C) and longer duration (230–250 s) are required. The pretreatment of mango slices before frying is also favorable for preserving their color; the common methodology to pretreat mango slices is OD using maltodextrin and citric acid (CA).

11.3.2.4 SWEET POTATOES AND PUMPKIN

Sweet potatoes and pumpkin are both popular vegetables known for their high β-carotene content. These vegetables have recently been reported to have color preservation from vacuum frying with one or more pretreatments. The ultrasound-pretreated sweet potato chips, fried using MVF, showed that cell integrity and color retention of the finished product were negatively affected by the strong application of ultrasound power as well as long processing time. Ultrasound pretreatment was investigated in the study at 300 W for 10, 20, and 30 minutes, and also at three power levels of 150, 300, and 450 W for a fixed 20-minute time length. The study showed that optimizing the pretreatment parameters based on the material chosen for frying was crucial for getting the best sensory quality of chips [22]. A similar study performed using pumpkin chips after ultrasound-assisted osmotic pretreatment in maltodextrin solution showed 70% enhanced stability of carotenoid in the chips compared to those that were not pretreated [21].

11.3.3 ANTHOCYANIN (ACN) COLORED FRUITS AND VEGETABLES

Anthocyanins (ACNs) are water-soluble flavonoid compounds that give purple, blue, and red color pigments to a variety of vegetables and fruits. Blueberry, blackberry, eggplant, purple yam, and purple potato are the primary sources for the dietary intake of this class of compounds. In recent years, an array of health benefits such as anticancer, anti-inflammatory, antioxidant, and anti-aging properties have been associated with ACNs, which has sparked interest among researchers for their use in food applications. Because of the vivid color they provide to any product, ACNs have been explored as an alternative to synthetic color pigments to enhance the visual appeal of new food products. However, like carotenoids, they are susceptible to degradation due to high temperature and hence have limited stability while processing. Figure 11.3 provides the color values of some of the carotenoid and ACN-containing fruits and vegetables.

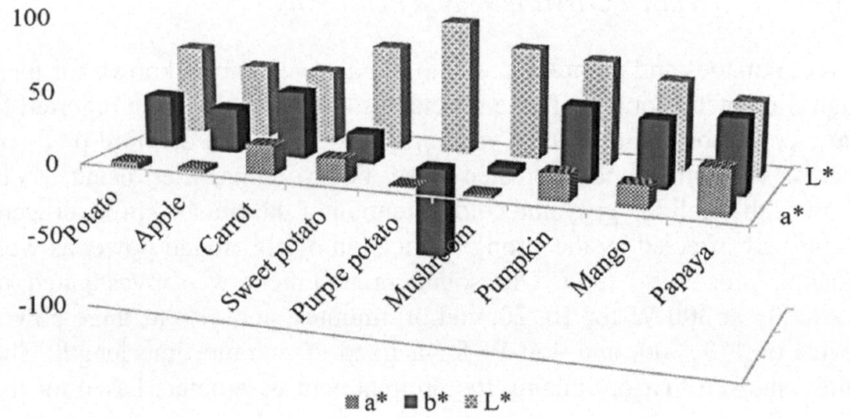

FIGURE 11.3 L* (lightness), a* (green-red chromaticity), b* (blue-yellow chromaticity) values of carotenoid/anthocyanin rich vegetables and fruits.

Structurally, ACNs comprise a core molecule of trihydroxyl-2-phenyl benzopyran cation with aglycones present at the third carbon site attached through glycosidic bonds [6]. Glycosylation of the anthocyanidin forms ACN; its stability is dependent on the molecular nature and type of substitution groups on the structure (methyl, hydroxyl) and the number and position of such groups. Previous studies have shown that the presence of a hydroxyl group increases blue color but reduces stability, while the presence of a methyl group increases both red color and stability [2]. Some of the predominant ACN compounds include glucoside derivatives of delphinidin, malvidin, petunidin, cyanidin, and pelargonidin. The heterogeneity in the chemical structure is mainly because of the variations in the quantity of hydroxyl groups, methylation of these hydroxyl substituents, and the type of sugar moiety attached to a phenolic molecule.

The mechanism of ACN degradation is relatively unknown, although several possibilities have been proposed based on the understanding of its chemical structure. The initial degradation step upon heating is noted as the hydrolysis of glycosides to aglycone. The aglycone further decomposes into chalcones, which then transform into phenolic acids. The extent of deterioration caused by heat depends on pH, co-pigmentation, and the type of vegetable/fruit. Chemical modifications such as acylation and methoxylation enhance the stability of ACN when heated, although these changes vary from one product to another.

11.3.3.1 PURPLE AND RED FLESHED POTATO

Purple and red-fleshed potatoes are newer varieties of fried chips that have drawn consumer interest due to their color pigments and health benefits. Blue potato chips, when vacuum fried at 120°C for 400 seconds compared to traditionally fried at 165°C for 180 seconds, were compared for ACN stability. Vacuum fried blue potato chips retained the characteristic blue color with a higher b* value and lower a* value, indicating less yellowing/browning of the product compared to atmospheric frying. The retention of monomeric ACNs was also 60% higher than the control for the vacuum fried chips [8]. The significant difference in ACN retention explains the 'blueness' of the final product and highlights the impact of frying temperature on color. A recent study by Su et al. [27] introduced an ultrasound microwave-assisted vacuum frying (UMVF) method. The fryer was set at 90°C and 10 ± 2 kPa for temperature and vacuum levels, respectively, and was compared against simple vacuum frying (VF) and microwave vacuum frying (MVF) without ultrasound. The combination of MW (800 W) and ultrasound technology (600 W) effectively reduced the frying time from 16 minutes to 10 minutes, significantly impacting the ΔE value. The retention of ACNs was also 9% higher in the hybrid method compared to vacuum fried chips. A closer examination of the microstructure revealed a disrupted, porous structure in UMVF products, which aided in water removal and improved quality.

In contrast to UMVF, the structure was found to be intact in VF chips. Hence, a longer dehydration time was needed, which deteriorated the color, along with causing higher oil uptake and the loss of crispy texture. Similar findings were reported for purple yam, where UMVF provided the best sensory quality of the low-fat fried chips compared to the one-step vacuum frying method.

11.3.3.2 STRAWBERRY

Strawberries are highly perishable fruits that are a good source of ACNs, vitamin C, and ellagic acid. The main ACN in strawberries is pelargonidin-3-glucoside, which contributes to their bright red color. In a study by Chauhan et al. [7], the impact of freezing pretreatment on vacuum frying of strawberries was evaluated. Cut strawberries were stored for 24 hours at –18°C before being vacuum fried at 90°C for 25 minutes at 13.33 kPa. The authors observed that the frozen strawberries had significantly higher color retention, as well as higher ACN and flavonoid content in the final product.

11.3.4 CHLOROPHYLL PIGMENTED VEGETABLES

The loss of green color in processed vegetables results from chlorophyll degradation. Chlorophyll a and b are pigments that give several vegetables their characteristic green color. Like other pigments discussed in this chapter, they are susceptible to chemical (temperature, pH changes) and enzymatic degradation. Both chlorophylls a and b contain magnesium atoms attached to the porphyrin ring in their structure. Heating causes the loss of magnesium, resulting in the formation of pheophytin. This reaction is accompanied by a color change from green to olive brown, which is considered a quality defect in processed products. Chlorophyll b is reported to have higher heat stability compared to chlorophyll a due to differences in their molecular structure. Some chlorophyll-rich vegetables, such as edamame, green peas, and beans, are processed into chips/crisps using high heat. While vacuum frying has shown to provide superior quality compared to traditional atmospheric frying, pretreatments, especially ultrasound and MW frying, have shown the best results in retaining the green color of these vegetables. The preservation of chlorophyll through vacuum frying is less documented compared to other color pigments like carotenoids. A recent study by Islam et al. [15] compared the use of VF, MVF, and UMVF for frying edamame. The best temperature/time combination reported using UMVF was 100°C for 12 minutes, resulting in a lighter L* value and a greener (a*) looking product compared to VF and MVF fried chips.

11.4 SUMMARY

Color is an essential parameter for consumer acceptance of fried fruit/vegetable snacks. The frying temperature above 150°C used in traditional atmospheric frying has been well-documented to have a negative impact on color, which affects product quality. Additionally, fried products also have high oil uptake, generate harmful trans fatty acids, lose nutrients, and form lipid oxidation products, which raises health concerns for consumers. Therefore, the evolution of vacuum frying offers a safe alternative by reducing the frying temperature under OHVCs, which mitigates undesirable heat-induced reactions. While several commonly consumed fried chips/crisps have shown benefits from vacuum frying, the use of pretreatments often has synergistic effects. Among the various pretreatment methods, ultrasound technology shows promise in achieving the best results for stabilizing various classes of color pigments, mainly due to its ability to shorten frying time. Ultrasound also accelerates dehydration kinetics by disrupting cell structure and increasing the porosity of the product being vacuum fried. The use of such hybrid technologies has also been explored further due to their potential in enhancing the nutritional and sensory aspects of vacuum-fried snacks. The studies published so far have primarily relied on analytical instrumental assays to assess color. While this is a reliable approach for accurately quantifying color differences, there is often a human threshold for consumers' visual perception of these differences. Therefore, future studies should focus on integrating sensory data with instrumental color analysis to understand the value of vacuum frying compared to deep frying, as well as the benefits of using one or more pretreatments with vacuum fryers (VFs). Additionally, future research should prioritize optimizing the process conditions for VFs to accommodate newer product varieties that are gaining consumer attention. Furthermore, it is crucial to investigate the storage stability of products manufactured using VFs, including the color stability throughout their shelf life, as well as any other physical and chemical changes that may negatively impact the final product.

KEYWORDS

- **anthocyanin**
- **browning**
- **carotene**
- **carotenoid**
- **chlorophyll**
- **color pigments**
- **frying**
- **microwave-assisted vacuum frying**
- **pulsed electric field**
- **ultrasound assisted vacuum frying**

REFERENCES

1. Al Faruq, A., Zhang, M., & Adhikari, B., (2019). A novel vacuum frying technology of apple slices combined with ultrasound and microwave. *Ultrasonics Sonochemistry, 52*, 522–529.
2. Albuquerque, B. R., Oliveira, M. B. P., Barros, L., & Ferreira, I. C., (2021). Could fruits be a reliable source of food colorants? Pros and cons of these natural additives. *Critical Reviews in Food Science and Nutrition, 61*(5), 805–835.
3. Andreou, V., Dimopoulos, G., Tsonas, T., Katsimichas, A., Limnaios, A., Katsaros, G., & Taoukis, P., (2021). Pulsed electric fields-assisted drying and frying of fresh zucchini. *Food and Bioprocess Technology, 14*(11), 2091–2106.
4. Asokapandian, S., Swamy, G. J., & Hajjul, H., (2020). Deep fat frying of foods: A critical review on process and product parameters. *Critical Reviews in Food Science and Nutrition, 60*(20), 3400–3413.
5. Ayustaningwarno, F., Fogliano, V., Verkerk, R., & Dekker, M., (2021). Surface color distribution analysis by computer vision compared to sensory testing: Vacuum fried fruits as a case study. *Food Research International, 143*, 110230.
6. Cai, D., Li, X., Chen, J., Jiang, X., Ma, X., Sun, J., Tian, L., et al., (2022). A comprehensive review on innovative and advanced stabilization approaches of anthocyanin by modifying structure and controlling environmental factors. *Food Chemistry, 366*, 130611.
7. Chauhan, O., Pandey, A., Ravi, N., Patki, P., & Semwal, A., (2019). Use of vacuum frying technology for the development of strawberry chips. *International Journal of Tropical Agriculture, 37*(2), 147–156.
8. Da Silva, P. F., & Moreira, R. G., (2008). Vacuum frying of high-quality fruit and vegetable-based snacks. *LWT-Food Science and Technology, 41*(10), 1758–1767.
9. Devi, S., Zhang, M., & Law, C. L., (2018). Effect of ultrasound and microwave assisted vacuum frying on mushroom (*Agaricus bisporus*) chips quality. *Food Bioscience, 25*, 111–117.
10. Dueik, V., Marzullo, C., & Bouchon, P., (2013). Effect of vacuum inclusion on the quality and the sensory attributes of carrot snacks. *LWT–Food Science and Technology, 50*(1), 361–365.
11. Dueik, V., Robert, P., & Bouchon, P., (2010). Vacuum frying reduces oil uptake and improves the quality parameters of carrot crisps. *Food Chemistry, 119*(3), 1143–1149.
12. Fan, K., Zhang, M., & Bhandari, B., (2019). Osmotic-ultrasound dehydration pretreatment improves moisture adsorption isotherms and water state of microwave-assisted vacuum fried purple-fleshed sweet potato slices. *Food and Bioproducts Processing, 115*, 154–164.
13. Faruq, A. A., Zhang, M., & Fan, D., (2019). Modeling the dehydration and analysis of dielectric properties of ultrasound and microwave combined vacuum frying apple slices. *Drying Technology, 37*(3), 409–423.
14. Garayo, J., & Moreira, R., (2002). Vacuum frying of potato chips. *Journal of Food Engineering, 55*(2), 181–191.

15. Islam, M., Zhang, M., & Fan, D., (2019). Ultrasonically enhanced low-temperature microwave-assisted vacuum frying of edamame: Effects on dehydration kinetics and improved quality attributes. *Drying Technology*, *37*(16), 2087–2104.
16. Lund, M. N., & Ray, C. A., (2017). Control of Maillard reactions in foods: Strategies and chemical mechanisms. *Journal of Agricultural and Food Chemistry*, *65*(23), 4537–4552.
17. Maity, T., Bawa, A., & Raju, P., (2018). Effect of preconditioning on physicochemical, microstructural, and sensory quality of vacuum-fried jackfruit chips. *Drying Technology*, *36*(1), 63–71.
18. Mariscal, M., & Bouchon, P., (2008). Comparison between atmospheric and vacuum frying of apple slices. *Food Chemistry*, *107*(4), 1561–1569.
19. Oladejo, A. O., Ma, H., Qu, W., Zhou, C., Wu, B., Yang, X., & Onwude, D. I., (2017). Effects of ultrasound pretreatments on the kinetics of moisture loss and oil uptake during deep fat frying of sweet potato (*Ipomea batatas*). *Innovative Food Science & Emerging Technologies*, *43*, 7–17.
20. Pankaj, S. K., & Keener, K. M., (2017). A review and research trends in alternate frying technologies. *Current Opinion in Food Science*, *16*, 74–79.
21. Piyalungka, P., Sadiq, M. B., Assavarachan, R., & Nguyen, L. T., (2019). Effects of osmotic pretreatment and frying conditions on quality and storage stability of vacuum-fried pumpkin chips. *International Journal of Food Science & Technology*, *54*(10), 2963–2972.
22. Qiu, L., Zhang, M., Wang, Y., & Bhandari, B., (2018). Effects of ultrasound pretreatments on the quality of fried sweet potato (*Ipomea batatas*) chips during microwave-assisted vacuum frying. *Journal of Food Process Engineering*, *41*(8), e12879.
23. Ren, A., Pan, S., Li, W., Chen, G., & Duan, X., (2018). Effect of various pretreatments on quality attributes of vacuum-fried shiitake mushroom chips. *Journal of Food Quality, 2018*(1), 4510126.
24. Sakooei-Vayghan, R., Peighambardoust, S. H., Hesari, J., & Peressini, D., (2020). Effects of osmotic dehydration (with and without sonication) and pectin-based coating pretreatments on functional properties and color of hot-air dried apricot cubes. *Food Chemistry*, *311*, 125978.
25. Shyu, S. L., & Hwang, L. S., (2001). Effects of processing conditions on the quality of vacuum fried apple chips. *Food Research International*, *34*(2, 3), 133–142.
26. Soto, M., Pérez, A. M., Servent, A., Vaillant, F., & Achir, N., (2021). Monitoring and modelling of physicochemical properties of papaya chips during vacuum frying to control their sensory attributes and nutritional value. *Journal of Food Engineering, 299*, 110514.
27. Su, Y., Zhang, M., Bhandari, B., & Zhang, W., (2018). Enhancement of water removing and the quality of fried purple-fleshed sweet potato in the vacuum frying by combined power ultrasound and microwave technology. *Ultrasonics Sonochemistry*, *44*, 368–379.
28. Su, Y., Zhang, M., Chitrakar, B., & Zhang, W., (2021). Reduction of oil uptake with osmotic dehydration and coating pre-treatment in microwave-assisted vacuum fried potato chips. *Food Bioscience*, *39*, 100825.

29. Su, Y., Zhang, M., Zhang, W., Adhikari, B., & Yang, Z., (2016). Application of novel microwave-assisted vacuum frying to reduce the oil uptake and improve the quality of potato chips. *LWT-Food Science and Technology*, *73*, 490–497.
30. Su, Y., Zhang, M., Zhang, W., Liu, C., & Adhikari, B., (2018). Ultrasonic microwave-assisted vacuum frying technique as a novel frying method for potato chips at low frying temperature. *Food and Bioproducts Processing, 108*, 95–104.
31. Yoo, K. S., Bang, H., Pike, L., Patil, B. S., & Lee, E. J., (2020). Comparing carotene, anthocyanins, and terpenoid concentrations in selected carrot lines of different colors. *Horticulture, Environment, and Biotechnology*, *61*(2), 385–393.
32. Zhang, X., Zhang, M., & Adhikari, B., (2020). Recent developments in frying technologies applied to fresh foods. *Trends in Food Science & Technology*, *98*, 68–81.

CHAPTER 12

EFFECT OF EXTRUSION PROCESSING PARAMETERS ON COLOR PROPERTIES OF EXTRUDATES

ANTO PRADEEP RAJA CHARLES,[1] RAMNARAIN RAMAKRISHNA,[1] PULIVARTHI MANOJ KUMAR,[2] MOHANA YOGANANDAN,[3] CHATRAPATI VEERA RAGHAVA KUMAR,[4] and JIAJIA RAO[1]

[1]*Food Ingredients and Biopolymer Laboratory, Department of Plant Sciences, North Dakota State University, Fargo, North Dakota, USA*

[2]*Department of Grain Science and Industry, Kansas State University, Manhattan, Kansas, USA*

[3]*The Mennel Milling Company, Fostoria, Ohio, USA*

[4]*Idaho Milk Products, Idaho, USA*

ABSTRACT

Extrusion technology is widely used to produce food products such as meat analogues, texturized proteins, breakfast cereals, and ready-to-eat (RTE) foods. The processing parameters involved in extrusion can influence the taste, color, nutritional value, and physicochemical properties of the food products. This chapter specifically focuses on the change in the color profile of the extrudates due to the impact of processing parameters. To thoroughly understand how extrusion parameters affect the color profile, the extrusion mechanism, operating parameters, and their influence on physical properties and composition are initially summarized. Models

Nonthermal Food Processing Technologies: Impact on Color Profile.
Kaavya Rathnakumar and R. Pandiselvam (Eds.)

of rice, pulses, and texturized protein extrudates are selected to discuss the influence of processing parameters on the color profile. Finally, the behavior of natural colorants, many of which are also potent bioactive molecules, during extrusion is reviewed.

12.1 INTRODUCTION

Extrusion is one of the most conventional food processing techniques used globally. Food extrusion originated in the early 1870s with the processing of meat for sausage production. It is believed that General Mills first introduced food extrusion in the early 1930s to produce dry pasta and breakfast cereals, which led to its popularity in other sectors of the food industry [1, 2]. Food extrusion involves the use of "feedstock," consisting of ingredients such as soy, corn, sorghum, and wheat, which undergo multiple unit operations including mixing, shearing, heating, and high-pressure processing (HPP) simultaneously before being forced out through a shaped die/hole at a programmed rate [3]. The extrusion process can be described as a short-time reaction at high temperatures [4]. Extrusion is used in the development of various food products, including instant flours, energy bars, breakfast cereals, snack products, baby food, texturized proteins, pet food, ready-to-eat (RTE) food, and confectioneries. One of the key advantages of food extrusion is its high turnover or production rate in a short period of time. Additional benefits include its versatility, ability to operate continuously, and energy and cost efficiency. However, extrusion processing does have certain disadvantages, such as the high initial investment required to procure the extrusion screw system, the need to optimize process parameters, and incompatibility with heat-sensitive foods and ingredients.

Color is the most important sensory attribute that attracts consumers and determines the success of a food product in the market. It serves as a potential indicator of the food's freshness, flavor, nutrient value, and safety [5]. The color of a food product is influenced by factors such as the characteristics of raw materials, operating conditions, packaging, and storage. However, the processing techniques used in manufacturing the product play a significant role in determining its final color profile. Extrusion, a thermomechanical operation involving high pressure, temperature, and shear, is commonly employed. The high temperatures used during extrusion have an impact on microorganisms and anti-nutrient enzymes, thereby improving the shelf-life of the products [6]. Additionally, extrusion leads to various chemical and structural changes in the ingredients, including enzymatic reactions, protein and vitamin degradation, starch gelatinization, and lipid oxidation. These changes can develop or alter the color profile of extrudates [7]. The extent of color development largely depends on processing factors such as barrel temperature, feed moisture, and screw speed. Therefore, it is crucial to understand how extrusion parameters affect the color profile of extrudates.

Food colorants have been used for decades to produce extrudates with the desired color [8]. Both natural and synthetic food colorants are widely used in food extrusion. Consumer preference for natural colorants has expanded due to negative perceptions of synthetic food dyes [9]. Among the natural colorants, anthocyanins (ACNs), carotenoids, betalains, phycocyanins, and chlorophyll are commonly used. Apart from acting as coloring agents, these compounds can also serve as potential bioactives due to their functional properties. However, the processing conditions (e.g., high temperature, acidic environment) involved during extrusion affect their color-promoting capacity and functional ability [10].

This review aims to provide an overview of the extrusion parameters and their impact on the color attributes of extrudates. Furthermore, changes in color retention and beneficial functionalities of pigments derived from natural sources during extrusion are also extensively discussed.

12.2 TYPES OF EXTRUDERS

Extruders are divided into single-screw and twin-screw categories based on the number of screws. As the name suggests, a single-screw extruder has one screw, and it is further categorized based on the heat source or screw design [11]. Based on the screw diameter, the single-screw extruders may be classified either as parallel (diameter remains constant throughout its length) or varying diameter (diameter increases towards the ends of the screw) extruders [6]. Single-screw systems are cost and energy efficient compared to twin-screw extruders and therefore are the most widely used type in producing simple food products such as pasta, noodles, and high carbohydrate products.

Twin-screw extruders consist of dual screws enclosed in a barrel. They are classified as co-rotating extruders or counter-rotating extruders based on the direction of screw rotation [12]. In co-rotating extruders, the screws turn in the same direction, while in counter-rotating extruders, they turn in opposite directions. Furthermore, co-rotating screw extruders can be further classified as non-intermeshing (material is moved by squeezing action between the two screws) or inter-meshing (material is moved by the gliding action of one screw on the threads of the other) systems, based on how the material is advanced through the barrel. Twin-screw extruders are highly adaptable and versatile for food applications, as they can achieve higher degrees of shear and heat generation due to increased surface area. This allows for better contact between materials and improved cooking [13].

12.3 EXTRUDER COMPONENTS AND PROCESS

A generic extruder consists of a motor that runs the screw enclosed in a barrel and a gearbox to control the screw speed. The feeding port is used to insert feed material into the screw for processing [14]. The feed material passes through three processing sections, namely the conveying section, mixing or compression section, and melting section. The conveying section moves the feed from the feeding port to the compression section through the screw. In the compression zone, the food substance is sheared within a compact region and generates heat due to the combined effect of continuous screw movement and heating elements surrounding the barrel. The melting section is the final section close to the die of the extruder [13]. The feed melts in this region owing to the heat generated by the screws' temperature, shear, and pressure. Due to the quick pressure drop, the melted feed material expands as it is pushed out of the die.

A single or twin-screw extruder typically has feed, compression, and cooking sections [15]. The feed section consists of a feed hopper into which pre-conditioned materials are added as hydrated particulates. The materials are generally added with water to increase the feed viscosity. The material density is low in this section due to the air trapped along the feed. Subsequently, the material moves towards the compression region in which it loses its particulate nature, and the feed density increases. The continuous spin of the screw compresses the feed material and generates extensive heat and pressure. Additional heat from heating segments contributes to maintaining a high temperature in this compartment to cook the extrudate properly. The materials reach the plasticized state at the end of the compression section [16]. The texturization of materials occurs in the cooking section. This section has the highest temperature, pressure, and shear rates, helping to force out the material through the die. Most physicochemical transformations initiated in the compression section end in the cooking section.

12.4 CHANGES DURING EXTRUSION COOKING

Food extruders are used in the food industry to perform various operations such as shearing, size reduction, shaping, dehydration, thermal cooking, texturization, pasteurization, homogenization, protein denaturation, starch gelatinization, lipid modification, and enzyme inactivation [17]. Extrusion cooking involves the use of different raw materials, including cereals, grains, tubers, oil seeds, and legumes. The physicochemical changes that occur during extrusion depend on the type of material, its physical characteristics, chemical composition (such as proteins, starch, fats, and sugar), moisture content, and pH. These physical and chemical changes lead to improved nutrient availability and organoleptic properties, as well as the inactivation of heat-labile enzyme inhibitors and toxic compounds [18].

12.4.1 INFLUENCE ON PHYSICAL PROPERTIES

12.4.1.1 COLOR

The color of extruded products is critical for consumer acceptance and is significantly impacted by the processing conditions and feed matrix composition. The physicochemical reactions occurring during extrusion are generally responsible for the color change of the extrudate, in which non-enzymatic browning and pigment degradation play a major role [19]. The Maillard reaction and caramelization are two major non-enzymatic browning reactions that cause the formation of colored compounds, imparting color to the extrudate. Low feed moisture, as well as high barrel temperature, favor the reaction between proteins and reducing sugar (Maillard reaction), thereby increasing the color intensity and antioxidant property of the final product [20]. In contrast, maintaining uniform feed moisture and increasing the protein content decreases the color intensity.

12.4.1.2 TEXTURE

The texture is a crucial factor influencing consumer acceptance, with the crispness of the extrudates frequently being a desirable attribute [21]. During the extrusion process, the ingredients play a critical part in determining the texture of the final products. Specifically, the composition of the flour acts as the main factor impacting the product texture. The combined effects of temperature, screw speed, pressure, and moisture levels also influence the texture during extrusion [22]. Research has indicated that the crispness of the RTE products is correlated with the moisture content when the critical water activity is above 0.5 [23]. In plant-based extrudates, the crispness varies depending on the thickness of the cell wall and the diameter of air cells.

12.4.1.3 EXPANSION RATIO

The expansion ratio is the range of expansion determined based on the die's cross-sectional diameter and the diameter of the extrudate. During the extrusion process, the extrudate can expand in size by two to six-fold, causing a change in volume and taste [24]. The expansion ratio varies based on the barrel temperature, feed moisture, and screw speed. Several studies indicated that the expansion ratio and feed moisture are inversely proportional [25, 26]. High feed moisture achieves a better foaming effect due to the transformation of moisture into steam [27]. Additionally, it increases viscosity and reduces melt elasticity. In contrast, the expansion ratio rises with the barrel temperature. However, if the temperature is too high, the expansion ratio of the extrudate decreases. The screw speed is another critical parameter closely related to the expansion ratio due to its impact on the viscosity. High screw speed results in a drop in viscosity, thereby leading to a rise in expansion ratio.

12.4.1.4 BULK DENSITY

The bulk density of the extrudates is influenced by the screw speed, temperature, and feed moisture level [28]. Interestingly, the impact of these parameters on the bulk density is opposite to that of the expansion ratio. Higher screw speed, temperature, and feed moisture result in decreased bulk density of the extrudates. In a previous study, the effect of barrel temperature and screw speed on the bulk density of oat and rice flour extrudates was examined [29]. The response surface analysis revealed that the screw speed had a negative quadratic effect, while the temperature had a linear effect on the extrudates.

12.4.1.5 WATER ABSORPTION INDEX (WAI)

The volume taken after the swelling of starch in an excess amount of water is known as the water absorption index (WAI) [30]. The WAI is influenced by molecular interactions that occur within the amorphous and crystalline structure. Similar to other physical properties, the WAI is affected by various processing variables, including temperature, moisture content, and screw speed. The WAI increases with higher screw speeds due to increased expansion and greater mechanical shear, with the expansion being related to starch gelatinization [31]. Elevated temperatures also lead to an increase in WAI, attributed to dextrinization that occurs at higher temperatures. Conversely, WAI decreases with higher moisture levels. In the case of the rice and carrot blend extrudate, a high WAI was observed at low moisture content and high barrel temperature [32].

12.4.1.6 WATER SOLUBILITY INDEX (WSI)

The water solubility index (WSI) is the quantity of soluble unbound from the granules when surplus water is incorporated. The WSI increases with temperature, as high temperatures promote starch gelatinization and enhance the quantity of soluble starch [23]. High screw speed increases the WSI of the extrudate due to its association with specific mechanical energy (SME), while an increase in moisture content reduces the WSI. The water solubility values of rice and oat extrudates decrease at low screw speed and increase at high screw speed, attributed to the fragmentation of longer starch chains into shorter chains [29].

12.4.2 IMPACT ON COMPOSITION

The principal components of foods are carbohydrates, proteins, lipids, and dietary fibers. Each of these components undergoes numerous chemical changes during extrusion, thereby enabling the production of desired food extrudates.

12.4.2.1 CARBOHYDRATE

Depending on the extrusion parameters, carbohydrates undergo several transformations. The extent of functional and structural change depends on the shearing force, moisture content, temperature, and pressure [33]. Starch undergoes chain fragmentation, gelatinization, and melting when subjected to these parameters [34, 35]. To produce food with enhanced sensory and nutritional quality, it is necessary to control the modifications of starch during extrusion. Sugar loss occurs due to the breakdown of sucrose into fructose and glucose. Additionally, reducing sugars (fructose) are lost due to interaction with proteins in the Maillard reaction. Fructose loss is preferred during extrusion due to its flatulence-causing behavior [36].

12.4.2.2 PROTEIN

The nutritional value of proteins is primarily impacted by temperature [37]. High temperatures during the extrusion process denature the proteins and inactivate enzymes and anti-nutritional factors, such as lectins and antitrypsin. This alteration affects the functionality of the protein, reducing its solubility and improving protein digestibility by exposing susceptible enzyme sites. However, applying very high temperatures during extrusion can decrease protein digestibility [38]. Depending on the raw ingredients and other processing conditions, there may be a change in the amino acid profile, which can make the food either nutritious or deleterious.

12.4.2.3 LIPIDS

The lipid content in most extruded materials is lower than 5%, as high content can reduce the torque and eventually affect material expansion. The increased lipid level causes slippage within the internal parts of the extruder [39]. Based on the lipids' lipophilic-hydrophilic balance, type, quantity, and source, the chemical changes during extrusion differ and are attributed explicitly to starch-lipid binding [40]. The high temperatures involved during the extrusion affect the activity of lipase and lipoxygenase, and hence the development and oxidation of free fatty acids is reduced. Additionally, the reduced free fatty acid content prevents off-flavor development.

12.4.2.4 DIETARY FIBER

The chemical properties of dietary fiber are modified during extrusion. The processing conditions affect the solubility and total dietary fiber (TDF) content. Increasing the SME augments the solubility of dietary fiber [41]. Studies have indicated that the solubilization of dietary fibers is more dependent on mechanical stress than thermal energy. The change in solubility and structural properties influences the viscosity of the dietary fiber [42]. Based on the literature, it is evident that the TDF content may increase or decrease after extrusion. The transformation of dietary fiber into soluble fiber and resistant starch into digestible starch during extrusion reduces the TDF content. In contrast, resistant starch formation during extrusion increases the TDF content. The modification in morphological properties of dietary fiber also occurs throughout the process. Increasing the SME has decreased the particle size of the citrus fiber [43, 44].

12.5 EXTRUSION PROCESSING PARAMETERS AND THEIR INFLUENCE ON THE COLOR PROFILE OF THE EXTRUDATE

The properties of extrudates rely on the independent and dependent variables of extrusion, which are summarized in Figure 12.1. The independent variables are parameters that can be controlled. Dependent variables are parameters that change as one or more independent variables shift [23]. The operating conditions (Figure 12.2) involved in an extrusion process are essential in altering various quality properties of food material. As the material passes through the extrusion barrel, it absorbs thermal and shear energies resulting in physicochemical changes to the processed material. Numerous studies indicate that extrusion parameters, including screw speed and geometry, moisture content, and composition of food material, greatly influence the quality of extrudate [45]. These parameters enhance the palatability and functionality of the food, as well as its texture, color, and other visual qualities. Although there are several parameters, they are all interconnected. For example, the rate at which raw material enters the barrel depends on the feed rate set at the feeder. As the feed rate increases, the remaining process parameters, namely temperature, pressure, and residence time, change and affect the final product quality [46].

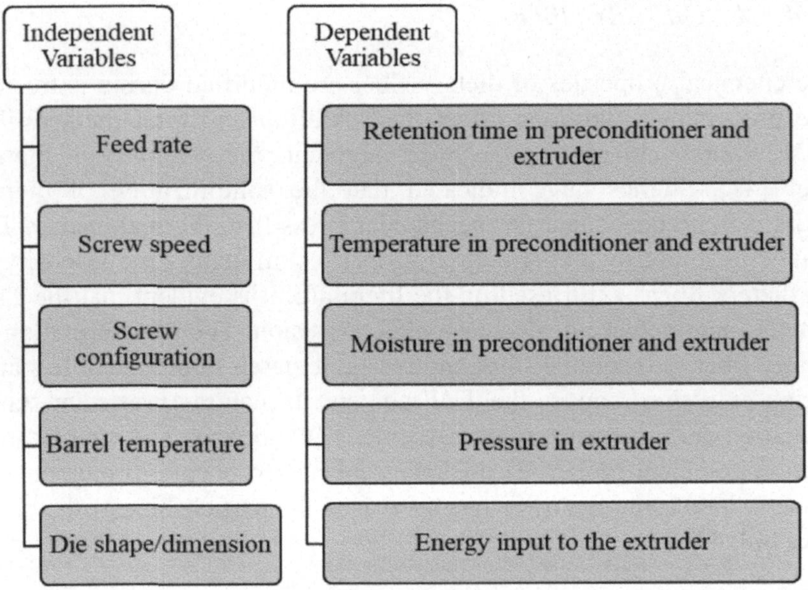

FIGURE 12.1 Independent and dependent variables involved in the extrusion.

Several researchers use color as a quality indicator to assess food. The hot extrusion process involves the addition of heat, causing non-enzymatic reactions in the food material. The Maillard reaction, caramelization, and pigment degradation are some of the critical chemical pathways that contribute to the color of the food material [47]. These reactions generally result in browning and redness of the extrudates. The color variation depends on the process conditions, as these reactions are highly influenced by temperature and moisture. Understanding the degree of color development in food material requires considering factors such as temperature, pressure, feed moisture, and residence time. Among these, residence time is directly related to the screw geometry and speed [48]. Additionally, the sheer force and thermal energy applied to the food material further accelerate these reaction rates due to their synergistic effect. Therefore, changes in color in food material can be primarily attributed to parameters such as moisture content, temperature, screw speed, and geometry.

Color, flavor, and texture changes due to extrusion are not always desirable and, therefore, need to be optimized based on the application. It is crucial to thoroughly investigate the role of these extrusion parameters on quality parameters to achieve desirable properties.

In a study on honey-enriched cereal flour extrudates, the lightness (L^*) value increased with the moisture content of raw materials [49]. However, the barrel temperature was inversely related to the L^* values. Additionally, increasing honey in the initial feed resulted in lower lightness values. Interestingly, the redness (a^*) and yellowness (b^*) values increased with an increase in temperature and amount of honey. Similarly, another research optimized the extrusion process to develop fiber-rich cereals from rice and fruits [50]. The increase in barrel temperature led to lower L^* values for the rice-based extrudates. The increased screw speed has led to lower L^* values. Jakkanwar et al. [51] indicated that the L^* values decreased with a rise in die temperature for the cowpea-based extrudates. The decrease in L^* values and increase in a^* and b^* values with increased temperature were attributed to the browning reaction due to caramelization and Maillard reactions. These studies indicate that the color profile of the extrudates is highly dependent on the processing parameters. Therefore, optimizing the process parameters to positively influence the color profile is important. Table 12.1 summarizes the color profile of various products influenced by extrusion parameters. The subsection discusses the color profile of rice-based extrudates, textured proteins, and pulse products affected by processing parameters.

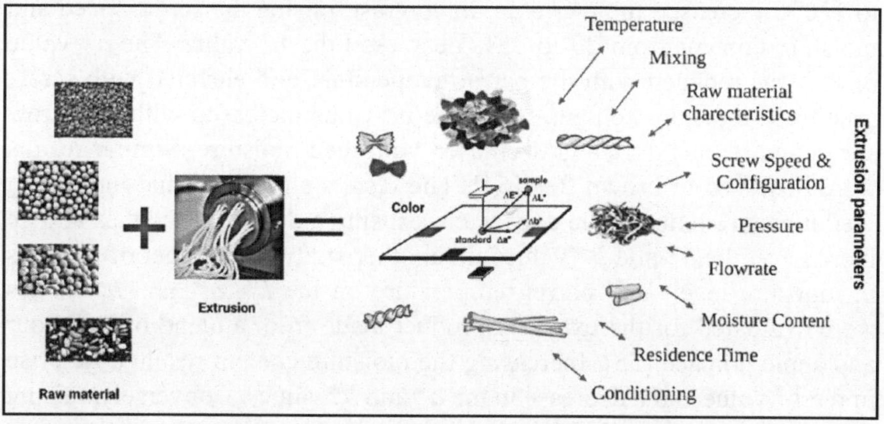

FIGURE 12.2 Extrusion processing parameters that influence the color profile of the extrudate.

12.5.1 RICE-BASED PRODUCTS

Rice is an excellent raw material widely used in the production of dietetic food, modified starch, breakfast cereals, and animal foods. Its extensive usage in the food industry is attributed to its color, bland taste, flavor, and suitability during processing. Among these characteristics, the color of rice is crucial for consumer acceptance [50]. Therefore, it is important to understand how extrusion parameters affect the color of rice extrudates (REs). The color varies depending on the raw materials, processing conditions, and chemical reactions. Non-enzymatic browning reactions, such as caramelization and Maillard reactions, form colored compounds that influence the color of the RE. For example, studies have shown that the b^* (yellowness), h^0 (hue angle), and C^* (chromaticity) can increase with non-enzymatic reactions in pigmented REs [30].

The total color (ΔE) of RE is mainly impacted by the moisture level and temperature. The ΔE increases with temperature due to the increase in the rate of browning reaction. The impact of temperature and moisture on the color of extrudates prepared using rice flour was previously examined [52]. The lightness value (L^*) of the extruded flour increased with temperature and decreased with feed moisture. Meanwhile, the redness (a^*) and yellowness (b^*) values increased with feed moisture and decreased with temperature. Gat & Ananthanarayan [53] studied the influence of barrel temperature, screw speed, and feed moisture on the colored REs added with paprika extract. The shift in barrel temperature from 100 to 120°C increased the L^* value. In contrast, raising the screw speed and moisture content from 20 to 25% decreased the L^* value. The a^* value of RE was reduced with the barrel temperature and elevated with screw speed and moisture content, while the b^* value increased with the screw speed and temperature and decreased with feed moisture. Similar results were observed in brown Res [54]. The results elucidated that increasing feed moisture reduced the expansion, causing a drop in L^* values and an increase in the a^* and b^* values. In another study, the impact of changes in moisture levels and barrel temperature on the L^*, a^*, and b^* values were evaluated for the extruded product made from a blend of rice flour and apple pomace [55]. Increasing the moisture content resulted in a rise in the L^* value and a decrease in the a^* and b^* values. Conversely, raising the barrel temperature led to a decrease in the L^* value and an increase in the a^* and b^* values. Fiber-rich cereals were produced by extrusion using rice flour, carambola pomace, and seeded banana [50]. As the temperature increased, the L^*, a^*, and b^* values decreased, which was attributed to the Maillard reaction. Overall, the literature on extrusion suggests that increasing the moisture content can reduce the L^* value and increase the a^* and b^* values. However, it is difficult to generalize as the color may vary depending on the composition of the raw materials, material properties, polymer concentrations, and operating conditions.

12.5.2 TEXTURIZED PROTEINS

In the process of texturization, the protein is stretched, layered, or cross-linked to form filamentous structures, crumbly surfaces, or other physical morphologies from its globular structure [56]. The proteins melt and denature in the extruder due to the high temperatures and pressures they encounter, hence losing their secondary and/or tertiary structure [7]. Depending on the quantity of water utilized in the extrusion, two distinct extrusion processes are used to texturize proteins, namely high-moisture extrusion (40–80%) and low-moisture extrusion (20–40%). Low-moisture textured proteins generally require rehydration with additional additives, which is not necessary for high-moisture extruded materials. The main functional characteristics of extruded textured protein include density, size, shape, and oil/water absorption. These qualities are influenced by the initial feed material, secondary cutting, die cutting, and extrusion conditions [57].

12.5.2.1 TEXTURIZED VEGETABLE PROTEINS

Extrusion is frequently used to transform protein-rich (50–70%) plant materials into fibrous products. It is a thermomechanical procedure that involves mechanical shear, heat, and pressure [58]. Pea protein concentrate/isolate, soy protein concentrate/isolate, defatted soybean meal, peanut protein, and wheat gluten are primary plant protein ingredients currently used in extrusion [58]. The texturization of plant proteins is the process of fabricating layered and cross-linked protein mass by rearranging protein complexes. The fabricated protein mass is resistant to disruption during processing conditions. The denatured proteins rearrange themselves in the flow direction as they move along the screw, resulting in different cross-linking due to exposed bonding sites. As a result of cross-linking, globular plant proteins become increasingly texturized and modified into a fibrous and laminar structure like meat. Water evaporation occurs due to high temperatures and pressure as the product leaves the extruder's outlet, causing material expansion and taking on the final puffed appearance [57]. The shape and texture of the finished product are substantially influenced by the design of the die. The volatile bioactives and trace nutrients are added post extrusion to the final products due to the potential denaturation of vitamins and minerals [57].

12.5.2.2 IMPACT OF PROTEIN TEXTURIZATION ON COLOR CHANGES

The Maillard reaction determines the color of the extrudate in a high moisture extrusion process [59]. The extrudate's color turns darker during the extrusion due to enhanced Maillard reactions occurring due to high temperature or synchronous shear [60]. The addition of sodium chloride promotes the formation of dark color by increasing the Maillard reaction rate and affecting water distribution by decreasing the materials' viscosity. Peng et al. [61] added sodium chloride in the texturization of wheat gluten to obtain a darker extrudate.

The impact of SME on the color profile of texturized soy protein during high moisture extrusion was explored [48]. The increase in SME raised redness (a^*) and significantly decreased the lightness (L^*), indicating that higher SME produces darker extrudates. This outcome is attributed to the high temperature and frictional heat produced at higher SME, which enhances the intensity of the Maillard reaction. Siaotong [62] investigated the impact of processing variables on the color profile of flax fiber-polyethylene composites. The study indicated that higher mechanical energy could result in darkened extrudates. Another study discovered that SME had a negative influence on the lightness (L^*) but a positive impact on redness (a^*) in a study on color development utilizing a rice-glucose-lysine mixture in an extrusion-cooked system [63]. Many studies observed similar results, demonstrating that higher SME results in darker extrudates due to the Maillard reaction [64, 65].

The moisture content is another crucial factor that impacts the color of the textured protein. In research, the combined impact of moisture content and temperature on texturized pea protein was investigated [66]. The lightness (L^*) increased with the increasing moisture content at a constant temperature, indicating that the lightness (L^*) of the extrudate is highly associated with the moisture content. In contrast, the increase in moisture content at a constant temperature decreased the redness (a^*) as well as the yellowness (b^*) of the texturized pea protein. This study indicates that the moisture content significantly impacts the color profile of texturized protein more than the extrusion temperature.

12.5.3 PULSE-BASED PRODUCTS

Pulses such as beans, chickpeas, and lentils are a good source of protein, containing high amounts of essential amino acids [67]. In addition to being a vital protein source, they are also rich in vitamin B, iron, zinc, and other micronutrients. Extrusion is widely used to produce pulse and pulse-based products, as it can enhance protein digestibility, solubility, and quality by modifying the tertiary and quaternary structure [16]. Furthermore, compared to traditional cooking and baking methods, extrusion helps preserve the protein content [68].

The color of pulse extrudates is influenced by the processing variables and composition of the raw material through the Maillard reaction [69]. Luo et al. [70] discovered that injecting nitrogen during extrusion increased the lightness (L^*) compared to standard extrusion, resulting in lighter extrudates with an increase in nitrogen injection pressure. Research on lentils and horse gram-based extrudates indicated that the L^* value increased with high extrusion temperature and low feed moisture [71]. However, for pea extrudates, the extrusion temperature did not affect the color, and lower feed moisture led to darker and more brownish extrudates [4]. These studies suggest that the influence of processing variables may vary depending on the material and its composition.

12.6 EFFECT OF EXTRUSION PROCESSING ON THE FUNCTIONALITY OF NATURAL COLORANTS

Evolving consumer preferences increasingly favor the adoption of natural ingredients, including colorants, by food manufacturers across a broad

TABLE 12.1 Studies on Extrusion Parameters Affecting the Color Profile of Extrudates

Extruded Product	Type of Extruder	Major Base Material Used	Determinant Parameters	Results	References
Ready to eat breakfast cereal.	Co-rotating twin screw extruder.	Whole wheat flour.	Feed moisture, temperature of barrel zones, and whole grain wheat flour:corn flour ratio.	Presence of whole wheat grain flour and bran reduced the L^* value. Degradation of pigments like carotene due to high temperature caused the L^* value to increase. High temperature and low moisture increased the redness (a^*) and increasing bran level negatively impacted the yellowness (b^*) of the extrudate.	[72]
Breakfast cereal	Counter-rotating twin-screw extruder.	Maize	Carotenoid content	High temperature and low moisture led to carotenoid (zeaxanthin and lutein) loss. The yellowness (b^*) of the extruded product had a strong positive correlation with carotenoid content.	[73]
Mango bars	Single screw extruder.	Dried mango powder	Ingredient formulation, barrel temperature, and feed moisture.	Higher feed moisture decreased the brightness (L^* value). Xanthan gum and barrel temperature displayed a positive impact on the redness (a^*). Moisture level and glycerin had a positive influence on the b^* value, while the barrel temperature had a negative quadratic effect.	[45]
Modified starch	Co-rotating twin screw extruder.	Corn and potato	Degree of cooking (temperature), pigment destruction. Ingredient formulation.	L^*, a^*, b^*, hue angle, and chroma values reduced with an increase in temperature. a^* values were higher for potato starch compared to corn starch.	[74]
Pulse based snack.	Co-rotating twin screw extruder.	Yellow pea	Feed moisture. Nitrogen injection pressure.	Increase in feed moisture from 14 to 18% had no significant impact on the color profile of extrudates. Increase in the pressure of blowing agent raised the L^*, a^*, b^*, and chroma values and decreased the hue value.	[75]

TABLE 12.1 *(Continued)*

Extruded Product	Type of Extruder	Major Base Material Used	Determinant Parameters	Results	References
Meat analog	Twin screw extruder.	Faba bean protein	Ingredient formulation and chemical composition.	Increase in Faba bean isolates resulted in dark-colored fibrous extrudates. The carbohydrate content was also found to be responsible for color development.	[76]
			Feed water content.	Increase in feed water content led to an increase in the L^* value.	
Quinoa based extrudates.	Twin screw extruder	Quinoa, oat, corn, rice, sweet potato.	Product formulation.	The quinoa formulation substituted with rice led to a higher L^* value.	[77]
			Barrel temperature.	The increase in barrel temperature resulted in darker extrudates.	
Meat analog	Intermeshing and co-rotating twin-screw extruder.	Soy protein	Screw profile-specific mechanical energy (SME)	Increase in SME led to the formation of darker extrudates, while the a^* value and total color difference value increased. However, the change in screw profile did not affect the b^* values of the fibrous extrudates.	[48]
Meat analog	Twin screw extruder	Pea protein	Amylose:amylopectin ratio	The quantity of amylose and amylopectin were responsible for obtaining higher (L^*) and yellowish extrudates (b^*). The change in amylose and amylopectin ratio showed no significant effect on the color profile of extrudates implying starch content to be a major contributing factor to color development.	[78]
Potato-maize snack.	Single screw extruder	Maize and potato	Ingredient formulation, and potato:maize ratio.	The increase in potato substitution caused the maize snacks to turn darker. The redness and yellowness of the extrudates were found to be lower compared to the control sample (without potato substitution).	[79]
Purple sweet potato snack.	Intermeshing co-rotating twin screw extrude.	Sweet potato, butterfly pea flower.	Barrel temperature	The variation in extrusion barrel temperature had no significant impact on the color profile, even though the phenolic content, antioxidant capacity, and anthocyanin contents varied significantly.	[80]

TABLE 12.1 *(Continued)*

Extruded Product	Type of Extruder	Major Base Material Used	Determinant Parameters	Results	References
Potato snack	Single screw extruder.	Potato	Potato variety and color (presence of anthocyanins).	Substitution with colored potatoes-red and purple, has given a novel visual profile.	[81]
				The raw material formulation with colored potato flours lowered the lightness L^* values.	
				The formulation from red color potato flour substitution gave the snacks an attractive and novel look.	
Rice and vegetable snack.	Single screw extruder.	Rice and vegetables	Screw speeds (80, 100, and 120 rpm).	Screw speed did not greatly affect the color properties of the snacks in comparison to the variation in additive type and level.	[82]
			Amount of additive	The darkening of extrudates was observed with the addition of vegetable pulps.	
				Color profile of the snacks was directly proportional to the natural vegetable colors.	
Breakfast cereal	Single screw extruder.	Wheat bran, oat bran, corn grits, malt.	Ingredient formulation (different levels of dehydrated naranjita pomace (DPC))	Increase in DPC content resulted in higher b^* values at all extrusion temperatures.	[83]
				Lightness (L^*) value for all the treatment conditions did not significantly differ. However, the addition of DPC has darkened the breakfast cereals.	

spectrum of food products. Food dyes derived from natural sources have garnered consumers' attention as alternatives to synthetic food dyes for a number of reasons. In recent years, natural colorants have been often utilized in the food industry. The natural colorants that are primarily utilized in food extrusion are summarized in Figure 12.3. Negative consumer perceptions of synthetic food dyes as being harmful to human and environmental health, in conjunction with trends that favor the inclusion of natural, organic, and clean-label foods, are key drivers of the demand for natural pigments [9]. Evidence of certain synthetic dyes involved in the development of allergic reactions and hyperactive behavior in children has also emerged [9, 84–86]. As a result, efforts to explore and utilize natural colorants have increased within the last decade.

Apart from providing color, pigments derived from natural sources are also potent bioactive functional compounds. They have demonstrated beneficial biological activities, especially in the context of preventing and/or managing chronic metabolic disorders [86]. The production and utilization of natural colorants may also be environmentally sustainable as they can be extracted through the valorization of agro-industrial byproducts, making them attractive for large-scale adoption [87]. However, natural colorants also present certain technological challenges that potentially limit their utility. Natural food colors often lack the color intensity and stability of synthetic dyes used in the food industry [9]. Their ability to provide the desired colors is highly sensitive to processing conditions. Their functionality is particularly diminished by neutral and acidic pH levels, exposure to radiant energy, and high temperatures [9, 84, 86]. Therefore, in order to successfully expand the application of natural colorants to extruded foods, it is essential to understand the effect of thermomechanical processing on such compounds. Thus, this subsection aims to outline the effect of extrusion on major natural colorants currently used in the food industry.

12.6.1 ANTHOCYANINS (ACNS)

Anthocyanins (ACNs) are the predominant group of vacuolar pigments that impart an array of colors ranging from orange to red, purple, pink, and blue to flowers, fruits, as well as certain dark-colored cereal grains, tubers, and legumes [88]. ACNs are phenolic compounds comprising a 15-carbon (C6-C3–C6) skeleton consisting of a heterocyclic chromane ring carrying

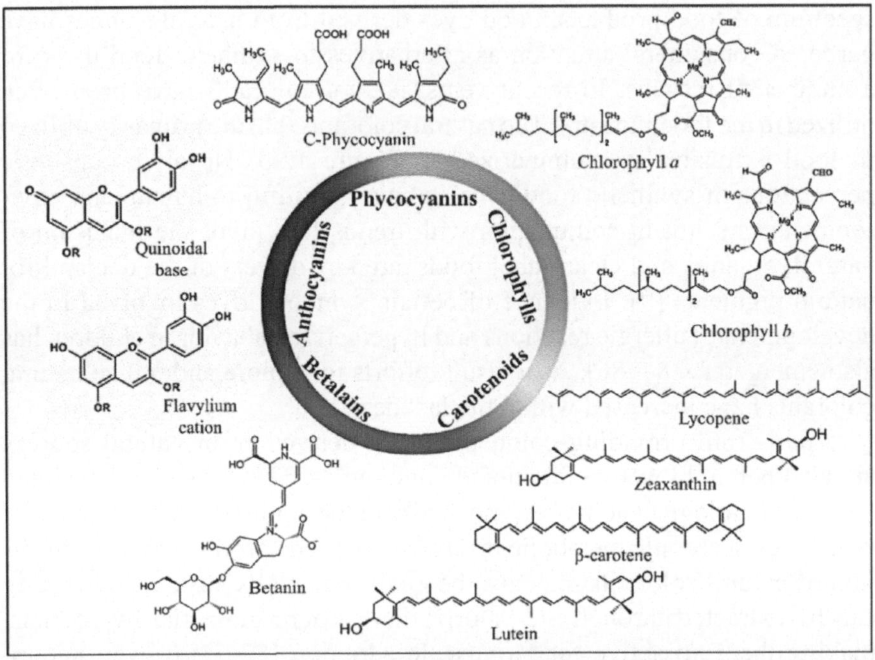

FIGURE 12.3 Major groups of natural colorants and their representative compounds.

an aromatic B-ring at the second carbon position and multiple sugar moieties attached at various hydroxylated positions. ACNs (glycones) are glycosidic forms of their corresponding anthocyanidins (aglycones). The color and hue imparted by these compounds are a function of the chemical combinations with glycosides and/or acyl groups within the basic C6–C3–C6 structure and its interaction with other molecules [9]. Variations in colors and hues of ACNs are determined by a multitude of chemical (substitution patterns, nature of esterification, and level of glycosylation) and physical factors (presence of co-pigments, pH, temperature, and solvents) [9, 84, 86, 88].

The pH of a food matrix is a crucial determinant of the suitability of ACNs as a potential food colorant. In aqueous solutions, under highly acidic conditions (pH 1 to 3), the ACNs exist as red-colored flavylium cations. In comparison, they undergo deprotonation to establish a colorless carbinol pseudo base at mildly acidic conditions (pH 5). In neutral to alkaline conditions (pH $\geq$ 6), they are transformed into a blue-purple quinoidal base [9, 88]. Sensitivity to other factors such as heat, light, and oxygen levels during processing could also limit the application of ACN-based colorants in food products [9, 86].

ACNs are known to be extremely unstable during extrusion, with reported losses ranging from 50% to 90%. The loss of color and activity is particularly exacerbated by high barrel temperatures and low feed moisture [88]. Escalante-Aburto et al. [89] reported that *a** values (red/green hues) are strongly influenced by processing temperature and feed moisture, with the highest values, especially for red, observed at low moisture contents and elevated temperature. This is remarkable, as these conditions usually correspond to higher levels of ACN degradation. In the same study, *b** values (yellow and blue hues) were unaffected by temperature, while lower *b** values (higher blue) were observed at higher levels of feed moisture. Notably, the receptiveness to degradation during extrusion and the color stability of the ACNs in the dry extrudates were relatively stable during storage [90, 91]. ACN degradation and loss of color during extrusion are largely mediated by oxidation, polymerization-browning reactions due to high temperatures, the inclusion of metal particles from screw wear, and the formation of Maillard-reaction products and sugar hydrolysis [90].

Thermomechanical degradation of ACN content and bioactivity can be minimized by reducing exposure to elevated temperature, shear stress, and increasing moisture levels. The total ACN and cyanidin-3-glucoside contents in the extrudate were only influenced by specific feed moisture levels (16% and 16.5%) and did not show any correlation with other extrusion parameters [89]. Similarly, Hirth et al. [92] discovered that higher water content (22%) improved ACN retention in high temperature and short time (HTST) extrusion of native cornstarch containing chokeberry extracts. Interestingly, the stability of other polyphenols, such as hydrocinnamic acids and proanthocyanins from the chokeberry extract, were not significantly affected by the stresses associated with extrusion. However, in the case of bilberry extract-incorporated corn starch, a combination of factors such as screw speed, barrel temperature, moisture content, and flow rate also influenced the color of the extrudates [91]. ACN retention improved with the use of high screw speeds and increased flow rates, as it resulted in reduced residence times of the dyed food material within the barrel.

Extrusion affects the retention of various anthocyanins (ACNs) differently. The content of cyanidin-3-glucoside was found to increase in extruded colored maize and corn, while this process was detrimental to the retention of pelargonidin-3-glucoside [89, 93]. Furthermore, large molecular weight procyanidin oligomers and polymers were broken down into monomers, dimers, and trimers during twin screw extrusion. The relative concentration of these low molecular weight products was influenced by barrel temperature and screw speeds [94]. Although the bioactive functionality of the extrudates improved compared to the non-extruded food material, the color was negatively impacted due to the loss of polymeric ACNs.

Optimizing process parameters to minimize shear and temperature stress during extrusion represents the simplest means of reducing thermo-mechanical degradation [95, 96]. Various other methods were investigated to improve ACN retention in extruded foods [97]. Co-pigmentation of ACNs via molecular or complex associations with compounds like organic acids, metals, polysaccharides, and proteins has shown significant color protective effects [97, 98]. Durge et al. [99] found that adding 1% CA raised the ACN retention to 18.2% in extruded rice flour, potentially reducing the amount of pre-extrusion coloring required by 25%, without detrimentally affecting the organoleptic properties of the extrudate. In the same research, adding sodium carbonate to the feed mixture reduced ACN retention in the RE. This can be ascribed to the increased stability of flavylium cation responsible for intense red coloration under acidic conditions. Similarly, in extruded and nixtamalized blue maize flours, color intensity, antioxidant activity, ACN, and total phenolic contents demonstrated a negative correlation with the concentration of calcium hydroxide, while the opposite was observed with the concentration of calcium lactate [93]. Extrusion resulted in greater retention of ACN, total phenolic contents, and antioxidant activity relative to nixtamalization.

The incorporation of ACNs in food composites containing protective biopolymers may also help enhance the stability and retention of functional bioactive compounds. For example, antioxidant activities and retention of total phenolic and ACN were significantly improved in value-added food composites (VAFC) containing rutin-enriched buckwheat and ACN-enriched potato, vitamin E, and soy lecithin during hot melt extrusion [100].

Multiple studies concurred that the stability of ACNs in dry extrudates was unaffected in long-term storage [90, 91]. Beneficial bioactive properties of phenolic compounds were unaffected significantly post-extrusion. However, ACN degradation and the resulting loss in color stability still pose a challenge to adopting these natural colorants in extruded foods [101].

12.6.2 CAROTENOIDS

Carotenoids are plastidial, lipophilic pigments found in higher plants, bacteria, fungi, micro- and macroalgae, yeasts, insects, and birds. Chemically, they are classified as tetraterpenoids and are further subdivided into two main groups: (i) carotenes, which contain exclusively polyunsaturated hydrocarbons (e.g., lycopene, α-carotene, and β-carotene); and (ii) xanthophylls, which contain oxygenated polyunsaturated hydrocarbon chains (e.g., β-cryptoxanthin, violaxanthin, astaxanthin, lutein, and zeaxanthin) [9, 84]. The long chains of isoprenoid units in carotenoids give rise to extensive π-electron-rich conjugation systems, which enable them to absorb radiation in the visible spectrum, resulting in red, orange, and yellow colors. The difference in colors results from varying sizes of the conjugation systems and the occurrence of cyclization within the molecular structure [9, 102].

Carotenoids provide functional health benefits to consumers while also playing an essential role in maintaining normal functions of human physiology. For instance, vitamin A (retinol), which is derived from dietary carotenoids and bio-transformed within the human body, is essential for various metabolic functions such as vision and immunity. Lycopene exhibits significant antioxidant activity and reduces the threat of chronic diseases linked to redox imbalance and oxidative stress [102, 103].

The high degree of unsaturation in their chemical structures makes carotenoids particularly prone to degradation by oxidation and isomerization during processing and storage. The oxidation degree is affected by factors such as exposure to light, temperature, moisture, enzymes, peroxides, metals, lipids, and antioxidants [9]. As an oil-soluble compound, carotenoids are more sensitive to thermal stress during extrusion than water-soluble vitamins [88].

The stability of vitamin A and its provitamins during extrusion has been studied extensively. The provitamin β-carotene is relatively more vulnerable to degradation than retinol, with retention rates in extruded feed mixtures reported at 25% and 50%, respectively [104]. The addition of quinoa or chia flours to corn flour was found to reduce the overall loss of naturally occurring carotenoids, regardless of whether the additives were significant sources of carotenoids [105]. This indicates the benefits of using composite flours or feeds as an effective means of resisting carotenoid degradation.

Carotenoids exhibit varying levels of stability or susceptibility to degradation. Xanthophylls (lutein and zeaxanthin) exhibited greater stability than carotenes during extrusion [106]. Likewise, in extrudates prepared from a blend of corn with pumpkin tissue flour, amaranth seeds, and Jerusalem artichoke, lutein was found to be more stable than lycopene and β-carotene [107].

The effect of extrusion temperature on carotenoid retention varies according to the applied shear force, moisture levels, and feed formulation. Carotenoids are vulnerable to the effects of temperature, and their stability and retention were found to decrease at elevated temperatures due to oxidation and isomerization from the more stable *trans*-forms to *cis*-forms [102]. Moisture content is a key determinant of pigment retention, with higher levels favoring free and encapsulated lycopene retention. At high moisture levels, the feed materials' viscosity and temperature are also reduced, thereby minimizing the thermomechanical degradation of carotenoids. The effect of shear on carotenoid stability also depends on the mode of extrusion under consideration; shear effects due to increased viscosity and friction during cold extrusion are expected to have a more pronounced detrimental effect compared to hot extrusion [102]. Also, oil media containing high levels of unsaturated fatty acids were found to favor carotenoid degradation via the generation of oxidizing species, including peroxides and free radicals [102].

Protective techniques, such as coacervation and microencapsulation using polymers like chitosan, liposomes, β-cyclodextrin matrices, maltodextrin, pullulans, as well as emulsification, have been shown to significantly increase the stability, retention, and bioactivity of carotenoids post-extrusion [102, 108]. The incorporation of oil-soluble bioactives into a protein-carbohydrate matrix has also demonstrated positive effects on their stability [109]. Including potent hydrophilic antioxidants (polyphenols, flavonoids, vitamin C) and lipophilic antioxidants (tocopherols, vitamins E) can also improve the stability and, therefore, the retention of carotenoids [102]. Additionally, using additives like gums and pectin to modulate the rheological properties of the extruded mixtures by reducing residence time and promoting fluidity may also reduce carotenoid loss [110].

While carotenoids are vulnerable to degradation during extrusion, this processing could also potentially improve their bioaccessibility. Ortak et al. [111] found a 48% decrease in lutein during the extrusion of a 1:4 w/w mixture of carrot pulp-corn grit, while the dialyzable lutein content concurrently increased by 70%. Furthermore, Tonyali et al. [112] discovered that the bioaccessibility of provitamin A is dependent on the barrel temperature during extrusion.

12.6.3 BETALAINS

Betalains are water-soluble, heterocyclic compounds containing nitrogen, which are ammonium derivatives of betalamic acid. Two major groups of betalains are recognized depending on the type of compound with which betalamic acid undergoes condensation reactions: (i) betacyanins (betalamic acid + *cyclo*-Dopa), which impart red-violet and deep crimson hues; and (ii) betaxanthins (betalamic acid + amines), which impart yellow to orange hues. They are predominantly seen in plants belonging to the order Caryophyllales and are relatively uncommon in nature [86, 102].

Between the pH range of 3–7, betalains are stable, making them suitable for neutral and mildly acidic foods such as yogurt and ice cream. They are often used in conjunction with ACNs [102]. Betanin, a natural colorant derived from red beetroot (*Beta vulgaris*), is the only betalain-based pigment approved for food applications by the Food and Drug Administration (FDA). It is mainly used as a replacement for the artificial pigment Red 40 [86]. Other potential sources of betalains include *Opuntia* species (prickly pear), *Hylocereus* (dragon fruit), *Eulychnia* species of cacti, *Ravina humilis* (pigeonberry plant), *Chenopodium quinoa* (quinoa), tubers of *Ullucustuberosus* (Ulluco), and certain fungi such as *Agaricus muscaria* (fly agaric) and certain genera of *Basidiomycetes* [9, 113].

Betalain retention and stability during processing are influenced by temperature, oxygen, light, and water activity. At high pH levels and temperatures, their aldimine bonds are hydrolyzed, yielding betalamic acid, which results in the formation of undesired yellow hues [102]. They impart an unappealing earthy taste to food products when used at high concentrations. However, they have advantages over ACNs due to their higher tinctorial strength, neutral pH stability, and water solubility [9, 86, 102]. Findings on the health benefits of betalains in animal and *in vitro* models have highlighted various biological effects, which include antioxidant, anticarcinogenic, hypolipidemic, hypotensive, and antimicrobial properties [114, 115]. Human studies, while being limited in number, have mainly focused on exploring the potential of nutritional supplements containing betalains to help improve athletic performance based on their nitrate content [114, 115].

The color of betalain changes from red to yellow under alkaline conditions and high temperatures [9]. Ruiz-Gutiérrez et al. [116] reported a loss of betalain content (36%–54%) during the extrusion of encapsulated red cactus pear powder (2.5%) and corn starch. However, increasing screw speeds minimized the residence time, improving the retention of red color (a^*). Additionally, higher temperatures increased the intensity of red color in the extrudate, primarily due to increased formation of Maillard reaction byproducts rather than the retention of betalains. While both groups of betalains are negatively affected by high temperatures, betaxanthins have been found to be more susceptible. Elevated temperatures reduce the stability and content of betalains, but changes in material flow properties under these conditions lower the exposure time, inhibiting further degradation [108].

Durge et al. [117] found that an increase in screw speed and moisture content of the feed materials could enhance the retention of red beetroot color. Conversely, higher die temperature had the opposite effect. Additionally, the addition of CA (1%) to the feed reduced the color requirement by 40% by acting as a pre-extrusion colorant. Furthermore, the use of additives such as antioxidants, chelating agents, preservatives, and gums has also been shown to improve betalain stability in foods [9].

12.6.4 PHYCOCYANINS

Phycocyanins are blue pigments found naturally in cyanobacteria, rhodophytes, and cryptophytes [118]. They are preferred over ACNs to impart vibrant blue hues to low acid and neutral foods and beverages within the pH range of 5–7.5. Phycocyanin derived from *Spirulina* (*Arthrospira platensis*) phycobiliproteins has been certified by the FDA as generally recognized as safe (GRAS) and is the only naturally occurring blue colorant permitted for use in the USA, Europe, and Asia [86].

Like other natural pigments, phycocyanins demonstrate free-radical scavenging, anti-inflammatory, antiviral, anticancer, antioxidant activity, and hypolipidemic effects. However, investigations into these effects have thus far been restricted to *in vitro* and animal models. In human clinical studies, dietary supplements containing phycocyanins have shown potential as functional therapeutic agents for alleviating chronic pain [86].

Thus far, studies focusing on the effect of extrusion on purified phycocyanins have been lacking, while extruded products containing whole *Spirulina* extracts have been studied. Tańska et al. [119] found extrusion to be detrimental to the retention of algal pigments, including phycocyanins, in extruded corn grit containing *Spirulina* powder, with only trace amounts of the pigment found in the extrudates. Notably, they found that the addition of baking powder to the feed mixture demonstrated a limited protective effect on the color and phycocyanin content of the extrudate. Microencapsulation of microalgal phycocyanins within carbohydrates like alginate and chitosan could potentially be used to improve their stability during extrusion [120, 121]. Other investigators noted that the use of *Spirulina* extracts in extruded foods resulted in dark green hues due to the combined effects of pigments phycocyanins, β-carotene, xanthophylls, and chlorophyll, and resulted in improved total antioxidant activity [122, 123].

12.6.5 CHLOROPHYLL

Chlorophylls are photosynthetic pigments that are found widely in algae, plants, and cyanobacteria. They consist of a characteristic porphyrin macrocycle, which includes four pyrrole rings and a central magnesium ligand, and are classified as tetrapyrroles. Based on the functional groups associated with carbon atoms, six types of chlorophyll (chlorophyll a, b, c1, c2, d, f) have been identified [8].

The role of chlorophyll as a functional bioactive ingredient has attracted interest, and it has been reported to have various health benefits such as anticancer properties, inhibition of hyperproliferative and heme-induced cytotoxic effects, as well as antimutagenic and anti-proliferative activities [103].

During food processing operations involving acidic conditions and high temperatures, chlorophyll molecules are susceptible to degradation, resulting in the loss of magnesium atoms and the formation of brown-colored pyropheophytins [103]. The understanding of the specific effects of extrusion on chlorophyll retention and bioactivity is far from adequate. An & Ryu [124] found that the contents of chlorophylls *a* and *b* in extruded rice containing green tea mix were significantly reduced as a result of both conventional and cold extrusion processing. On the other hand, extrusion was found to be significantly more efficient in high pressure in terms of retaining pigments, including chlorophyll *a* and *b,* in hemp seed oil [125].

12.7 SUMMARY

Understanding the influence of extrusion parameters on the color profile is crucial, as color serves as a primary quality indicator for extrudates. The Maillard reaction, caramelization, and pigment degradation are significant chemical pathways that determine the color of food products. Since these reactions are particularly influenced by temperature and moisture, the processing conditions during extrusion have a significant impact on the color of the extrudate. Overall, studies on REs suggest that increasing moisture content can reduce the L^* value and increase the a^* and b^* values. In texturized proteins, an increase in SME can reduce the L^* value and increase redness, indicating that higher SME produces darker extrudates. However, it is difficult to generalize as the color can vary based on the composition of raw materials, material properties, polymer concentrations, and operating conditions.

With the expected increase in consumer preference and demand for clean-label food products in this decade and beyond, the use of naturally sourced pigments in food products is likely to become more widely adopted across the food industry. Specifically, their potential in extruded foods is significant, as indicated by various studies mentioned in the previous section. Certain natural colorants derived from vegetables and fruits have already been utilized to dye commercially available extruded products such as pasta and breakfast cereal. Moreover, extrusion, being an extremely versatile processing technique, can be extended to create innovative functional foods that contain natural pigments. These pigments have the potential to provide health benefits by enhancing the bioavailability and bioactivity of these compounds. However, despite their advantages, natural pigments also present technical challenges that may limit their utility. Therefore, it is crucial to make efforts to optimize extrusion parameters and ingredient formulation to maximize color retention and the health-improving properties of these pigments in extruded foods.

KEYWORDS

- **color**
- **extrusion parameters**
- **generally recognized as safe**
- **natural colorants**
- **pulses**
- **rice**
- **rice extrudate**
- **texturized proteins**
- **value-added food composites**
- **water absorption index**
- **water solubility index**

REFERENCES

1. Pichmony, E., & Ganjyal, G. M., (2020). Basics of extrusion processing. In: Ganjyal, G. M., (ed.), *Extrusion Cooking* (pp. 1–28). Wood head Publishing. https://doi.org/10.1016/B978-0-12-815360-4.00001-8.
2. Shurtleff, W., & Aoyagi, A., (2020). *History of Extrusion Cooking and Extruders (1938–2020): Extensively Annotated Bibliography and Sourcebook.* Soy info Center.
3. Zhang, B., Liu, G., Ying, D., Sanguansri, L., & Augustin, M. A., (2017). Effect of extrusion conditions on the physico-chemical properties and in vitro protein digestibility of canola meal. *Food Research International*, *100*, 658–664. https://doi.org/10.1016/J.FOODRES.2017.07.060.
4. Saldanha Do, C. C., Varela, P., Poudroux, C., Dessev, T., Myhrer, K., Rieder, A., Zobel, H., et al., (2019). The impact of extrusion parameters on physicochemical, nutritional, and sensorial properties of expanded snacks from pea and oat fractions. *LWT*, *112*, 108252. https://doi.org/10.1016/J.LWT.2019.108252.
5. Martins, N., Roriz, C. L., Morales, P., Barros, L., & Ferreira, I. C. F. R., (2016). Food colorants: Challenges, opportunities, and current desires of agro-industries to ensure

consumer expectations and regulatory practices. *Trends Food Sci. Technol.*, *52*, 1–15. https://doi.org/10.1016/J.TIFS.2016.03.009.

6. Rokey, G. J., Plattner, B., & De Souza, E. M., (2010). Feed extrusion process description. *Revista Brasileira de Zootecnia*, *39*(Suppl. 1), 510–518. https://doi.org/10.1590/S1516-35982010001300055.
7. Zhang, J., Liu, L., Liu, H., Yoon, A., Rizvi, S. S., & Wang, Q., (2019). Changes in conformation and quality of vegetable protein during texturization process by extrusion. *Crit. Rev. Food Sci. Nutr.*, *59*(20), 3267–3280.
8. Ghosh, S., Sarkar, T., Das, A., & Chakraborty, R., (2022). Natural colorants from plant pigments and their encapsulation: An emerging window for the food industry. *LWT*, *153*, 112527. https://doi.org/10.1016/J.LWT.2021.112527.
9. Sigurdson, G. T., Tang, P., & Giusti, M. M., (2017). Natural colorants: Food colorants from natural sources. *Annu. Rev. Food Sci. Technol.*, *8*, 261–280. https://doi.org/10.1146/ANNUREV-FOOD-030216-025923.
10. Neder-Suárez, D., Lardizabal-Gutiérrez, D., Zazueta-Morales, J. De J., Meléndez-Pizarro, C. O., Delgado-Nieblas, C. I., Wong, B. R., Gutiérrez-Méndez, N., et al., (2021). Anthocyanins and functional compounds change in a third-generation snacks prepared using extruded blue maize, black bean, and chard: An optimization. *Antioxidants*, *10*(9), 1368. https://doi.org/10.3390/ANTIOX10091368.
11. Navale, S. A., Thakor, N. J., & Swami, S. B., (2015). Extrusion cooking technology for foods: A review. *Journal of Ready to Eat Food*, *2*(3), 66–80.
12. Uitterhaegen, E., & Evon, P., (2017). Twin-screw extrusion technology for vegetable oil extraction: A review. *J. Food Eng.*, *212*, 190–200. https://doi.org/10.1016/J.JFOODENG.2017.06.006.
13. Adekola, K. A., (2016). Engineering review food extrusion technology and its applications some of the authors of this publication are also working on these related projects: Studies in food extrusion die view project engineering review food extrusion technology and its application. *J. Food Sci. Eng.*, *6*, 149–168. https://doi.org/10.17265/2159-5828/2016.03.005.
14. Riaz, M. N., (2019). Food extruders. In: Kutz, M., (ed.), *Handbook of Farm, Dairy and Food Machinery Engineering* (pp. 483–497). Academic Press. https://doi.org/10.1016/B978-0-12-814803-7.00019-1.
15. Huber, G. R., (2000). Twin-screw extruders. In: *Extruders in Food Applications* (pp. 81–114).
16. Pasqualone, A., Costantini, M., Coldea, T. E., & Summo, C., (2020). Use of legumes in extrusion cooking: A review. *Foods*, *9*(7), 958. https://doi.org/10.3390/FOODS9070958.
17. Choton, S., Gupta, N., Bandral, J. D., Anjum, N., & Choudary, A., (2020). Extrusion technology and its application in food processing: A review. *Pharma Innov.*, *9*(2), 162–168. https://doi.org/10.22271/tpi.2020.v9.i2d.4367.
18. Ragaee, S., Seetharaman, K., El-Sayed, & Abdel-Aal, M., & Abdel-Aal, E. S. M., (2014). The impact of milling and thermal processing on phenolic compounds in cereal grains. *Crit. Rev. Food Sci. Nutr.*, *54*(7), 837–849. https://doi.org/10.1080/10408398.2011.610906.

19. Gümüşay, Ö. A., & Şeker, M., (2021). Effects of extrusion parameters on physicochemical properties of third generation corn snacks expanded by microwave heating. *J. Food Process Preserv.*, *45*(7), 45. https://doi.org/10.1111/JFPP.15630.
20. Prabha, K., Ghosh, P., Abdullah, S., & Joseph, R. M., (2021). Recent development, challenges, and prospects of extrusion technology. *Future Foods*, *3*, 100019.
21. Starowicz, M., & Zieliński, H., (2019). How Maillard reaction influences sensorial properties (color, flavor and texture) of food products? *Food Reviews International*, *35*(8), 707–725. https://doi.org/10.1080/87559129.2019.1600538.
22. Leyva-Corral, J., Quintero-Ramos, A., Camacho-Dávila, A., Zazueta-Morales, J. De. J., Aguilar-Palazuelos, E., Ruiz-Gutiérrez, M. G., Meléndez-Pizarro, C. O., et al., (2016). Polyphenolic compound stability and antioxidant capacity of apple pomace in an extruded cereal. *LWT – Food Science and Technology, 65*, 228–236. https://doi.org/10.1016/J.LWT.2015.07.073.
23. Alam, M. S., Kaur, J., Khaira, H., & Gupta, K., (2016). Extrusion and extruded products: Changes in quality attributes as affected by extrusion process parameters: A review. *Crit. Rev. Food Sci. Nutr.*, *56*(3), 445–473. https://doi.org/10.1080/10408398.2013.779568.
24. Ye, J., Hu, X., Luo, S., Liu, W., Chen, J., Zeng, Z., & Liu, C., (2018). Properties of starch after extrusion: A review. *Starch-Stärke*, *70*(11, 12), 1700110. https://doi.org/10.1002/STAR.201700110.
25. Waramboi, J., Gidley, M., & Sopade, P., (2014). Influence of extrusion on expansion, functional and digestibility properties of whole sweet potato flour. *LWT – Food Science and Technology*, *59*(2), 1136–1145.
26. Tas, A., & Shah, A., (2021). The replacement of cereals by legumes in extruded snack foods: Science, technology, and challenges. *Trends Food Sci. Technol.*, *116*, 701–711.
27. Martin, A., Osen, R., Karbstein, H. P., & Emin, M. A., (2021). Impact of rapeseed press cake on the rheological properties and expansion dynamics of extruded maize starch. *Foods*, *10*(3), 616.
28. Verma, T., & Subbiah, J., (2020). Use of residence time versus screw speed in the response surface model for microbial inactivation during single-screw extrusion of low-moisture food. *Food Control*, *115*, 107293.
29. Sandrin, R., Caon, T., Zibetti, A. W., & De Francisco, A., (2018). Effect of extrusion temperature and screw speed on properties of oat and rice flour extrudates. *J. Sci. Food Agric.*, *98*(9), 3427–3436. https://doi.org/10.1002/JSFA.8855.
30. Dalbhagat, C. G., Mahato, D. K., & Mishra, H. N., (2019). Effect of extrusion processing on physicochemical, functional and nutritional characteristics of rice and rice-based products: A review. *Trends Food Sci. Technol.*, *85*, 226–240. https://doi.org/10.1016/J.TIFS.2019.01.001.
31. Hirunyophat, P., Chalermchaiwat, P., On-nom, N., & Prinyawiwatkul, W., (2022). Selected physicochemical properties and sensory acceptability as affected by addition of lecithin and calcium carbonate in extruded breakfast cereals made with silkworm pupae powder and rice flour. *Int. J. Food Sci. Technol.*, *57*(1), 631–642. https://doi.org/10.1111/IJFS.15356.
32. Yousf, N., Nazir, F., Salim, R., Ahsan, H., Sirwal, A., Yousf, N., Nazir, F., et al., (2017). Water solubility index and water absorption index of extruded product from rice and carrot blend. *J. Pharmacogn. Phytochem.*,.*6*(6), 2165–2168.

33. Pezzali, J. G., & Aldrich, C. G., (2019). Effect of ancient grains and grain-free carbohydrate sources on extrusion parameters and nutrient utilization by dogs. *J. Anim. Sci.*, *97*(9), 3758–3767. https://doi.org/10.1093/JAS/SKZ237.
34. Charles, A. P. R., Jin, T. Z., Mu, R., & Wu, Y., (2021). Electro hydrodynamic processing of natural polymers for active food packaging: A comprehensive review. *Compr. Rev. Food Sci. Food Saf.*, *20*(6), 6027–6056. https://doi.org/10.1111/1541–4337.12827.
35. Charles, A. P. R., Mu, R., Jin, T. Z., Li, D., Pan, Z., Rakshit, S., Cui, S. W., & Wu, Y., (2022). Application of yellow mustard mucilage and starch in nanoencapsulation of thymol and carvacrol by emulsion electrospray. *Carbohydr Polym.*, 120148. https://doi.org/10.1016/j.carbpol.2022.120148.
36. Hussain, S. Z., Singh, B., Rather, A. H., & Qazi, N. A., (2014). Study on color coordinates and Maillard precursors of re-fabricated rice. *Appl Biol Res.*, *16*(2), 176. https://doi.org/10.5958/0974-4517.2014.00008.1.
37. Nikmaram, N., Leong, S. Y., Koubaa, M., Zhu, Z., Barba, F. J., Greiner, R., Oey, I., & Roohinejad, S., (2017). Effect of extrusion on the anti-nutritional factors of food products: An overview. *Food Control*, *79*, 62–73. https://doi.org/10.1016/J.FOODCONT.2017.03.027.
38. Singh, S., Gamlath, S., & Wakeling, L., (2007). Nutritional aspects of food extrusion: A review. *Int. J. Food Sci. Technol.*, *42*(8), 916–929. https://doi.org/10.1111/J.1365-2621.2006.01309.X.
39. Zhang, Z., Zhang, L., He, S., Li, X., Jin, R., Liu, Q., Chen, S., & Sun, H., (2022). High-moisture extrusion technology application in the processing of textured plant protein meat analogues: A review. *Food Reviews International*. https://doi.org/10.1080/87559129.2021.2024223.
40. Agarwal, S., Singh, C. E., & Ekta, S. C. C., (2019). *Extrusion Processing: The Effect on Nutrients and Based Products,8*(4), 464–470. thepharmajournal.com (accessed on 23 January 2024).
41. Robin, F., Schuchmann, H. P., & Palzer, S., (2012). Dietary fiber in extruded cereals: Limitations and opportunities. *Trends Food Sci. Technol.*, *28*(1), 23–32. https://doi.org/10.1016/J.TIFS.2012.06.008.
42. Chen, H., Zhao, C., Li, J., Hussain, S., Yan, S., & Wang, Q., (2018). Effects of extrusion on structural and physicochemical properties of soluble dietary fiber from nodes of lotus root. *LWT*, 204–211.
43. Gajula, H., Alavi, S., Adhikari, K., & Herald, T., (2008). Precooked bran-enriched wheat flour using extrusion: Dietary fiber profile and sensory characteristics. *J. Food Sci*, *73*(4). https://doi.org/10.1111/J.1750-3841.2008.00715.X.
44. Redgwell, R. J., Curti, D., Robin, F., Donato, L., & Pineau, N., (2011). Extrusion-induced changes to the chemical profile and viscosity generating properties of citrus fiber. *J. Agric. Food Chem.*, *59*(15), 8272–8279. https://doi.org/10.1021/JF201845B.
45. Salgado, N., Giraldo, G. I., & Orrego, C. E., (2017). Influence of the extrusion operating conditions on the antioxidant, hardness, and color properties of extruded mango. *LWT*, *86*, 209–218. https://doi.org/10.1016/J.LWT.2017.07.049.
46. Zhang, T., Dou, W., Zhang, X., Zhao, Y., Zhang, Y., Jiang, L., & Sui, X., (2021). The development history and recent updates on soy protein-based meat alternatives. *Trends Food Sci. Technol.*, *109*, 702–710. https://doi.org/10.1016/J.TIFS.2021.01.060.

47. Lei, H., Fulcher, R. G., Ruan, R., & Van, L. B., (2007). Assessment of color development due to twin-screw extrusion of rice–glucose–lysine blend using image analysis. *LWT – Food Science and Technology,40*(7), 1224–1231. https://doi.org/10.1016/J.LWT.2006.08.016.
48. Fang, Y., Zhang, B., & Wei, Y., (2014). Effects of the specific mechanical energy on the physicochemical properties of texturized soy protein during high-moisture extrusion cooking. *J. Food Eng.*, *121*(1), 32–38. https://doi.org/10.1016/J.JFOODENG.2013.08.002.
49. Bobade, H., & Sharma, S., (2017). Effect of extrusion on color characteristics of honey enriched whole grain cereal flour extrudates. *International Journal of Agriculture Engineering*, *10*(1), 37–42. https://doi.org/10.15740/HAS/IJAE/10.1/37-42.
50. Borah, A., Lata, M. C., & Kalita, D., (2016). Optimization of process parameters for extrusion cooking of low amylose rice flour blended with seeded banana and carambola pomace for development of minerals and fiber rich breakfast cereal. *J. Food Sci. Technol.*, *53*(1), 221–232. https://doi.org/10.1007/S13197-015-1772-9.
51. Jakkanwar, S. A., Rathod, R. P., & Annapure, U. S., (2018). Development of cowpea-based (*Vigna unguiculata*) extruded snacks with improved in vitro protein digestibility. *Int. Food Res. J.*, *25*(2), 804–813.
52. Da Silva, T. C., Da Silva, E. M. M., Chávez, D. W. H., De Carvalho, C. W. P., & Ascheri, J. L. R., (2017). Effects of whey protein concentrate, feed moisture and temperature on the physicochemical characteristics of a rice-based extruded flour. *Food Chem.*, *228*, 287–296.
53. Gat, Y., & Ananthanarayan, L., (2016). Use of paprika oily extract as pre-extrusion coloring of rice extrudates: Impact of processing and storage on color stability. *J. Food Sci. Technol.*, *53*(6), 2887–2894. https://doi.org/10.1007/S13197-016-2271-3.
54. Tepsongkroh, B., Jangchud, K., Jangchud, A., Charunuch, C., & Prinyawiwatkul, W., (2019). Healthy brown rice-based extrudates containing straw mushrooms: Effect of feed moisture and mushroom powder contents. *J. Food Process. Preserv.*, *43*(9), e14089. https://doi.org/10.1111/JFPP.14089.
55. Mehraj, M., Naik, H. R., Reshi, M., & Mir, S. A., (2018). Development and EV development and evaluation of extruded production of extruded product of rice flour and apple Pomace. *Development,13*(1), 21–26.
56. Kinsella, J. E., & Franzen, K. L., (1978). Texturized proteins: Fabrication, flavoring, and nutrition. *Crit. Rev. Food Sci. Nutr.,10*(2), 147–207. https://doi.org/10.1080/10408397809527248.
57. Pam, I. B., Senaratne-Lenagala, L., Stube, A., Brackenridge, A., Ismail, B. P., Senaratne-Lenagala, L., Stube, A., et al., (2020). Protein demand: Review of plant and animal proteins used in alternative protein product development and production. *Animal Frontiers*, *10*(4), 53–63. https://doi.org/10.1093/AF/VFAA040.
58. Kyriakopoulou, K., Dekkers, B., & Van, D. G. A. J., (2019). Plant-based meat analogues. *Sustainable Meat Production and Processing*, 103–126.
59. Zahari, I., Ferawati, F., Helstad, A., Ahlström, C., Östbring, K., Rayner, M., & Purhagen, J. K., (2020). Development of high-moisture meat analogues with hemp and soy protein using extrusion cooking. *Foods*, *9*(6), 772. https://doi.org/10.3390/FOODS9060772.

60. Maurya, A. K., & Said, P. P., (2014). Extrusion processing on physical and chemical properties of protein rich products: An overview. *Journal of Bioresource Engineering and Technology*, *2*(4), 61–67.
61. Peng, J., Zhu, K., Guo, X., Peng, W., Chen, Y., Li, Q., & Zhou, H., (2021). NaCl mediated physicochemical and structural changes of textured wheat gluten. *LWT*, *140*, 110704. https://doi.org/10.1016/J.LWT.2020.110704.
62. Siaotong, B. A. C., (2006). *Effects of Fiber Content and Extrusion Parameters of Falx Fiber – Polyethylene Composites*. University of Saskatchewan, Saskatoon.
63. Lei, H., Ruan, R., Fulcher, R. G., & Van, L. B., (2008). Color development in an extrusion-cooked model system. *International Journal of Agricultural and Biological Engineering*, *1*(2), 55–63. https://doi.org/10.3965/j.issn.1934-6344.2008.02.055-063.
64. Kang, L. N., (2007). *Texturization Technology and Mechanism of Soybean Protein by High Moisture Extrusion*. Northwestern University of Agricultural and Forestry Science and Technology.
65. Rausch, D. T., (2009). *Influence of Extrusion Parameters and Recipe Compounds on Flavor Formation and Its Detection and Quantification*. Technical University of Berlin, Berlin.
66. Zhang, J., Liu, L., Jiang, Y., Faisal, S., & Wang, Q., (2020). A new insight into the high-moisture extrusion process of peanut protein: From the aspect of the orders and amount of energy input. *J. Food Eng.*, *264*, 109668. https://doi.org/10.1016/J.JFOODENG.2019.07.015.
67. Boye, J., Zare, F., & Pletch, A., (2010). Pulse proteins: Processing, characterization, functional properties and applications in food and feed. *Food Research International*, *43*(2), 414–431. https://doi.org/10.1016/J.FOODRES.2009.09.003.
68. Cotacallapa-Sucapuca, M., Vega, E. N., Maieves, H. A., Berrios, J. D. J., Morales, P., Fernández-Ruiz, V., & Cámara, M., (2021). Extrusion process as an alternative to improve pulses products consumption. A review. *Foods*, *10*(5), 1096. https://doi.org/10.3390/FOODS10051096.
69. Parmar, P., Bobade, H., Singh, B., & Pathania, S., (2021). Extrusion technologies for cereal–pulses blends. In: Tiwari, B. K., Gowen, A., & McKenna, B., (eds.), *Pulse Foods: Processing, Quality and Nutraceutical Applications* (pp. 393–421). Academic Press. https://doi.org/10.1016/B978-0-12-818184-3.00016-7.
70. Luo, S., Chan, E., Masatcioglu, M. T., Erkinbaev, C., Paliwal, J., & Koksel, F., (2020). Effects of extrusion conditions and nitrogen injection on physical, mechanical, and microstructural properties of red lentil puffed snacks. *Food and Bioproducts Processing*, *121*, 143–153. https://doi.org/10.1016/J.FBP.2020.02.002.
71. Ghumman, A., Kaur, A., Singh, N., & Singh, B., (2016). Effect of feed moisture and extrusion temperature on protein digestibility and extrusion behavior of lentil and horsegram. *LWT*, *70*, 349–357. https://doi.org/10.1016/J.LWT.2016.02.032.
72. Oliveira, L. C., Schmiele, M., & Steel, C. J., (2017). Development of whole grain wheat flour extruded cereal and process impacts on color, expansion, and dry and bowl-life texture. *LWT*, *75*, 261–270. https://doi.org/10.1016/J.LWT.2016.08.064.
73. Cueto, M., Farroni, A., Schoenlechner, R., Schleining, G., & Buera, P., (2017). Carotenoid and color changes in traditionally flaked and extruded products. *Food Chem*, *229*, 640–645. https://doi.org/10.1016/J.FOODCHEM.2017.02.138.

74. Gandhi, N., Singh, B., Singh, P., & Sharma, S., (2021). Functional, rheological, morphological, and micro-structural properties of extrusion-processed corn and potato starches. *Starch-Stärke*, *73*(1, 2), 2000140. https://doi.org/10.1002/STAR.202000140.
75. Koksel, F., & Masatcioglu, M. T., (2018). Physical properties of puffed yellow pea snacks produced by nitrogen gas assisted extrusion cooking. *LWT*, *93*, 592–598. https://doi.org/10.1016/J.LWT.2018.04.011.
76. Kantanen, K., Oksanen, A., Edelmann, M., Suhonen, H., Sontag-Strohm, T., Piironen, V., Diaz, J. M. R., & Jouppila, K., (2022). Physical properties of extrudates with fibrous structures made of Faba bean protein ingredients using high moisture extrusion. *Foods*, *11*(9), 1280. https://doi.org/10.3390/FOODS11091280.
77. Alajil, O., Hymavathi, T. V., Robert, P., & Deepika, L. V. S. B., (2018). Effect of flour composition and temperature on physico-chemical and sensory properties of quinoa based extrudates. *J. Pharm. Res. Int.*, *24*(5), 1–13. https://doi.org/10.9734/JPRI/2018/45346.
78. Chen, Q., Zhang, J., Zhang, Y., Meng, S., & Wang, Q., (2021). Rheological properties of pea protein isolate-amylose/amylopectin mixtures and the application in the high-moisture extruded meat substitutes. *Food Hydrocoll.*, *117*, 106732. https://doi.org/10.1016/J.FOODHYD.2021.106732.
79. Gumul, D., Areczuk, A., Ziobro, R., Ivanišová, E., & Zieba, T., (2018). The influence of freeze-dried red and purple potatoes on content of bioactive polyphenolic compounds and quality properties of extruded maize snacks. *Quality Assurance and Safety of Crops & Foods*, *10*(1), 51–60. https://doi.org/10.3920/QAS2016.1055.
80. Kantrong, H., Klongdee, S., Jantapirak, S., Limsangouan, N., & Pengpinit, W., (2022). Effects of extrusion temperature and puffing technique on physical and functional properties of purpled third-generation snack after heat treatment. *J. Food Sci. Technol.*, *59*(6), 2209–2219. https://doi.org/10.1007/S13197-021-05234-X/FIGURES/4.
81. Nemš, A., Pęksa, A., Kucharska, A. Z., Sokół-ŁĘtowska, A., Kita, A., Drozdz, W., & Hamouz, K., (2015). Anthocyanin and antioxidant activity of snacks with colored potato. *Food Chem.*, *172*, 175–182. https://doi.org/10.1016/J.FOODCHEM.2014.09.033.
82. Lisiecka, K., Wójtowicz, A., Bouasla, A., & Kasprzak, K., (2021). Design of new gluten-free extruded rice snack products supplemented with fresh vegetable pulps: The effect on processing and functional properties. *Int. Agrophys*, *35*(1), 41–60. https://doi.org/10.31545/INTAGR/132121.
83. Delgado-Nieblas, C., Ruiz-Beltrán, K., Sánchez-Lizárraga, J., Zazueta-Morales, J. De J., Aguilar-Palazuelos, E., Carrillo-López, A., Camacho-Hernández, I. L., & Quintero-Ramos, A., (2019). Effect of extrusion on physicochemical, nutritional and antioxidant properties of breakfast cereals produced from bran and dehydrated naranjita pomace. *CyTA – Journal of Food*, *17*(1), 240–250. https://doi.org/10.1080/19476337.2019.1566276.
84. Luzardo-Ocampo, I., Ramírez-Jiménez, A. K., Yañez, J., Mojica, L., & Luna-Vital, D. A., (2021). Technological applications of natural colorants in food systems: A review. *Foods*, *10*(3), 634. https://doi.org/10.3390/FOODS10030634.

85. Oplatowska-Stachowiak, M., & Elliott, C. T. (2017). Food colors: Existing and emerging food safety concerns. *Crit. Rev. Food Sci. Nutr., 57*(3), 524–548. https://doi.org/10.1080/10408398.2014.889652.
86. De Mejia, E. G., Zhang, Q., Penta, K., Eroglu, A., & Lila, M. A., (2020). The colors of health: Chemistry, bioactivity, and market demand for colorful foods and natural food sources of colorants. *Annu. Rev. Food Sci. Technol., 11*(1), 145–182. https://doi.org/10.1146/ANNUREV-FOOD-032519-051729.
87. Mansoor, M., Singh, J., Gani, A., & Noor, N., (2021). Valorization of natural colors as health-promoting bioactive compounds: Phytochemical profile, extraction techniques, and pharmacological perspectives. *Food Chem., 362*, 130141.
88. Gulati, P., Brahma, S., & Rose, D. J., (2020). Extrusion cooking: Cereal grains processing. In: Ganjyal, G. M., (ed.), *Extrusion Cooking – Cereal Grains Processing* (pp. 433–435). Woodhead Publishing.
89. Escalante-Aburto, A., Ramírez-Wong, B., Torres-Chávez, P., Figueroa-Cárdenas, J., López-Cervantes, J., Barrón-Hoyos, J., Morales-Rosas, I., et al., (2013). Effect of extrusion processing parameters on anthocyanin content and physicochemical properties of nixtamalized blue corn expanded extrudates. *CyTA-Journal of Food, 11*(SUPPL.1), 29–37. https://doi.org/10.1080/19476337.2013.764929.
90. Camire, M. E., Chaovanalikit, A., Dougherty, M. P., & Briggs, J., (2002). Blueberry and grape anthocyanins as breakfast cereal colorants. *J. Food Sci., 67*(1), 438–441. https://doi.org/10.1111/J.1365-2621.2002.TB11425.X.
91. Hirth, M., Leiter, A., Beck, S. M., & Schuchmann, H. P., (2014). Effect of extrusion cooking process parameters on the retention of bilberry anthocyanins in starch-based food. *J. Food Eng., 125*(1), 139–146. https://doi.org/10.1016/J.JFOODENG.2013.10.034.
92. Hirth, M., Preiß, R., Mayer-Miebach, E., & Schuchmann, H. P., (2015). Influence of HTST extrusion cooking process parameters on the stability of anthocyanins, procyanidins and hydroxycinnamic acids as the main bioactive chokeberry polyphenols. *LWT – Food Science and Technology, 62*(1), 511–516. https://doi.org/10.1016/J.LWT.2014.08.032.
93. Sánchez-Madrigal, M. Á., Quintero-Ramos, A., Martínez-Bustos, F., Meléndez-Pizarro, C. O., Ruiz-Gutiérrez, M. G., Camacho-Dávila, A., Torres-Chávez, P. I., & Ramírez-Wong, B., (2015). Effect of different calcium sources on the bioactive compounds stability of extruded and nixtamalized blue maize flours. *J. Food Sci. Technol., 52*(5), 2701–2710. https://doi.org/10.1007/S13197-014-1307-9/FIGURES/4.
94. Khanal, R. C., Howard, L. R., Brownmiller, C. R., & Prior, R. L., (2009). Influence of extrusion processing on procyanidin composition and total anthocyanin contents of blueberry pomace. *J. Food Sci., 74*(2), H52–H58. https://doi.org/10.1111/J.1750-3841.2009.01063.X.
95. Killeit, U., (1994). Vitamin retention in extrusion cooking. *Food Chem., 49*(2), 149–155. https://doi.org/10.1016/0308-8146(94)90151-1.
96. Yajnik, M. M., Dolan, K. D., & Ng, P. K. W., (2010). Thermal and mechanical effects on retention of food-grade β-carotene during extrusion processing. *International Journal of Food Engineering, 6*(3). https://doi.org/10.2202/1556-3758.1568/MACHINEREADABLECITATION/RIS.

97. Fan, L., Wang, Y., Xie, P., Zhang, L., Li, Y., & Zhou, J., (2019). Copigmentation effects of phenolics on color enhancement and stability of blackberry wine residue anthocyanins: Chromaticity, kinetics and structural simulation. *Food Chem.*, *275*, 299–308. https://doi.org/10.1016/J.FOODCHEM.2018.09.103.
98. Castañeda-Ovando, A., Pacheco-Hernández, M. De. L., Páez-Hernández, M. E., Rodríguez, J. A., & Galán-Vidal, C. A., (2009). Chemical studies of anthocyanins: A review. *Food Chem.*, *113*(4), 859–871. https://doi.org/10.1016/J.FOODCHEM.2008.09.001.
99. Durge, A. V., Sarkar, S., & Singhal, R. S., (2013). Stability of anthocyanins as pre-extrusion coloring of rice extrudates. *Food Research International*, *50*(2), 641–646. https://doi.org/10.1016/J.FOODRES.2011.05.017.
100. Azad, M. O. K., Adnan, M., Sung, I. J., Lim, J. D., Baek, J. S., Lim, Y. S., & Park, C. H., (2022). Development of value-added functional food by fusion of colored potato and buckwheat flour through hot-melt extrusion. *J. Food Process Preserv.*, *46*(5), e15312. https://doi.org/10.1111/JFPP.15312.
101. Nayak, B., Berrios, J. D. J., Powers, J. R., & Tang, J., (2011). Effect of extrusion on the antioxidant capacity and color attributes of expanded extrudates prepared from purple potato and yellow pea flour mixes. *J. Food Sci.*, *76*(6), C874–C883. https://doi.org/10.1111/J.1750-3841.2011.02279.X.
102. Pinho, L. S., Rodrigues, C. E. C., Favaro-Trindade, C. S., & Campanella, O. H., (2021). The incorporation of carotenoids on ready to eat foods studied through their stability during extrusion processing. *Food Engineering Reviews*, *13*(4), 902–915.
103. Rodriguez-Amaya, D. B., (2019). Natural food pigments and colorants. In: Mérillon, J. M., Ramawat, K. G., (eds.), *Reference Series in Phytochemistry* (Vol. 2, p. 902). Springer Science and Business Media B.V. https://doi.org/10.1007/978-3-319-78030-6_12/COVER.
104. Camire, M. E., (2001). Extrusion and nutritional quality. In: *Extrusion Cooking – Technology and Applications* (pp. 108–129). Wood head Publishing.
105. Ortiz, D., Ponrajan, A., Bonnet, J. P., Rocheford, T., & Ferruzzi, M. G., (2018). *Carotenoid Stability During Dry Milling, Storage, and Extrusion Processing of Biofortified Maize Genotypes* (Vol. 66, No. 18, pp. 4683–4691). ACS Publications. https://doi.org/10.1021/acs.jafc.7b05706.
106. Obradović, V., Babić, J., Šubarić, D., Jozinović, A., Ačkar, Đ., & Klarić, I., (2015). Influence of dried Hokkaido pumpkin and ascorbic acid addition on chemical properties and color of corn extrudates. *Food Chem.*, *183*, 136–143.
107. Kolniak-Ostek, J., Kita, A., Pęksa, A., Wawrzyniak, A., Hamułka, J., Jeznach, M., Danilčenko, H., & Jariene, E., (2017). Analysis of the content of bioactive compounds in selected flours and enriched extruded corn products. *Journal of Food Composition and Analysis*, *64*, 147–155.
108. Ruiz-Gutiérrez, M. G., Quintero-Ramos, A., Ruiz-Gutiérrez, M. G., & Quintero-Ramos, A., (2021). Prickly pear (*Opuntia ficus-indica*) processing by extrusion-cooking. *Opuntia spp.: Chemistry, Bioactivity, and Industrial Applications*, 657–677. https://doi.org/10.1007/978-3-030-78444-7_34.
109. Ying, D. Y., Cheng, L. J., Chibracq, G., Sanguansri, L., Oiseth, S. K., & Augustin, M. A., (2015). The format of β-carotene delivery affects its stability during extrusion.

LWT – Food Science and Technology, *60*(1), 1–7. https://doi.org/10.1016/J.LWT.2014.09.034.

110. Ruiz-Armenta, X. A., Zazueta-Morales, J. De. J., Aguilar-Palazuelos, E., Delgado-Nieblas, C. I., López-Diaz, A., Camacho-Hernández, I. L., Gutiérrez-Dorado, R., & Martínez-Bustos, F., (2017). Effect of extrusion on the carotenoid content, physical and sensory properties of snacks added with bagasse of naranjita fruit: Optimization process. *CyTA – Journal of Food*, *16*(1), 172–180. https://doi.org/10.1080/19476337.2017.1368717.
111. Ortak, M., Caltinoglu, C., Sensoy, I., Karakaya, S., & Mert, B., (2017). Changes in functional properties and in vitro bioaccessibilities of β-carotene and lutein after extrusion processing. *J. Food Sci. Technol.*, *54*(11), 3543–3551. https://doi.org/10.1007/S13197-017-2812-4.
112. Tonyali, B., Sensoy, I., & Karakaya, S., (2016). The effect of extrusion on the functional components and in vitro lycopene bioaccessibility of tomato pulp added corn extrudates. *Food Funct.*, *7*(2), 855–860.
113. Akbar, H. E., Sadiq, Z., & Zia-Ul-Haq, M., (2018). Sources of betalains. *Betalains: Bimolecular Aspects*, 15–32. https://doi.org/10.1007/978-3-319-95624-4_2.
114. Choo, W. S., (2019). Betalains: Application in functional foods. *Reference Series in Phytochemistry*, 1471–1498. https://doi.org/10.1007/978-3-319-78030-6_38.
115. Fernández-López, J. A., Fernández-Lledó, V., & Angosto, J. M., (2020). New insights into red plant pigments: More than just natural colorants. *RSC Adv.*, *10*(41), 24669–24682. https://doi.org/10.1039/D0RA03514A.
116. Ruiz-Gutiérrez, M. G., Amaya-Guerra, C. A., Quintero-Ramos, A., Pérez-Carrillo, E., Ruiz-Anchondo, T. D. J., Báez-González, J. G., & Meléndez-Pizarro, C. O., (2015). Effect of extrusion cooking on bioactive compounds in encapsulated red cactus pear powder. *Molecules*, *20*(5), 8875–8892. https://doi.org/10.3390/MOLECULES20058875.
117. Durge, A. V., Sarkar, S., Survase, S. A., & Singhal, R. S., (2013). Impact of extrusion on red beetroot color used as pre-extrusion coloring of rice flour. *Food Bioproc Tech.*, *6*(2), 570–575. https://doi.org/10.1007/S11947-011-0723-3/FIGURES/1.
118. Martelli, G., Claudia, F., Visai, L., Daglia, M., & Ferrari, D., (2014). Thermal stability improvement of blue colorant c-phycocyanin from *spirulina platensis* for food industry applications. *Process Biochemistry*, *49*(1), 154–159.
119. Tańska, M., Konopka, I., & Ruszkowska, M., (2017). Sensory, physico-chemical and water sorption properties of corn extrudates enriched with *Spirulina*. *Plant Foods for Human Nutrition*, *72*(3), 250–257. https://doi.org/10.1007/S11130–017–0628-Z/TABLES/5.
120. Suzery, M., Hadiyanto, Sutanto, H., Soetrisnanto, D., Majid, D., Setyawan, D., & Azizah, N., (2015). The improvement of phycocyanin stability extracted from *Spirulina* sp using extrusion encapsulation technique. *AIP Conf Proc.*, *1699*(1), 030011. https://doi.org/10.1063/1.4938296.
121. Pradeep, H. N., & Nayak, C. A., (2019). Enhanced stability of c-phycocyanin colorant by extrusion encapsulation. *J. Food Sci. Technol.*, *56*(10), 4526–4534. https://doi.org/10.1007/S13197-019-03955-8/FIGURES/6.
122. Castellari, M., Boukid, F., Tork, M. B., Vazifedoost, M., Hesarinejad, M. A., Didar, Z., & Zenoozian, M. S., (2022). Fabrication of Dragee containing *Spirulina platensis*

microalgae to enrich corn snack and evaluate its sensorial, physicochemical, and nutritional properties. *Foods, 11*(13), 1909. https://doi.org/10.3390/FOODS11131909.

123. Rodríguez De, M. E., Steffolani, M. E., Martínez, C. S., & León, A. E., (2014). Effects of *Spirulina* biomass on the technological and nutritional quality of bread wheat pasta. *LWT – Food Science and Technology, 58*(1), 102–108. https://doi.org/10.1016/J.LWT.2014.02.054.
124. An, S. H., & Ryu, G. H., (2015). Comparison of antioxidant activities of extruded rice with vegetables by cold and conventional extrusion. *Journal of the Korean Society of Food Science and Nutrition, 44*(8), 1212–1218. https://doi.org/10.3746/jkfn.2015.44.8.1212.
125. Moon, Y. H., Song, Y. S., Kim, K. S., Lee, J. E., Yu, G. D., Lee, Y. H., Lee, K. B., et al., (2017). Comparison of color and major components of hempseed oils extracted with pressuring and extruding methods. *Journal of the Korean Applied Science and Technology, 34*(3), 666–672. https://doi.org/10.12925/JKOCS.2017.34.3.666.

CHAPTER 13

NONTHERMAL TECHNIQUES IN THE FOOD INDUSTRY: SAFETY AND REGULATORY ASPECTS

ANANDU CHANDRA KHANASHYAM,[1]
RAMNARAIN RAMAKRISHNA,[2] PRIYAMVADA THORAKKATTU,[3]
ANTO PRADEEP RAJA CHARLES,[2] KARTIK SHAH,[4]
KARTHIK SAJITH BABU,[3] and KAAVYA RATHNAKUMAR[5]

[1]*Department of Food Science and Technology, Kasetsart University, Ladyao, Chatuchak, Bangkok, Thailand*

[2]*Food Ingredients and Biopolymer Laboratory, Department of Plant Science, North Dakota State University, Fargo, North Dakota, USA*

[3]*Department of Animal Sciences and Industry, Food Science Institute, Kansas State University, Manhattan, Kansas, USA*

[4]*Sargento Foods, Elkhart Lake, Wisconsin, USA*

[5]*Department of Food Science, University of Wisconsin, Madison, USA*

ABSTRACT

The food industry widely employs traditional thermal processing technologies for shelf-life extension. However, the desirable organoleptic qualities may be lost as a result of such treatments, and minerals and vitamins that are temperature-sensitive may also be affected. As a result, cutting-edge nonthermal techniques are employed to achieve shelf-life extension and food safety while reducing the influence on a product's nutritive and

Nonthermal Food Processing Technologies: Impact on Color Profile.
Kaavya Rathnakumar and R. Pandiselvam (Eds.)

qualitative qualities. Regulatory organizations generally agree that novel nonthermal technology is advantageous to both food processors and consumers. Validating procedures are a challenge for the industry due to the complexity of food matrices and the variety of foods produced. Proper validation of comparable goods for a specific technology, in terms of shelf-life extension, nutritive and sensory characteristics, along with functional properties, customer acceptability, and environmental effects, is necessary for developing proper regulatory guidelines for an intended technology. Regulations occasionally serve as a potent catalyst for innovation as the ability to innovate is better encouraged when regulation is more flexible with reduced compliance requirements and bureaucratic hassles.

13.1 INTRODUCTION

The innovation in the food processing industry is relatively consumer-centered, and consumer needs have historically been a major driving force for food scientists to develop novel technologies that meet consumer demands. Moreover, foods with exceptional quality, free from preservatives/additives, healthier, microbiologically acceptable, and with extended shelf life are becoming increasingly popular with consumers. The majority of techniques available for maintaining the food safety qualities of food products involve thermal treatments, which negatively impact their nutrition and sensory qualities. Recently, there has been a significant increase in the investigation of nonthermal techniques that guarantee the quality and safety of food products [68, 91]. Indeed, these technologies provide significant benefits to improve the preservation of food products, as well as to overcome the mounting difficulties brought on by heightened regulatory and customer demands. Interestingly, there is no single novel technology that is more suitable for broad use and general purposes than another; each available technology has its benefits and drawbacks [114]. The use of these available technologies should be targeted towards specific purposes such as monitoring food safety, ensuring food quality, and assessing overall quality while being cost-effective, energy-efficient, and possibly environmentally friendly. Adopting new food process technologies includes difficulties and other variables that affect consumer and stakeholder approval, and must be taken into consideration when making a choice to adopt an emerging technology. The way risks are perceived by consumers and how risks are evaluated by food scientists must be taken into consideration to successfully adapt an existing emerging technology.

The loss of aroma, color, flavor, texture, and nutritional content caused by traditional food industry processing techniques has been the subject of debate over optimization processes. However, the acceptance of these novel nonthermal techniques into food processing is currently being explored, and factors that affect the acceptance need further testing. The industry-level adoption of an emerging technology is influenced by factors such as market share, absorptive ability of individual companies' innovativeness, and technological expertise. New processing technologies are now deemed essential to deliver products that meet specifications while preserving other key quality characteristics, including nutritional and sensory qualities. Nonthermal technologies can be successfully adopted in the food industry to lower energy costs, reduce energy consumption, and benefit society at large by lowering the usage of energy resources. It has become crucial in the food processing industry to increase energy efficiency and replace current inefficient operations with new efficient ones. The demand for food, energy, and water has skyrocketed due to the population explosion, bringing up the need for more innovative and sustainable technologies in the food industry. To integrate these emerging technologies into the food processing lines, it is required to create new food regulations and modify the existing ones. Therefore, before a product can be commercialized, a thorough assessment and validation of the total quality/safety of food products processed by emerging technologies is deemed essential [12, 47]. This chapter discusses the safety, quality, and regulatory aspects of nonthermal technology, along with its environmental impacts and quality aspects.

13.2 NONTHERMAL TECHNOLOGIES: AN HISTORICAL OVERVIEW

While providing a safe and nourishing food supply is the goal of food processing, there are numerous ways to achieve this, and they have all seen significant development in recent years. Throughout the course of human evolution, it was difficult for people and their ancestors to access food or consume it uncooked [70]. Humans have sought and, through innovation, created ways to enhance the availability, digestibility, transportability, and storage of food products. Beginning with the establishment of agriculture, food processing sparked a wide-ranging expansion of activities and procedures that included all edible food sources. Processing food has always focused on making it safe to eat, transportable, and preserving its nutritional content through storage. In the 19th century, thermal processing methods such as sterilization and pasteurization increased the range and effectiveness of food preservation. Ancient Romans and Chinese people used cooling techniques such as ice and snow to preserve delicacies [70].

Some of the nonthermal techniques, such as high hydrostatic pressure (HHP), cold plasma (CP) technology, UV radiation, pulsed electric field (PEF), and ultrasound, are discussed in this section. It is noteworthy that some of these technologies trace their origin back to 1800 [74]. For instance, the first use of HPP was reported in 1985 to kill pathogenic bacteria [87]. Similarly, the pasteurization of milk using HPP was reported to increase its shelf life in 1899 [53, 121]. Another study by Bridgman [18] reported the use of HPP treatment on egg whites, which resulted in coagulation attributed to protein denaturation [10, 18]. In 1974, the treatment of low-acid foods in a container using HPP and temperatures (82°C–103°C) was reported to have sterilization-like effects [87]. The first high-pressure processed foods, such as sauce, jellies, and jams, were introduced to the Japanese market in the early 1990s [140]. On April 23, 1990, Meijiya Food Company introduced HHP-treated jam to the market for the first time [43], and they were characterized by their natural color and taste after HPP processing. Other foods, including fruit juices in Portugal and France, oysters, and guacamole in the USA, followed the trend [57].

The application of PEF is gaining importance. Implementing this technology helps reduce the negative impact that biological materials may undergo during enzymatic, chemical, and thermal processes [37]. The PEF equipment and techniques were patented (DE 1237541 B) by a German researcher named Doevenspeck, often referred to as the father of PEF technology [34]. The author later conducted several studies to investigate the impact of PEF on cell disruption in food matrices, such as *Salmonella* sp. inactivation in marination brine [35]. In 1967, a group of researchers adopted a detailed approach to study microbial inactivation and cell repair. The authors noted that treatment time (TT), electric field intensity, and pulse width/count are crucial aspects of PEF for inactivating microorganisms [49]. In the 1980s, Germany's Krupp Maschinentechnik GmbH evaluated the effects of electric pulse treatments on the "Elcrack process," which uses electricity to crack and disintegrate animal and plant cells, as well as on the "Elsteril process," which uses electricity to pasteurize or sterilize pumpable liquid media [134]. The Dutch food company Hoogsteger is credited with being the first in Europe to process fresh juices using PEF technology. By using PEF processing technology, this company increased the shelf life of refrigerated fruit juices by two weeks. Despite the extensive studies conducted on PEF technology, there are still very few industrial and commercial systems that utilize it. In recent years, the potato business has been one of the largest markets for PEF, as the technology is capable of increasing drying effectiveness and sugar leaching while obtaining French fries of outstanding quality.

Among the nonthermal technologies, cold plasma (CP) has also generated some interest as a decontamination technique. The word "plasma" means "formed" or "molded" in the Greek language. Remarkable advancements in plasma physics and engineering since 1928 have made it possible to create numerous atmospheric pressure plasma sources without the need for expensive mechanical parts. In recent years, CP has been evaluated for various food applications, including: (i) modifying the hydrophobicity of food surfaces [89] and the structural characteristics of macromolecules [86]; (ii) altering packaging materials used in food applications [103]; (iii) microbial inactivation of food packaging materials and a variety of foods [90]; (iv) inactivation of enzymes in food applications [88]; and (v) improving grain germination, to name a few [113].

The history of some of the other nonthermal technologies also dates back to the early 1900s. Evidence suggests that UV radiation had been studied on bacteria as early as 1903 [115]. Additionally, the application of UV radiation for disinfection of drinking water dates back to France in 1906 [85]. A study conducted in 1932 reported that when milk and other liquids were subjected to audible sound waves with an 8,900-cycle frequency, both the total bacterial count and the *E. coli* count were decreased [24]. The acceptance of this technology in the food business is somewhat gradual since UV light is thought to be a more suitable application for disinfecting water, surfaces of materials, and air. While milk was first sterilized with radiation in 1960 [153], a new technique for irradiating milk and other food products was developed in 1976 [4]. In some other examples of nonthermal history, Barnett described many forms of cavitation that could be used in the food sector in the year 1998. In 2003, it was shown that using a single nonthermal technology alone is insufficient to promote food safety. Therefore, researchers also recommend the combination of various other nonthermal technologies to improve the food product shelf-life [4].

Prior to and even during the implementation of nonthermal food technologies, it is crucial for the food industry to have a thorough understanding of their individual mechanisms, as well as their advantages and disadvantages. Some of the advantages and disadvantages of the principles of nonthermal technology are outlined in Table 13.1.

13.3 SAFETY ASPECTS OF NONTHERMAL TECHNOLOGY

Consumer demand for foods that have undergone the least amount of processing, as well as food processors' interest in new, energy-efficient methods to enhance food safety and nutritional quality, have both contributed to the rise of nonthermal food processing technology over the past 10 years. This section will provide a brief overview of the role of nonthermal processes in improving microbiological and chemical food safety.

13.3.1 MICROBIOLOGICAL FOOD SAFETY

In their natural state, almost all food matrices tend to contain a plethora of microflora, which may also include spoilage and pathogenic strains. Traditional microbial mitigation processes invariably involve the application of heat in some form, which, while effective, may compromise nutritional and quality parameters in some foods. Consequently, nonthermal processing methods are uniquely positioned to overcome these challenges while ensuring microbial food safety and quality.

HPP is among the most widely studied methods for inactivating pathogenic microbes, including yeasts, molds, and bacteria such as *C. perfringens, B. cereus, E. coli,* and *S. aureus* [46, 160]. The mode of action of HPP is outlined in Table 13.2. While HPP can reduce the viable cell count of vegetative pathogens at both ambient and cold conditions, treated foods require refrigeration post-processing. Spores of pathogens such as *Clostridium botulinum* are not inactivated by HPP alone and therefore present a significant risk, especially in low acid foods. Furthermore, this method is ineffective on powders and flours, low moisture content foods, and products containing air bubbles [55].

TABLE 13.1 The Advantages and Disadvantages of Several Non-thermal Technologies

Nonthermal Technology	Principle	Advantages	Disadvantage	References
High-pressure processing (HPP)	Le Chatelier's law (pressure-favored structural changes and responses leading to a volume reduction) and the isostatic principle where each food atom and molecule undergo the same pressure at the same moment during HPP treatment.	Post Packaging application. Homogeneous treatment of the food metrics. Destruction of bacteria and enzymes. Commercially available units.	Unavailability of continuous process. Increased cost of processing. Packaging to be conducive to the treatment. Not effective for spores' inactivation.	[40, 52, 78, 96, 108]
Cold plasma (CP)	Microbial inactivation through synergistic action such as UV radiation, surface erosion, oxidation by ozone and reactive oxygen and nitrogen species, and charged particles.	Capability to inactivate bacterial and fungal spores. Short treatment times. Relatively low cost. In-package treatment.	Varied effects depend on the physicochemical properties of food. Impact on quality attributes. Early-stage technology.	[52, 74, 78, 108]
Ultraviolet light (UV)	Inactivation of bacteria or inability to reproduce as a result of thymine dimers formation in the bacterial chromosome.	Low cost. Effective against bacterial spores. Applicable to food contact surfaces. Health-promoting compounds synthesis.	Transmittance through product critical. Effectiveness relates to parameters such as target microorganisms and concentration, food composition. Not very effective against all spores.	[78, 108, 128]

TABLE 13.1 *(Continued)*

Nonthermal Technology	Principle	Advantages	Disadvantage	References
Irradiation	Destruction of the DNA of target bacteria, inhibition inhibits DNA synthesis, hence preventing cell division.	Less processing time. Sustainable process: no chemicals or residue generated. No subsequent processing is required post-treatment.	Stringent safety requirement. Inactivation of spores required high doses. High doses not permitted. Efficacy varies depending on the product properties, e.g., turbidity. Negative consumer perception.	[52, 63, 74, 78, 94]
Pulsed electric field (PEF)	Short bursts of high voltage electricity (sub-microseconds–milliseconds) into foods.	Minimum effect on physical and sensory properties. Efficient use of energy. Shorter treatment times.	The increased cost of the treatment. The presence of bubbles in the food matrices can result in non-uniform treatment. Limited availability of industrial equipment.	[11, 52, 63, 105]
Pulsed light	Photochemical effect resulting in DNA denaturation and pyramid formation in dimmers; disintegration of cell due to overheating of its constituents.	Rapid infection in a short period. Sustainable; very less energy requirement. No nutritional and sensorial effect on product.	Low penetration power in certain foods. Selection of packaging for improved penetration. Limited control of food heating.	[133]

Exposure to ionizing radiation is a commonly used strategy for mitigating undesirable microbes from foods. The types of ionizing radiation approved for this purpose include γ-rays, X-rays, and accelerated electron beams, arranged in decreasing order of energy and penetrative power [22]. The antimicrobial activity of ionizing radiation can be attributed to the following phenomena: (i) radicidation (occurs at 1–10 kGy), (ii) radappertization (> 10 kGy), and (iii) radurization (> 20 kGy) [4]. These forms of radiation share similar modes of action as outlined in Table 13.2. While these methods are commonly used to process low moisture foods like spices, herbs, and condiments, their utility has been extended to ready-to-eat meat products (high doses, > 15 kGy) and fresh mushrooms (up to 3 kGy) as well [2]. Notably, food products meant for space flight, such as frozen, packaged meats, are required to be irradiated with a dose of 44 kGy or above. On the contrary, the use of high irradiation dosages on fresh meat may degrade myoglobin and polyunsaturated fatty acids, resulting in the development of unpleasant colors and odors, while also negatively impacting the functional properties of whole wheat flour and rice [22, 160]. Additionally, it is expensive and is associated with consumer concerns regarding the safety of the technique. As a result, the use of irradiation is limited to specific applications where an exceedingly high level of sanitation is required, which may not be otherwise achieved using conventional processing methods.

Application of pulsed light (PL) processing to ensure microbiological safety of food products has been extensively reviewed by multiple authors in recent years [14, 21, 65]. Among the different types of PL treatments, pulsed UV treatments were found to have the highest antimicrobial effect. It is recognized as an effective and commercially viable means of reducing microbial loads of foods in the United States. The mode of action by which pulsed UV light inactivates or damages microbial cells is outlined in Table 13.2. The antimicrobial efficacy of pulsed visible and infrared lights is significant only at high peak power settings, potentially limiting their application due to higher energy consumption [83].

Ultrasonication represents another viable avenue for subjecting foods to high energy levels, resulting in the inactivation and death of microbes. For applications requiring sterilization, sonication using frequencies and energy density greater than 20 kHz and 1 W cm^{-2}, respectively, is most suitable. For instance, pathogenic strains including *S. aureus, E. coli, Staphylococcus enteritidis,* and *Listeria innocua* in fresh produce (lettuce and strawberries) underwent a 2-log reduction when treated with an ultrasonic frequency of 37 kHz and 30 W cm^{-2} for 45 minutes [15]. Furthermore, high-energy ultrasonic treatment (30 kHz, 100 W, 15 minutes) was able to successfully sterilize fruit juices [67]. In both studies, sensory attributes of the sonicated produce were reported to be on par with the respective controls.

CP is produced around ambient temperatures and doesn't utilize heat treatment for incapacitate microorganisms [160]. Liao et al. [74] have provided an extensive review of the types of foods and microorganisms for which CP may be effectively utilized [75, 77]. Mechanisms of generation, including dielectric barrier discharge plasma, gliding arc plasma, atmospheric plasma jets (PJs), etc., can be used to categorize the CP [75, 77, 163]. The deleterious effect of CP on microbes is a result of interactions between plasma ions and cells and is outlined in Table 13.2.

PEF processing is characterized by the use of high-intensity electrical currents in pulses on the nanoseconds to milliseconds scale. This technology has garnered significant attention due to the short processing times and high energy efficiency involved [62]. This method is suitable for pumpable fluid and semi-fluid foods and can be operated in batch and continuous formats. The mode of action of PEF is described in Table 13.2. In addition to reducing microbial loads, undesirable enzymes may also be inactivated when subjected to high-intensity electric fields, thereby enhancing safety, quality, and shelf-life [62].

Ozonation differs from the methods discussed thus far in that it utilizes a chemical agent, ozone (O_3), to exert oxidative stress to mitigate microbes in food products [101]. This approach is especially advantageous due to the rapid action of ozone, even at low concentrations, and the lack of harmful reaction intermediates. Furthermore, it is generally regarded as safe (GRAS) and can exert a broad spectrum of action against bacterial, viral, parasitic, and fungal pathogens. Pandiselvam et al. [99] have reviewed the application of ozone as a disinfectant under various processing conditions in great detail [101]. It is important to note that ozonation by itself cannot guarantee the complete eradication of microbial loads in foods without compromising their sensory and nutritional properties.

13.3.2 CHEMICAL FOOD SAFETY

13.3.2.1 MYCOTOXIN DECONTAMINATION

A group of hazardous, low-molecular-weight secondary metabolites generated by fungi and molds are known as mycotoxins. Mycotoxins

TABLE 13.2 Summary of Description and Modes of Action of Commonly Employed Nonthermal Processes

Process	Brief Description	Determinants of Efficacy	Mechanism of Action	Intended Applications	References
High pressure processing (HPP)	200–700 MPa pressure with temperatures less than 50°C, is applied on food products, placed within a fluid (water) filled chamber.	Chamber dimensions, magnitude of pressure, treatment duration, temperature.	Direct: Pressure induced phase transitions in membrane leading to permeabilization and leakage of cell components. Damage to molecular structures of biomolecules, complex cellular mechanisms such as efflux pump proteins, ATPase activity and ribosomes. Indirect: Generation of reactive species and subsequent oxidative stress.	Pre-packaged foods and bulk foodstuffs.	[46, 55, 62, 163]
Ionizing radiation	Beams of energy are directed from a source towards the target food matrix. Gamma rays are emitted from 60Co or 137Cs and applied in doses of between 1 and 50 kGy (typically between 1 and 10 kGy). X-rays produced emitted from linear and Rhodotron-type accelerators. Electron beams are produced by particle accelerators.	Target food matrix characteristics (density, moisture content, antioxidant levels), temperature, oxygen levels, subsequent storage conditions, radiation dosage and source.	Direct action: Radiation induced chromosomal DNA damage and/or disruption of cytoplasmic membrane. Indirect action: Damage to cellular components by radiolytic products such as H^+, OH^-, free radicals.	Pre-packaged foods and bulk foodstuffs.	[22, 62, 127, 163]

TABLE 13.2 *(Continued)*

Process	Brief Description	Determinants of Efficacy	Mechanism of Action	Intended Applications	References
Pulsed light (PL)	High frequency (1–20 flashes per second) pulses of electromagnetic radiation, with an energy density between 0.01 and 50 $Jcm^{-2}$are directed onto the target food product.	Pulse frequency, distance between light source and target, product thickness.	Direct action: Absorption of UV radiation by microbial DNA results in mutations and severe damage. Indirect action: Damage to cellular components by free radical generation and subsequent oxidative stress.	Surface sanitization/ sterilization of food, processing equipment and food packaging materials.	[14, 21, 62, 163]
Cold plasma (CP)	Target food products are exposed to the plasma discharge between two parallelly placed electrodes.	Gas type/pressure/ composition/flow rate and power and plasma excitation frequency.	Damage and permeabilization of cell membrane by reactive species-mediated oxidative stress, and localized heating effects.	Pre-packaged foods and bulk foodstuffs.	[44, 62, 141]
Pulsed electric field (PEF)	Target food products are subjected to extremely short electric pulses of intensities ranging between 20 and 80 kV cm^{-1}, for treatment times lasting less than 1s.	The wave form type, frequency, and width of electric pulse, properties of target food matrix (electrical conductivity and pH).	Electroporation and disruption of cell membranes, resulting in cell lysis and leakage of cell contents.	Liquid foods	[22, 62, 163]
Ultrasound	Target food product is immersed in a fluid (water) filled chamber and exposed to sound energy (100 kHz: < 1 W cm^{-2} or 18–100 kHz: > 1 Wcm^{-2}).	Sound energy density, viscosity of food matrix, size of particulate matter in food matrix.	Direct: Cavitation induced membrane thinning and permeabilization, localized heating effects. Indirect: generation of lethal free radicals and subsequent oxidative stress.	Sterilization, emulsification, extraction, and freezing fresh foods.	[15, 62, 163]

TABLE 13.2 *(Continued)*

Process	Brief Description	Determinants of Efficacy	Mechanism of Action	Intended Applications	References
Ozone treatment	Target food product is exposed to ozone directly or through aqueous media containing ozone (washing/soaking).	Geometry and pH of food matrix, size of ozone bubbles, humidity of gas phase, temperature of water.	Direct: oxidation of biomolecules primarily on the surface of cell membrane and subsequent lysis and leakage of cell contents. Indirect: Hydroxyl free radical-mediated oxidative stress.	Fresh produce and food processing equipment.	[66, 86, 131, 135]

pose a serious threat to the safety of food products, particularly cereals, nuts, legumes, dried fruits, spices, milk, and their derivatives. While more than 100 fungal species are known to produce an estimated 400 toxic compounds, the ones of greater concern are the aflatoxins (AF) synthesized by members of the genera *Aspergillus, Penicillium, Fusarium,* and *Alternaria* [19]. The chemical diversity and thermostability of mycotoxins make it challenging to mitigate their presence in agro-food products using traditional processing methods [98]. Currently, the most effective means of reducing mycotoxin contamination in food is through preventive measures at pre- and post-harvest stages. However, lapses in implementing these measures, combined with environmental factors, can lead to mycotoxin contamination, thus requiring efficient, reliable, and innovative mitigation strategies [19, 31, 98]. In recent years, numerous novel technologies have been investigated for their potential and applicability in the decontamination of mycotoxins, with varying degrees of success. The efficacy and modes of action of these processes are outlined below.

Ozonation can effectively result in the decomposition of mycotoxins, especially in cereal grains, legumes, dried fruits, dairy, and poultry products [98]. Ozone has been reported to show effective decontamination of AF in corn grits and corn flour, AFM1 in milk containing probiotic *Lactobacillus acidophilus,* zearalenone (ZEN) – wheat bran, corn flour, and kernels, AFB1-peanuts, ochratoxin A (OTA) – corn kernels, vomitoxin or deoxynivalenol (DON) – whole wheat grains, semolina, wheat bran, and corn kernels [2, 33, 68, 73, 80, 107, 109, 125]. The molecular oxygen generated during ozone treatment reacts with the furan ring, forming derivatives of monoazizone such as aldehydes, ketones, and organic acid [68, 156]. The susceptibility of mycotoxins to ozonation is highly variable, with compounds containing unsaturated bonds being more sensitive to ozonolysis [98]. Furthermore, among mycotoxin classes, AFB1 and AFG1 were shown to be more susceptible compared to AFB2 and AFG2, while ZEN was more sensitive than OTA [80, 112]. Additionally, mycotoxin degradation efficacy of ozonation varies directly in accordance with the moisture content of the target food matrix [31].

Many recent studies have indicated the potential of cold plasma (CP) as an effective tool in mitigating *Alternaria* mycotoxins, such as DON, AFB1, OTA, and fumonisin B1 (FB1) [31, 64, 149, 154]. CP exists in a partially ionized gaseous state, consisting of various anions and cations, free radicals and electrons, ultraviolet (UV) photons, and O_3, which collectively disrupt the unsaturated bonds in the mycotoxins. For example, AFB1 undergoes degradation during CP treatment due to the oxidation of covalent bonds in the lactone and furan rings, as well as alterations in methoxyl structures and cyclopentenone [31]. The relative susceptibility of mycotoxins to CP, in descending order, follows the pattern: AFB1 > enniatin B > ZEN > enniatin A > sterigmatocystin. The high susceptibility of AFB1 may be attributed to the presence of long, exposed aliphatic chains, while the more condensed aromatic rings of sterigmatocystin provide it with a more compact structure, resulting in increased resilience to degradation [139].

Irradiating foods results in the degradation of mycotoxins through direct or indirect oxidative action. This leads to the cleavage of chemical bonds in the target compound and the formation of harmless byproducts [150]. Ionizing radiations, such as γ-radiation and EBI, can degrade mycotoxins through direct contact and indirectly by generating protons, hydroxyl radicals (OHs), and hydrated electrons via the radiolysis of water in the food matrix. At a dosage of 50 kGy, EBI resulted in a significant reduction (ZEN – 71.1% and OTA – 67.9%) [42]. Notably, even milder doses of γ-radiation (up to 3.11 mGy) were sufficient to degrade AFM1 by up to 99%, suggesting its comparative superiority to EBI [51]. Similar to ozonation, the efficacy of mycotoxin degradation by ionizing radiation may be improved by increasing the moisture content of the target food matrix [31]. These methods could potentially be adapted for mycotoxin decontamination with relative ease due to their existing commercial usage. Furthermore, they are applicable to a wide range of foods and can deeply penetrate food matrices, even when packaged [98, 135, 149].

UV exposure has been shown to effectively cause the photodegradation of AFs. AFB1 is capable of absorbing UV light, specifically at 222, 265, and 362 nm (absorption maximum), resulting in damage to the terminal furan ring and disruption of the active binding sites responsible for toxicity, including the C8–C9 and C1–C14 bonds [54, 103]. Additionally, AFs may be further damaged by hydroxyl-radical mediated photolysis. Sharareh et al. [127] observed that UV treatment significantly reduced concentrations of AFB_1 (98%), AFB_2 (99.5%), AFG_1 (99.8%), AFG_2 (100%), and total AF (99.1%) from their initial levels [129]. This method has also been found to be effective in other foods such as peanut oil and almonds [33, 61].

Since PL treatments primarily utilize electromagnetic (EM) radiation in the range of 200–400 nm, their mode of action resembles that of

UV radiation [150]. Reductions in ZEN, DON, OT, AFB1, and AFB_2 in response to PL treatments have been reported [92, 100]. In one study, PL exposure, in the presence of Fe^{2+} catalyst, resulted in the conversion of patulin to the relatively less toxic patulin-glutathione conjugates [120]. However, the exact mechanism by which mycotoxins are affected by PL treatments is unclear and warrants further investigation.

Ultrasonic treatments deviate from the aforementioned methods in that mycotoxins are decontaminated as a result of sonic cavitation, which occurs during the rarefaction cycle. Cavitation bubbles implode and rupture the covalent bonds in water molecules upon attaining a specific critical radius, resulting in the production of active free radicals that react with macromolecules and cleave them into low molecular weight compounds. For instance, this method has been discovered to alter the lactone ring and methoxy group of AFB_1, disrupt the unsaturated connection between the C8–C9 atoms in the furan ring, and detoxify samples of apple juice as a result [1].

In contrast to the methods discussed above, the mode of action of HPP with regard to mycotoxin degradation varies drastically. The exertion of physical pressure in a uniform manner during HPP is posited to favor the formation of sulfhydryl bonds, along with electrostatic and hydrophobic interactions, resulting in the structural alteration and thereby inactivation of mycotoxins [3, 155]. Currently, there is relatively limited information about the effectiveness and use of HPP in mitigating mycotoxins from food.

The impact of PEF treatment on mycotoxins has been studied to a limited extent. The concentration of AFM_1 in milk containing probiotic *Lactobacillus acidophilus* strains was reduced by up to 96% as a result of PEF application [68]. The authors attributed the observed effects to the ability of the electric pulse to alter the chemical nature of the toxin. However, in the same study, a simultaneous decrease in the viability of the *L. acidophilus* cells was also observed, which could potentially restrict the use of this method in fermented or probiotic foods. In another study, PEF treatment combined with pH modification was found to degrade AFB_1 between 77% and 97% in potato dextrose agar (PDA) [146].

13.3.2.2 PESTICIDE DEGRADATION

Pesticides are an indispensable category of chemical inputs in modern agriculture. However, the retention and carryover of pesticide residues from farm to plate is of great concern for consumer health. It is also a major limiting factor in the trade of agro-food products. The diversity in their chemical structures and the unpredictability of residual concentrations make it challenging to completely eliminate these compounds from food products. Additionally, if pesticides are not adequately decontaminated, their concentrations may be significantly increased by certain traditional downstream processing methods such as drying or dehydration. Therefore, efficient and innovative processes to address pesticide residues are crucial for maintaining a sustainable global food supply chain.

HHP successfully lowers the concentration of hydrophobic pesticide residues. Chlorpyrifos, a pesticide commonly used in vegetable cultivation, was found to be reduced by up to 75% in cherry tomatoes and 89% in Brussels sprouts through HPP treatment [59, 60]. Additionally, longer holding times and ethanol treatments were found to further improve chlorpyrifos decontamination from cherry tomatoes [58]. This method has been commercially adopted and, being a physical process, it does not involve the formation of toxic reactionary intermediates, making it an attractive decontamination strategy. However, data on the efficacy and exact mechanism of pesticide degradation, as well as the final fate of the residues during HPP, is limited and requires further investigation.

The application of PEF has been shown to successfully degrade various pesticides, such as methamidophos, chlorpyrifos, dimethoate, diazinon, cyprodinil, vinclozolin, procymidone, and pyrimethanil, in liquid matrices like apple juice and wine [26, 30, 159]. According to these studies, the use of electric fields to alter the rotation and vibration of polar molecules resulted in significantly lower pesticide concentrations compared to the initial levels [45]. Furthermore, increasing specific parameters such as electric field intensity, time, and applied energy was found to greatly enhance the decontamination efficacy of PEF. The effectiveness of PEF varies depending on the chemical structure of the pesticide, thus requiring more detailed kinetic modeling to determine the specific degradative mechanism involved in each case.

Gavahian & Khaneghah [44] have provided an extensive review on the application of cold plasma (CP) to mitigate contaminants in food, including pesticides. They observed significant reductions in the levels of cypermethrin (63%) and chlorpyrifos (74%) on the external surface of mangoes, as well as organophosphorus (99.6%) from wolfberry samples, without adversely affecting the quality of the test samples. Additionally, the available carotenoid content was enhanced in the case of the mango samples [106, 162]. The degradation of pesticide molecules is achieved through their oxidation by reactive free radicals generated when food matrices are exposed to plasma. While CP could potentially aid in pesticide degradation with minimal damage to food quality, the retention and impact of free radicals generated during plasma treatment on consumer health are still unknown and require further investigation [44].

Ultrasonication effectively degrades 16 different pesticide residues in strawberries to varying degrees (45.1%–91.2%) under the following conditions: 40 kHz, 2×240 W peak/period, and up to 5 minutes of exposure time. The action of the shock waves and agitation produced during sonication results in the removal of pesticides from the food matrices [44]. In another study, sonication was found to remove carbosulfan from vegetables such as coriander, beans, kale, and red chili compared to soaking in a water bath [79]. More recently, Zhu et al. [162] achieved reduction rates of up to 74%, 24% and 45% of chlorothalonil, carbendazim, and pyrazophos residues, respectively in pakchoi using ultrasound sonication. Furthermore, ultrasound processing has also been extended to fruit juices, resulting in the degradation of malathion (41.7%), chlorpyrifos (82.0%), and diazinon (51.3%) [158, 159]. From the above studies, it can be inferred that the initial concentrations of pesticide residues and the amount of sound energy applied are key determinants of the success of this method.

The oxidation and subsequent decontamination of pesticide residues by ozone follows first-order reaction kinetics, in addition to the indirect attack of these compounds through the formation of hydroxyl species. The significance of ozonation in decontaminating pesticide residues has been extensively reviewed by Misra [85]. As a process where the absorption of ozone is accompanied by a chemical reaction, the key determinants of pesticide removal efficiency would include all factors that influence the reaction kinetics and mass transfer.

Supercritical carbon dioxide, being a potent extraction method, has also garnered considerable attention as a means of removing pesticides from foods [44]. This method is especially advantageous as it can effectively extract or remove a wide spectrum of pesticide residues concurrently from a diverse array of foods, including vegetables, grains, honey, and banana flour [120, 123, 126, 142, 144]. Furthermore, Tao et al. [135] demonstrated the potential specificity of SC-CO2, whereby it was utilized to specifically mitigate fenbuconazole and its harmful chiral metabolites in foods through stereoselective degradation and conversion.

13.3.2.3 REMOVAL OF ALLERGENS

Food allergens are a widely prevalent risk that can cause severe immune responses and potentially death in susceptible consumers. Food allergens are predominantly water-soluble proteins (often glycoproteins) with molecular weights up to 70 kDa [36]. Mitigation of food allergens is challenging due to their resistance to gastrointestinal digestion and even harsh traditional processing methods. Nonthermal processes have shown interesting results regarding their ability to potentially decrease the risk of food immunoreactivity.

The immunoreactivity of soy protein extracts, β-conglycinin, walnut proteins, and gingko seed protein extract was significantly reduced in response to various HPP regimes. The reduction in allergenicity in these cases is due to the changes in protein structures induced by high pressure, potentially altering the IgE binding sites of the proteins and preventing their recognition by IgE antibodies [36]. However, the converse was also demonstrated, whereby HPP treatments enhanced the allergenicity of β-lactoglobulin and bullfrog skin native collagen. Since the effects of HPP depended on the protein type under consideration and its molecular structure, this method requires extensive evaluation when dealing with potentially allergenic foods and ingredients.

The allergenicity of soy, peanut, and shrimp proteins was found to decrease in response to PUV treatments [36]. Exposure to PUV can alter the structural confirmation of food allergens, triggering protein aggregation and/or degradation, reduced solubility, and a loss of conformational epitopes, via photochemical, photothermal, or photophysical effects [36].

PEFs are known to alter the molecular structure of proteins and thus their functional properties [50]. Exposure to high-intensity electric fields results in ionization, disruption of electrostatic interactions, unfolding of the 3D structure, and finally aggregation of proteins [152]. The influence of PEF on allergenicity, however, is unique to the type of food matrix, allergen, and molecular arrangement, necessitating further research and process parameter optimization, just like with HPP [36].

In CP application, functional/structural alteration of allergenic proteins is mediated via the action of reactive species that constitute plasma, leading to the masking or disruption of the antibody binding site [39]. While CP treatments are capable of potentially reducing the risk from major allergy-inducing foods including peanuts, soy, shrimp, and wheat, dairy allergens do not seem to be significantly affected, indicating possible matrix specificity [36]. Similarly, the cavitation effect produced by ultrasonication was found to structurally alter and potentially decrease the allergenicity of soy, peanut, and certain shellfish proteins, while having limited to no impact on dairy proteins like α-casein, β-lactoglobulin, and α-lactalbumin [36].

13.3.3 COMBINED APPLICATION OF NONTHERMAL PROCESSES FOR FOOD SAFETY

While nonthermal processes hold tremendous potential in improving overall food safety, high treatment intensities may be necessary to yield reasonable levels of efficacy when used individually. From the information presented in the previous sections, it is apparent that each type of nonthermal technology has its own unique set of limitations, which may be overcome by synergistic combinations of multiple methods. Combining these methods based on the principle of hurdle technologies and suitably optimizing process parameters could potentially result in vastly improved mitigation of microbial and chemical food safety hazards. Such strategies could circumvent the ability of safety hazards to survive individual processes, resulting in ensuring comprehensive levels of food safety. Numerous recent publications have reviewed various combinations of processing methods to improve the mitigation efficacy of a wide array of microbial and chemical hazards in foods [1, 3, 31, 36, 98].

13.4 MQUALITY ASPECTS AND ENVIRONMENT IMPACTS

Emerging nonthermal technologies have been incorporated into food products and have attracted the interest of industries and consumers, particularly those seeking freshness in food products. These technologies have extended shelf life with minimal negative effects [38]. This can be attributed to the improved flavor, color, aroma, and texture compared to thermal technologies, when used under suitable process parameters. Furthermore, these nonthermal technologies have been found to improve functional properties, maintain higher concentrations of bioactive compounds, increase the amount of volatile compounds, and overall make the food products more appealing [42].

The influence of emerging processing techniques on consumers' acceptability of foods has been documented in numerous studies. HPP is the most widely reported nonthermal technique, with research focusing on sensory acceptance. This could be attributed to the availability of this technology and its suitability for both solid and liquid foods. Shen et al. [128] observed that sonicated samples exhibited enhanced color, odor, and cloudiness, complete inactivation of microorganisms (<1 log CFU/mL), improved rheological properties, and suspension stability. It also showed increased antioxidant activity. Another study investigated the impact of CP (40 kV for 90s) and found that it prevented the decline in color, vitamin C, firmness, and soluble solid contents in fresh-cut cantaloupe, maintaining its quality during 10 days of refrigerated storage [161]. Furthermore, another study examined the effect of HPP on the shelf life of coconut water compared to HTST thermal treatment. The results showed that HPP significantly reduced nutrient losses (including proteins, total amino acids, vitamin C, sugar, phenols, and antioxidant potential) in coconut water, outperforming HTST. HPP also better preserved the natural flavor and color of coconut water, extending its shelf life from 15 days to 25 days under refrigerated conditions [82]. Mu et al. [93] investigated the effect of ultrasonic treatments on soybean milk (4°C). The $D_{4,3}$ value in non-sonicated soymilk was 2.47 ± 0.47 μm, whereas it was reduced to 0.44 ± 0.01 μm after 9 minutes of high-intensity ultrasound treatment. The overall number of microorganisms in soymilk was also reduced from 4.51 to 3.95 Log (CFU/mL) after 9 minutes of high-intensity ultrasound treatment. Another interesting study reported the addition of pasteurized orange juice (1–4%) to CP-treated (18 kV for 1.75 min) tender coconut water. The authors determined that treating tender coconut water containing 1% orange juice with CP had a positive effect on its sensory characteristics. The investigation results showed that CP effectively reduced the microbiological load in tender coconut water, and blending it with 1% orange juice produced a beverage with acceptable sensory qualities, while also masking the chemical odor [28]. Ricciardi, Plazzotta, Conte, & Manzocco [116] investigated the effect of PL treatment on ricotta cheese (4°C), and the results demonstrated an effective surface decontamination efficacy of PL on perishable food surfaces, thereby increasing the shelf life with negligible effect on color and texture. However, there was a negative effect on the odor when the PL was applied above 3.1 J/cm^2 and significant changes in the product when the PL was increased above 7.5 J/cm^2 [118]. Martins et al. [83] compared tropical mixed juice (acerola, cashew apple, and melon) exposed to four treatments: control (unprocessed), pasteurized (90°C for 1 min), and two HPP treatments (500 MPa: 5 min; 520 MPa: 2 min) for 28 days under refrigerated storage. Consumer perception of the HPP1 and HPP2 juices during the 28 storage days was similar to that of the control juice on day zero, according to the results. The juices from the control and high-pressure processed (HPP1 and HPP2) batches had an orange color, a fruity aroma, a fruity flavor, a sweet aroma, a fresh juice consistency, and were both sweet and aromatic, unlike the pasteurized sample which was characterized by a yellow color and unacceptable aroma and flavor with increasing storage. Vega-Gálvez et al. [143] explored the application of HPP (300/400/500 MPa: 1, 3, and 5 min) to aloe vera gel and observed that the antioxidant activity showed a maximum value at 500 MPa and influenced the gel firmness. The authors also reported that HPP at 500 MPa (5 min) could effectively preserve the quality attributes of aloe vera gel. Zhou et al. [161] studied the influence of HPP treatment at 200 MPa on sensory quality in chicken sausages with reduced sodium that were made with three different potassium salts (potassium lactate, potassium chloride, and potassium citrate). The results showed that sausages treated with HPP had reduced cook loss, higher firmness, and lower color characteristics (L*, a*, and b*). Additionally, the color and texture qualities were preserved throughout a 40-day cold storage period. It was also shown that employing HPP in conjunction with potassium salts can reduce the need for sodium chloride without sacrificing the textural qualities of sausage to consumer choice. It has been reported that the emerging techniques are able to extract compounds from foods and could be attributed to the enhanced flavor, aroma, and color of the products due to cavitation and electroporation [5, 28]. Overall, the nonthermal techniques were found to have improved flavor, color, aroma, and texture when compared to untreated or thermally processed products.

In this section, we discuss how nonthermal technologies have gained consumer interest and acceptance by improving the quality of foods. These technologies offer better nutritive value and enhanced sensory properties compared to conventional heat treatment (HT) methods. Additionally, emerging nonthermal technologies have positive effects on environmental impacts, such as improved energy utilization and cost economics, which contribute to sustainability. Studies show that agricultural production is responsible for approximately a quarter of all greenhouse gas emissions caused by human activity [13]. Therefore, the food industry is highly interested in newer technologies that offer better product characteristics while being environmentally friendly [9, 23]. It is worth noting that food processing accounts for around 25% of water usage worldwide and is the largest source of emissions of organic water pollutants [99]. Fruits and vegetables, as well as cereals and cereal products, contribute to 30% to 50% of all produced foods that are wasted, making them the leading causes of food loss throughout the entire food supply chain [17]. To effectively address these issues and reduce environmental impacts, it is essential to identify the hotspots where the environmental impact is greatest. Arnal et al. [6] consider the life cycle assessment approach as an innovative strategy. In the case of tomatoes, from farm to table, the processing steps include washing, sorting, thermophysical peeling, canning, sterilizing, and packaging. The processing technique that had the highest environmental impact in tomato processing was canning, contributing to 90% of the overall impact, followed by peeling. The combination of mild PEF treatment and steam was found to be effective in peeling tomatoes, while steam peeling alone requires 100 kPa, resulting in a 20% reduction in energy consumption. It was demonstrated that using PEF for thermophysical peeling reduced CO_2 equivalent by 18.72% [23]. Similarly, a life cycle assessment was conducted for HPP and traditional thermal processing in orange juice production by [20]. Economically, HPP-treated orange juice is 1.78 and 1.40 times more expensive than orange juice treated with indirect thermal processing or retort thermal processing, respectively. However, from a sustainability perspective, HPP was shown to be more effective, as thermal processing exhibits higher numerical values when considering environmental indicators. This could be explained by the fact that HPP uses electricity to create steam, while thermal processing requires fossil fuels. This also indicates that environmentally friendly technology may not always be the economically desirable solution. Thus, HPP has been reported to improve energy efficiency and help in sustainable food production. CP has been studied at different phases of food production and processing, including the treatment of raw materials, the use of packaging materials, the utilization of process equipment, and its impact on the environment. The sustainability of CP depends on its ability to function at low temperatures with minimal energy use, reducing processing time, and microorganism count [16, 138]. Ozone is another important nonthermal technology with GRAS status and is considered a green technology.

Paulikienė, Venslauskas, Raila, Žvirdauskienė, and Naujokienė [102] investigated the environmental impact of ozone in carrot processing. The authors reported that ozone technology and properly managed storage could reduce CO_2 equivalents and SO_2 equivalents emissions by 45.20% and product loss by up to 44%. Comparative production resources were also found to decrease, excluding the total loss of carrot: electricity and fuel were down to 42.72% and 48.54%.

13.5 COMMERCIAL AND REGULATORY REQUIREMENTS

The demand for nutritious food products from natural resources, free of additives and preservatives, has been increasing tremendously [160]. The advancement and commercialization of nonthermal processing technologies can produce safe and minimally processed foods with high nutritional value. These advantages have undoubtedly led to an increased demand for nonthermal processed foods. Nonthermal technology has also been extended to the food product packaging industry. For instance, nonthermal techniques such as electrospraying and electrospinning, collectively known as electrohydrodynamic processing, are increasingly being utilized for food packaging and nutraceutical applications [25].

Numerous governmental regulations and global standards have been implemented in nonthermal technologies to prevent food safety hazards [8]. Despite the advantages of nonthermal technologies in producing high-quality foods, certain limitations in regulations and standards have hampered their applications on an industrial scale. The most critical part of the commercialization is to obtain an operating license and permission from the regulatory authorities and government [8]. The procedure involved in obtaining a license for commercialization is political and time-consuming, owing to the complexity of food legislation. The opinion of interested parties such as consumers, equipment manufacturers, food safety officers, and regulatory organizations varies depending on commercialization, complicating the process. In the case of a nonthermal technique previously commissioned for other applications besides food, it is easy to adapt and commercialize it in the food industry [56]. Indeed, PL was commissioned by the Food and Drug Administration (FDA) in 1995 (food packaging applications), especially aimed at decontaminating bottle caps. Hence, it could be easily adapted by the food industry on an industrial scale.

The FDA regulations are closely related to the food regulations of the United States. However, these regulations are not only followed within the United States but also in several other countries. Currently, more than 150 countries supply food materials to the United States, accounting for 10-15% of the total food supply [4]. In order to ensure proper handling of these imported products, the FDA has developed new regulatory tools and implemented new enforcement measures. These steps include conducting foreign inspections, collaborating and educating foreign regulators and industries, and increasing supply chain transparency and accountability. The FDA's main focus is to deliver safe and nutritious food to the global population. When filing a novel process or technology with the FDA, a sequence of steps must be followed before approval [81]. It is crucial to maintain continuous and detailed communication with the FDA throughout the process development [8]. The technology or process should be described to the FDA, and its panel members should be invited to the site of operation. During the visit, the proposal outline should be provided, and the panel will review the equipment design and evaluate the novelty and efficiency. Data related to nutrition, toxicology, and microbial safety will also be assessed. Microbial studies are primarily conducted by comparing the impact of the process on resistant microorganisms. The applicant is required to submit information regarding food quality and physicochemical changes that occur in the food, such as the production of toxins and nutrition loss. Additionally, the applicant should only utilize approved materials from the government for manufacturing the processing unit.

Overall, the regulations are the most critical part that influences the commercialization of the nonthermal process. It is crucial to convince the regulatory officials regarding the system design, process efficiency, and food safety for approval. Commercializing nonthermal technologies helps to produce novel food products that are minimally processed and safe for consumers. The subsections discuss the commercialization and regulations of various nonthermal processing techniques.

13.5.1 PULSED-ELECTRIC FIELD

Despite several studies being performed using PEF, it is currently not commercialized at an industrial level and is still being used at an experimental stage or pilot plant. The Genesis Juice Corporation was the first to implement this processing technique in the United States for pasteurizing fruit juices at a volume of 200 $L.h^{-1}$. Electric field-based techniques are mainly used for food preservation and were previously commissioned by the FDA [48]. Currently, the PEF technique is not commercialized due to food safety concerns, particularly the physical hazards that arise from the chipping of electrode materials from the processing chamber and their migration into foods [143]. For these reasons, the FDA evaluates the PEF processing unit before approving the technique. After the evaluation, an edict is issued and added to the codex of regulations. Furthermore, the FDA requires a premarket approval application (PMA) that indicates the effectiveness and safety aspects of the technique, operational guidelines, a detailed description of the methodology, and the environmental impacts of PEF. Currently, several investigations have been performed regarding the processing efficiency and safety aspects of PEF. It is possible to address the queries raised by the FDA and implement PEF technology on a large scale.

13.5.2 OZONE TREATMENT

Ozone is a colorless gas formed by combining molecular oxygen with singlet oxygen. Ozone (O_3) is relatively denser than air and exhibits high instability. Due to its instability, ozone is produced whenever needed and not stored [62]. The Federal Food, Drug, and Cosmetic Act, passed in 1950, regulates the application of ozone for food purposes. In 1957, ozone was first used to store and preserve meat in gaseous form. Later, it was accepted in the bottled water industry at a concentration of 0.1 mg/L_{water}. In 1997, ozone was recognized as safe for food applications and categorized under the GRAS (Generally Recognized as Safe) status.

Ozonized water was initially used in Norway for fish preservation. It was employed to freeze the fish and prevent microbial contamination, thereby extending their shelf-life. Germany and Japan also use ionized water for similar purposes [119]. Additionally, the Canadian Food Inspection Agency (CFIA) has approved the usage of ozonized water, particularly for cleaning food surfaces. Following the regulations of several countries, the FDA approved ozone treatment for handling food and food contact surfaces on June 26th, 2001. By the end of 2001, the Ministry of Agriculture accepted ozone application for direct use in fish, meat, poultry, and crustaceans. Currently, the Food Safety and Inspection Service (FSIS) Directive 712.01 considers ozone a safe ingredient, and it is widely used for processing meat products. The CFR 173,368 points out that there is no limitation in the dosage level of ozone for food application and categorizes it as a secondary direct food additive capable of controlling microbial growth [136].

13.5.3 FOOD IRRADIATION

Food irradiation is a nonthermal process primarily used for microbial mitigation and food preservation. Compared to high temperature (HT) methods, food irradiation has a lesser impact on the physiochemical properties of food [93]. The FDA regulates food irradiation as a food additive under the Federal Food, Drug, and Cosmetic Act (*Section 201(s)*). According to the Act (*Section 409*), food additives require premarket approval, and it is illegal to market irradiated foods without complying with the conditions outlined in *Section 402(a)(7)*. Since the foods are irradiated before packaging, the packaging materials are also subjected to irradiation. Therefore, obtaining approval for both irradiated food and packaging materials is necessary [72].

Before 1999, the food additive petition (FAP) process was responsible for regulating irradiated foods. Later, it was added to Title 21 CFR Part 179, which includes the production, handling, and processing of irradiated foods. The 21 CFR 179 comprises Subpart B and C that cover the radiation and radiation sources [141]. Subpart B focuses on the radiation sources used to inspect food, food radiation requirements, and the regulations to use ionizing radiation, UV radiation, radiofrequency, PL, and carbon dioxide laser for food treatment. The irradiated food product should bear a specific logo (Figure 13.1) with the declaration "treated by irradiation" or "treated with irradiation" before distributing them. Additionally, manufacturers are encouraged to add information such as "The treatment does not include radioactive compounds" for consumer knowledge.

FIGURE 13.1 Labeling logo for irradiated foods according to 21 CFR 179 regulation.

The irradiation technology for food applications became famous after the FDA accepted the concept of "chemiclearance." Currently, over 40 countries have authorized food irradiation for different products [114]. According to the National Health Authority, each country has its own set of regulations for approving new technology, and few countries create codes for supporting novel technology based on case analysis. The regulations are generated either based on scientific research or by recommendations from organizations such as the FAO and WHO.

In Europe, the Council Directive 1999/2/EEC regulates irradiated food products. Its primary aim is to protect the public, as the word "irradiation" alone is sufficient to create unease among the public, despite scientific evidence indicating that the technology is effective and safe [151]. Food irradiation is predicted to be safe for food applications by the UK Food Standards Agency and the World Health Organization (WHO). However, Directive 1992/2/EEC emphasizes the need to treat foods within certain irradiation limits to protect public health. It also highlights that irradiation should only be used when there is an acceptable technological need, with no adverse effects on public health, benefiting consumers, and not as an alternative to good manufacturing or health practices [148]. Additionally, it should be noted that irradiation should only be used to treat pathogenic bacteria, prevent food spoilage, and delay premature ripening.

The main objective of the Council Directive is to create a list of approved foods and determine which foods can undergo radioactive treatment. Currently, only a few product categories, such as vegetable seasonings, spices, and aromatic herbs, are authorized for irradiation within Europe. Before being added to the list of authorized foods, a food product must obtain scientific opinions, including dosage levels, from the Council Directive [110]. Several ingredients and foods, including fish, cereals, vegetables, fruits, fresh meat, casein, caseinates, and gum Arabic, have received scientific opinions. For instance, microbial studies on dehydrated rennet casein, prepared according to European regulations, revealed the presence of *Salmonella* spp., *Escherichia coli,* and *Klebsiella pneumoniae.* Therefore, the Council Directive mandates subjecting rennet casein to a radiation dosage of 6 KGy to eliminate microbes. At this dosage level, the physicochemical properties of rennet casein remain unchanged [81].

The Council Directive defines the acceptable dose of radiation as 10 kGy or less for food applications, as a higher dose could degrade the quality and nutritional properties. The regulation also requires irradiating the packaging materials [141]. The labeling requirements defined are similar to those mentioned in the 21 CFR 179, stating "treated with ionizing radiation." Additionally, the label should include the facility name, address, and reference number. If only an ingredient is subjected to radiation instead of the whole product, similar labeling information should be indicated next to the ingredient on the label [110]. The same regulation applies to compound ingredients. Facilities that use irradiation technologies must be registered and approved by the national authority. After approval, information such as the nature of the irradiated foods, treatment date, batch number, packaging material, and personnel involved in the irradiation treatment is forwarded to the commission.

Before the list of authorized foods for irradiation from the Council Directive came into effect, Article 4(6) was responsible for regulating and granting access to treat food using radiation [71]. This chapter included a maximum dosage level for each product. In the case of fruits produced in the Czech Republic and Belgium, the maximum dosage level was 2 kGy; for potatoes produced in the Czech Republic and the UK, it was 0.2 kGy, and for potatoes in Poland, it was 0.1 kGy. Due to the inconsistency in the dosage level across borders, the regulation was ineffective. The following sections discuss the regulations of UV radiation, PL, and carbon dioxide laser for food applications.

In UV radiation, the source should consist of a low-pressure mercury lamp (radiation: 253.7 nm) [132]. The radiation intensity of 1 W for 5–10 ft^2 is employed for controlling surface microorganisms in food. For sterilizing potable water, the absorption coefficient is equal to 0.19/cm or less, water depth equal to 1 cm or less, flow rate of 100 $gallons.h^{-1}$ per watt of 2,537 Angstroms and lamp temperature of 36–6°C is employed. Xenon flash lamps are utilized for PL, and they emit broadband radiation with wavelengths between 200 and 1,100 nm [81]. The pulse duration should not exceed two milliseconds to control surface microorganisms in food, and merely a minimum treatment should be applied to the food to achieve the intended effects. Additionally, the total cumulative energy must not exceed 12.0 J/cm^2. The carbon dioxide laser for food treatment is devised to emit pulsed infrared radiation at a wavelength of 10.6 μm, and its maximum energy should not surpass 9.8 x 10^{-3} J/cm^2 [63]. The carbon dioxide laser is employed chiefly for etching the fruit surface. The maximum energy and surface area employed in etching citrus fruits should not exceed 1.5 x 10^{-3} J and 0.122 cm^2, respectively.

13.5.4 HIGH-PRESSURE PROCESSING

HPP is an efficient tool for inactivating vegetative spoilage microorganisms and harmful pathogens in liquid foods and solid foods with high moisture content [97]. Food products treated with HPP are commonly distributed under refrigerated conditions. The processed food must adhere to specific regulations, such as GMP conditions or HACCP, depending on the product type. In 1990, kiwi and apple jams developed in Japan were first subjected to pressure waves. Later in 1997, guacamole treated using HPP was commercialized in the United States. A decade later, around 120 HPP plants were installed worldwide for processing food products on an industrial scale. HPP offers the opportunity to develop value-added products with high quality. The intense pressure of 400–600 MPa and temperature below 45°C are used in HPP for food preservation [55]. This process has minimal impact on the nutritional value, texture, and taste of the foods. HPP is an efficient food processing technique for eliminating microorganisms like *Listeria monocytogenes, Salmonella*, and *E. coli* from processed meat. Due to its greater pressure resistance, *E. coli* O157:H7 has been suggested by the USDA as an indicator strain for HPP. The HPP condition capable of achieving a 5-log reduction is sufficient for producing microbially safe food products [25]. In Europe, the fruits-based products were initially pasteurized using HPP in 2001 and were categorized under novel ingredients after an application submitted to the regulatory authority in France by the Danone group. The French authorities stated that the high pressure of 8 kbar, temperature of 20°C, and processing time of 6 min is sufficient for food applications [160]. The test conducted on HPP indicated that it could preserve the functionality of food and not affect it to a greater extent as heat-based pasteurization. Since then, the application of HPP has tremendously increased in Europe, and currently, the HPP treated products are regulated under the Novel Food Regulations. The Pressure Equipment Directive regulates the container utilized for holding the treated food.

13.6 SUMMARY

In order to meet consumer needs, ensure food safety, and preserve nutritional value, food industries currently employ a variety of process technologies. Thermal processing techniques are frequently applied to increase the shelf life of foods; however, these technologies have been linked to long-term changes in the sensory and nutritional profiles of the treated foods. The main focus of food scientists and engineers in recent decades has been the development of alternative processing and preservation methods that can maintain the nutritional characteristics of food products while also being affordable and environmentally friendly. To meet consumer demands for convenience, safety, and freshness while reducing the use of additives and preservatives, a variety of nonthermal techniques have been developed.

Compared to other products, consumers' perceptions and acceptance of food are particularly conservative. Consumers are familiar with and generally accept traditional food processing methods like cooking and freezing. However, it is critical to comprehend how customers create impressions regarding "less known" cutting-edge emerging technologies. The assessment of available literature demonstrates the change in food culture resulting from the change in lifestyles and consumers' perception of environmental and personal well-being. Emerging technologies offer considerable potential for treating these types of foods while preserving their quality and beneficial ingredients.

However, introducing new technologies to the sector remains a significant challenge. Even though novel treatments have great potential to eliminate unwanted bacteria in food, achieving complete control over the variables involved in novel process operations is currently challenging. Nevertheless, there have been substantial improvements in the development of techniques that demonstrate commercial potential in recent years. The critical issue lies not in the classification or characterization of the treatment process, but in the validation and control of the specified critical control parameters. The food industry is also interested in technology with low environmental impact. There are extensive regulations surrounding nonthermal technology for foods, including product-specific criteria for various foods such as fruit juice and milk, as well as regulations governing novel foods, processed foods, and food contact materials that contain these foods. However, the lack of a single global set of laws in food manufacturing and the food safety aspects of nonthermal technologies still act as obstacles preventing wider industrial-scale adoption of these technologies. While there are some international standards in place, such as the Codex Alimentarius and reports from organizations like the WHO, which serve as a framework for most of the laws in Europe and around the world, these "international" standards are not enforceable. However, cutting-edge studies on the interactions between food and technology, applicable to a versatile range of food matrices, and the improvements in food engineering that can lead to reduced energy consumption, improved product quality, and reduced environmental impacts, may help in developing a regulatory framework for the majority of operations. Moreover, there are significant concerns about how consumers would react to these unique novel technologies, and it is impossible to ignore the skeptical attitude caused by consumers' unawareness about the technology. By establishing regulatory frameworks and regulations, these consumer perceptions and concerns can be fairly managed.

KEYWORDS

- **cold plasma**
- **generally recognized as safe**
- **high temperature**
- **high temperature short time**
- **high-pressure processing**
- **potato dextrose agar**
- **pulsed electric field**
- **pulsed light**
- **ultraviolet light**

REFERENCES

1. Adebo, O. A., Molelekoa, T., Makhuvele, R., Adebiyi, J. A., Oyedeji, A. B., Gbashi, S., & Njobeh, P. B., (2021). A review on novel non-thermal food processing techniques for mycotoxin reduction. *International Journal of Food Science and Technology, 56*(1), 13–27.
2. Alexandre, A. P. S., Castanha, N., Costa, N. S., Santos, A. S., Badiale-Furlong, E., Augusto, P. E. D., & Calori-Domingues, M. A., (2019). Ozone technology to reduce zearalenone contamination in whole maize flour: Degradation kinetics and impact on quality. *Journal of the Science of Food and Agriculture, 99*(15), 6814–6821.
3. Alizadeh, A. M., Hashempour-Baltork, F., Khaneghah, A. M., & Hosseini, H., (2021). New perspective approaches in controlling fungi and mycotoxins in food using emerging and green technologies. *Current Opinion in Food Science, 39*, 7–15.
4. Alsaleem, K. A., Hammam, A. R., & Awasti, N., (2021). Safety, regulatory aspects and environmental impacts of using non-thermal processing techniques for dairy industries. In: *Non-Thermal Processing Technologies for the Dairy Industry* (pp. 157–172): CRC Press.
5. Arcena, M. R., Leong, S. Y., Then, S., Hochberg, M., Sack, M., Mueller, G., & Oey, I., (2021). The effect of pulsed electric fields pre-treatment on the volatile and phenolic profiles of Merlot grape musts at different winemaking stages and the sensory characteristics of the finished wines. *Innovative Food Science and Emerging Technologies, 70*, 102698.
6. Arnal, Á. J., Royo, P., Pataro, G., Ferrari, G., Ferreira, V. J., López-Sabirón, A. M., & Ferreira, G. A., (2018). Implementation of PEF treatment at real-scale tomatoes processing considering LCA methodology as an innovation strategy in the agri-food sector. *Sustainability, 10*(4), 979.
7. Aronsson, A. K., Landquist, B., Esturo, A., Olafsdottir, G., Ramos, S., Pardo, G., & Bogason, S., (2014). *The Applicability of LCA to Evaluate the Key Environmental Challenges in Food Supply Chains.* Paper presented at the Proceedings of the 9th International Conference on Life Cycle Assessment in the Agri-Food Sector (LCA Food 2014), San Francisco, California, USA.
8. Awasti, N., Sunkesula, V., & Hammam, A. R., (2022). Food safety management systems in the food processing industry. In: *Biological and Chemical Hazards in Food and Food Products* (pp. 241–262). Apple Academic Press.
9. Babu, K. S., & Amamcharla, J. K., (2022). Generation methods, stability, detection techniques, and applications of bulk nanobubbles in agro-food industries: A review and future perspective. *Critical Reviews in Food Science and Nutrition*, 1–20.
10. Balny, C., (2006). What lies in the future of high-pressure bioscience? *Biochimica et Biophysica Acta – Proteins and Proteomics, 1764*(3), 632–639.
11. Barba, F. J., Parniakov, O., Pereira, S. A., Wiktor, A., Grimi, N., Boussetta, N., & Witrowa-Rajchert, D., (2015). Current applications and new opportunities for the use of pulsed electric fields in food science and industry. *Food Research International, 77*, 773–798.

12. Barbosa-Cánovas, G. V., Bermúdez-Aguirre, D., Franco, B. G., Candoğan, K., & Shin, G. Y., (2022). Novel food processing technologies and regulatory hurdles. In: *Ensuring Global Food Safety* (pp. 221–228). Elsevier.
13. Bennetzen, E. H., Smith, P., & Porter, J. R., (2016). Agricultural production and greenhouse gas emissions from world regions—The major trends over 40 years. *Global Environmental Change, 37*, 43–55.
14. Bhavya, M., & Umesh, H. H., (2017). Pulsed light processing of foods for microbial safety. *Food Quality and Safety, 1*(3), 187–202.
15. Birmpa, A., Sfika, V., & Vantarakis, A., (2013). Ultraviolet light and ultrasound as non-thermal treatments for the inactivation of microorganisms in fresh ready-to-eat foods. *International Journal of Food Microbiology, 167*(1), 96–102.
16. Bourke, P., Ziuzina, D., Boehm, D., Cullen, P. J., & Keener, K., (2018). The potential of cold plasma for safe and sustainable food production. *Trends in Biotechnology, 36*(6), 615–626.
17. Boye, J. I., & Arcand, Y., (2013). Current trends in green technologies in food production and processing. *Food Engineering Reviews, 5*(1), 1–17.
18. Bridgman, P., (2013). The coagulation of albumen by pressure. In: *Collected Experimental Papers* (Vol. II, pp. 735–736): Harvard University Press.
19. Buszewska-Forajta, M., (2020). Mycotoxins, invisible danger of feedstuff with toxic effect on animals. *Toxicon, 182*, 34–53.
20. Cacace, F., Bottani, E., Rizzi, A., & Vignali, G., (2020). Evaluation of the economic and environmental sustainability of high-pressure processing of foods. *Innovative Food Science and Emerging Technologies, 60*, 102281.
21. Cassar, J. R., Ouyang, B., Krishnamurthy, K., & Demirci, A., (2020). Microbial decontamination of food by light-based technologies: Ultraviolet (UV) light, pulsed UV light (PUV), and UV light-emitting diodes (UV-LED). In: *Food Safety Engineering* (pp. 493–521). Springer.
22. Chacha, J. S., Zhang, L., Ofoedu, C. E., Suleiman, R. A., Dotto, J. M., Roobab, U., & Hossaini, S. M., (2021). Revisiting non-thermal food processing and preservation methods—Action mechanisms, pros and cons: A technological update (2016–2021). *Foods, 10*(6), 1430.
23. Chakka, A. K., Sriraksha, M., & Ravishankar, C., (2021). Sustainability of emerging green non-thermal technologies in the food industry with food safety perspective: A review. *LWT, 151*, 112140.
24. Chambers, L. A., & Gaines, N., (1932). Some effects of intense audible sound on living organisms and cells. *Journal of Cellular and Comparative Physiology, 1*(3), 451–473.
25. Charles, A. P. R., Jin, T. Z., Mu, R., & Wu, Y., (2021). Electrohydrodynamic processing of natural polymers for active food packaging: A comprehensive review. *Comprehensive Reviews in Food Science and Food Safety, 20*(6), 6027–6056.
26. Chen, F., Zeng, L., Zhang, Y., Liao, X., Ge, Y., Hu, X., & Jiang, L., (2009). Degradation behavior of methamidophos and chlorpyrifos in apple juice treated with pulsed electric fields. *Food Chemistry, 112*(4), 956–961.
27. Chiozzi, V., Agriopoulou, S., & Varzakas, T., (2022). Advances, applications, and comparison of thermal (pasteurization, sterilization, and aseptic packaging) against

non-thermal (ultrasounds, UV radiation, ozonation, high hydrostatic pressure) technologies in food processing. *Applied Sciences, 12*(4), 2202.

28. Chutia, H., Mahanta, C. L., Ojah, N., & Choudhury, A. J., (2020). Fuzzy logic approach for optimization of blended beverage of cold plasma treated TCW and orange juice. *Journal of Food Measurement and Characterization, 14*(4), 1926–1938.
29. Crookes, W., (1879). V. the Bakerian lecture. —On the illumination of lines of molecular pressure, and the trajectory of molecules. *Philosophical Transactions of the Royal Society of London,* (170), 135–164.
30. Delsart, C., Franc, C., Grimi, N., Revel, G. D., Vorobiev, E., & Peuchot, M. M., (2016). *Effects of Pulsed Electric Fields on Four Residual Fungicides in White Wines.* Paper presented at the 1st world congress on electroporation and pulsed electric fields in biology, medicine, and Food & Environmental Technologies.
31. Deng, L. Z., Tao, Y., Mujumdar, A. S., Pan, Z., Chen, C., Yang, X. H., & Xiao, H. W., (2020). Recent advances in non-thermal decontamination technologies for microorganisms and mycotoxins in low-moisture foods. *Trends in Food Science and Technology, 106*, 104–112.
32. Diao, E., Hou, H., Chen, B., Shan, C., & Dong, H., (2013). Ozonolysis efficiency and safety evaluation of aflatoxin B1 in peanuts. *Food and Chemical Toxicology, 55*, 519–525.
33. Diao, E., Shen, X., Zhang, Z., Ji, N., Ma, W., & Dong, H., (2015). Safety evaluation of aflatoxin B1 in peanut oil after ultraviolet irradiation detoxification in a photodegradation reactor. *International Journal of Food Science and Technology, 50*(1), 41–47.
34. Doevenspeck, H., (1960). *Verfahren und Vorrichtungzur Gewinnung der Einzelnen Phasenaus Dispersen Systemen* (Vol. 1, pp. 237–541). DE.
35. Doevenspeck, H., (1961). Influencing cells and cell walls by electrostatic impulses. *Fleischwirtschaft, 13*(12), 968–987.
36. Dong, X., Wang, J., & Raghavan, V., (2021). Critical reviews and recent advances of novel non-thermal processing techniques on the modification of food allergens. *Critical Reviews in Food Science and Nutrition, 61*(2), 196–210.
37. Donsì, F., Ferrari, G., & Pataro, G., (2010). Applications of pulsed electric field treatments for the enhancement of mass transfer from vegetable tissue. *Food Eng. Rev., 2*, 109–130.
38. Dos Santos, R. C., Magnani, M., Ramos, G. L. D. P. A., Bezerril, F. F., De Freitas, M. Q., Cruz, A. G., & Pimentel, T. C., (2022). Emerging technologies in food processing: Impacts on sensory characteristics and consumer perception. *Current Opinion in Food Science*, 100892.
39. Ekezie, F. G. C., Cheng, J. H., & Sun, D. W., (2018). Effects of nonthermal food processing technologies on food allergens: A review of recent research advances. *Trends in Food Science and Technology, 74*, 12–25.
40. Elamin, W. M., Endan, J. B., Yosuf, Y. A., Shamsudin, R., & Ahmedov, A., (2015). High pressure processing technology and equipment evolution: A review. *Journal of Engineering Science and Technology Review, 8*(5).
41. Feizollahi, E., Iqdiam, B., Vasanthan, T., Thilakarathna, M. S., & Roopesh, M., (2020). Effects of atmospheric-pressure cold plasma treatment on deoxynivalenol

degradation, quality parameters, and germination of barley grains. *Applied Sciences, 10*(10), 3530.
42. Galanakis, C. M., (2021). Functionality of food components and emerging technologies. *Foods, 10*(1), 128.
43. Gaonkar, A. G., (1995). *Food Processing: Recent Developments.*
44. Gavahian, M., & Khaneghah, A. M., (2020). Cold plasma as a tool for the elimination of food contaminants: Recent advances and future trends. *Critical Reviews in Food Science and Nutrition, 60*(9), 1581–1592.
45. Gavahian, M., Pallares, N., Al Khawli, F., Ferrer, E., & Barba, F. J., (2020). Recent advances in the application of innovative food processing technologies for mycotoxins and pesticide reduction in foods. *Trends in Food Science and Technology, 106*, 209–218.
46. Gayán, E., Govers, S. K., & Aertsen, A., (2017). Impact of high hydrostatic pressure on bacterial proteostasis. *Biophysical Chemistry, 231*, 3–9.
47. Gomez-Lopez, V. M., Pataro, G., Tiwari, B., Gozzi, M., Meireles, M. A. A., Wang, S., & Sastry, S., (2022). Guidelines on reporting treatment conditions for emerging technologies in food processing. *Critical Reviews in Food Science and Nutrition, 62*(21), 5925–5949.
48. Góngora-Nieto, M., Sepúlveda, D., Pedrow, P., Barbosa-Cánovas, G., & Swanson, B., (2002). Food processing by pulsed electric fields: Treatment delivery, inactivation level, and regulatory aspects. *LWT–Food Science and Technology, 35*(5), 375–388.
49. Hamilton, W., & Sale, A., (1967). Effects of high electric fields on microorganisms: II. Mechanism of action of the lethal effect. *Biochimica et Biophysica Acta – General Subjects, 148*(3), 789–800.
50. Han, Z., Cai, M. J., Cheng, J. H., & Sun, D. W., (2018). Effects of electric fields and electromagnetic wave on food protein structure and functionality: A review. *Trends in Food Science and Technology, 75*, 1–9.
51. Hassanpour, M., Rezaie, M. R., & Baghizadeh, A., (2019). Practical analysis of aflatoxin M1 reduction in pasteurized milk using low dose gamma irradiation. *Journal of Environmental Health Science and Engineering, 17*(2), 863–872.
52. Hirneisen, K. A., Black, E. P., Cascarino, J. L., Fino, V. R., Hoover, D. G., & Kniel, K. E., (2010). Viral inactivation in foods: A review of traditional and novel food-processing technologies. *Comprehensive Reviews in Food Science and Food Safety, 9*(1), 3–20.
53. Hite, B. H., (1899). *The Effect of Pressure in the Preservation of Milk: A Preliminary Report* (Vol. 58). The Station.
54. Hojnik, N., Cvelbar, U., Tavčar-Kalcher, G., Walsh, J. L., & Križaj, I., (2017). Mycotoxin decontamination of food: Cold atmospheric pressure plasma versus "classic" decontamination. *Toxins, 9*(5), 151.
55. Huang, H. W., Wu, S. J., Lu, J. K., Shyu, Y. T., & Wang, C. Y., (2017). Current status and future trends of high-pressure processing in food industry. *Food Control, 72*, 1–8.
56. Huang, R., & Chen, H., (2019). Sanitation of tomatoes based on a combined approach of washing process and pulsed light in conjunction with selected disinfectants. *Food Research International, 116*, 778–785.
57. Hugas, M., Garriga, M., & Monfort, J., (2002). New mild technologies in meat processing: High pressure as a model technology. *Meat Science, 62*(3), 359–371.

58. Iizuka, T., Maeda, S., & Shimizu, A., (2013). Removal of pesticide residue in cherry tomato by hydrostatic pressure. *Journal of Food Engineering, 116*(4), 796–800.
59. Iizuka, T., & Shimizu, A., (2014a). Removal of pesticide residue from brussels sprouts by hydrostatic pressure. *Innovative Food Science and Emerging Technologies, 22*, 70–75.
60. Iizuka, T., & Shimizu, A., (2014b). Removal of pesticide residue from cherry tomatoes by hydrostatic pressure (part 2). *Innovative Food Science and Emerging Technologies, 26*, 34–39.
61. Jabłońska, J., & Mańkowska, D., (2014). The influence of UV, X, and microwave radiation on the aflatoxin B1 concentration in nuts. *Biotechnology and Food Science, 78*(2), 111–119.
62. Jadhav, H. B., Annapure, U. S., & Deshmukh, R. R., (2021). Non-thermal technologies for food processing. *Frontiers in Nutrition*, 248.
63. Jan, A., Sood, M., Sofi, S., & Norzom, T., (2017). Non-thermal processing in food applications: A review. *International Journal of Food Science and Nutrition, 2*(6), 171–180.
64. Janić, H. E., Vukić, M., Pezo, L., Orčić, D., Puač, N., Škoro, N., & Šoronja, S. D., (2019). Effect of atmospheric cold plasma treatments on reduction of *Alternaria* toxins content in wheat flour. *Toxins, 11*(12), 704.
65. John, D., & Ramaswamy, H. S., (2018). Pulsed light technology to enhance food safety and quality: A mini-review. *Current Opinion in Food Science, 23*, 70–79.
66. Khanashyam, A. C., Shanker, M. A., Kothakota, A., Mahanti, N. K., & Pandiselvam, R., (2021). Ozone applications in milk and meat industry. *Ozone: Science and Engineering*, 1–16.
67. Khandpur, P., & Gogate, P. R., (2015). Effect of novel ultrasound-based processing on the nutrition quality of different fruit and vegetable juices. *Ultrasonics Sonochemistry, 27*, 125–136.
68. Khoori, E., Hakimzadeh, V., Mohammadi, S. A., & Rashidi, H., (2020). Effect of ozonation, UV light radiation, and pulsed electric field processes on the reduction of total aflatoxin and aflatoxin M1 in acidophilus milk. *Journal of Food Processing and Preservation, 44*(10), e14729.
69. Knorr, D., & Watzke, H., (2019). Food processing at a crossroad. *Frontiers in Nutrition, 6*, 85.
70. Komolprasert, V., (2016). Packaging food for radiation processing. *Radiation Physics and Chemistry, 129*, 35–38.
71. Koutchma, T., Keener, L., & Kotilainen, H., (2018). Global harmonization initiative (GHI) consensus document on food irradiation.*Discordant International Regulations of Food Irradiation are a Public Health Impediment and a Barrier to Global Trade*, pp. 1–18.
72. Krstović, S., Krulj, J., Jakšić, S., Bočarov-Stančić, A., & Jajić, I., (2021). Ozone as decontaminating agent for ground corn containing deoxynivalenol, zearalenone, and ochratoxin A. *Cereal Chemistry, 98*(1), 135–143.
73. Li, X., & Farid, M., (2016). A review on recent development in non-conventional food sterilization technologies. *Journal of Food Engineering, 182*, 33–45.
74. Liao, X., Liu, D., Xiang, Q., Ahn, J., Chen, S., Ye, X., & Ding, T., (2017). Inactivation mechanisms of non-thermal plasma on microbes: A review. *Food Control, 75*, 83–91.

75. Liu, Y., Li, M., Liu, Y., & Bian, K., (2019). Structures of reaction products and degradation pathways of aflatoxin B1 by ultrasound treatment. *Toxins, 11*(9), 526.
76. López, M., Calvo, T., Prieto, M., Múgica-Vidal, R., Muro-Fraguas, I., Alba-Elías, F., & Alvarez-Ordóñez, A., (2019). A review on non-thermal atmospheric plasma for food preservation: Mode of action, determinants of effectiveness, and applications. *Frontiers in Microbiology, 10*, 622.
77. Lou, F., DiCaprio, E., Li, X., Dai, X., Ma, Y., Hughes, J., & Li, J., (2016). Variable high-pressure-processing sensitivities for genogroup II human noroviruses. *Applied and Environmental Microbiology, 82*(19), 6037–6045.
78. Lozowicka, B., Jankowska, M., Hrynko, I., & Kaczynski, P., (2016). Removal of 16 pesticide residues from strawberries by washing with tap and ozone water, ultrasonic cleaning and boiling. *Environmental Monitoring and Assessment, 188*(1), 1–19.
79. Luo, X., Wang, R., Wang, L., Li, Y., Wang, Y., & Chen, Z., (2014). Detoxification of aflatoxin in corn flour by ozone. *Journal of the Science of Food and Agriculture, 94*(11), 2253–2258.
80. Lyng, J. G., Lalor, F., Quinn, G., & Pedrós-Garrido, S., (2022). Regulation of foods processed by pulsed or moderate electric fields (PEF or MEF). In: *Pulsed Electric Fields Technology for the Food Industry* (pp. 541–561). Springer.
81. Ma, Y., Xu, L., Wang, S., Xu, Z., Liao, X., & Cheng, Y., (2019). Comparison of the quality attributes of coconut waters by high-pressure processing and high-temperature short time during the refrigerated storage. *Food Science and Nutrition, 7*(4), 1512–1519.
82. Mahendran, R., Ramanan, K. R., Barba, F. J., Lorenzo, J. M., López-Fernández, O., Munekata, P. E., & Tiwari, B. K., (2019). Recent advances in the application of pulsed light processing for improving food safety and increasing shelf life. *Trends in Food Science and Technology, 88*, 67–79.
83. Martins, I. B. A., De Souza, C. R., De Alcantara, M., Rosenthal, A., Ares, G., & Deliza, R., (2022). How are the sensory properties perceived by consumers? A case study with pressurized tropical mixed juice. *Food Research International, 152*, 110940.
84. Masschelein, W. J., & Rice, R. G., (2016). *Ultraviolet Light in Water and Wastewater Sanitation*. CRC press.
85. Misra, N., (2015). The contribution of non-thermal and advanced oxidation technologies towards dissipation of pesticide residues. *Trends in Food Science and Technology, 45*(2), 229–244.
86. Misra, N., Koubaa, M., Roohinejad, S., Juliano, P., Alpas, H., Inácio, R. S., & Barba, F., (2017). Landmarks in the historical development of twenty first century food processing technologies. *Food Research International, 97*, 318–339.
87. Misra, N., Schlüter, O., & Cullen, P., (2016). *Fundamentals and Applications*. In: Elsevier.
88. Misra, N., Sullivan, C., Pankaj, S., Alvarez-Jubete, L., Cama, R., Jacoby, F., & Cullen, P., (2014). Enhancement of oil spreadability of biscuit surface by nonthermal barrier discharge plasma. *Innovative Food Science and Emerging Technologies, 26*, 456–461.
89. Misra, N., Tiwari, B., Raghavarao, K., & Cullen, P., (2011). Nonthermal plasma inactivation of food-borne pathogens. *Food Engineering Reviews, 3*(3), 159–170.

90. Moreau, M., Lescure, G., Agoulon, A., Svinareff, P., Orange, N., & Feuilloley, M., (2013). Application of the pulsed light technology to mycotoxin degradation and inactivation. *Journal of Applied Toxicology, 33*(5), 357–363.
91. Morehouse, K. M., & Komolprasert, V., (2004). *Irradiation of Food and Packaging: An Overview.*
92. Motarjemi, Y., Moy, G., & Todd, E., (2013). *Encyclopedia of Food Safety.* Academic Press.
93. Mu, Q., Su, H., Zhou, Q., Xiao, S., Zhu, L., Xu, X., & Hu, H., (2022). Effect of ultrasound on functional properties, flavor characteristics, and storage stability of soybean milk. *Food Chemistry, 381*, 132158.
94. Muntean, M. V., Marian, O., Barbieru, V., Cătunescu, G. M., Ranta, O., Drocas, I., & Terhes, S., (2016). High pressure processing in food industry–characteristics and applications. *Agriculture and Agricultural Science Procedia, 10*, 377–383.
95. Nabi, B. G., Mukhtar, K., Arshad, R. N., Radicetti, E., Tedeschi, P., Shahbaz, M. U., & Aadil, R. M., (2021). High-pressure processing for sustainable food supply. *Sustainability, 13*(24), 13908.
96. Nunes, V. M., Moosavi, M., Khaneghah, A. M., & Oliveira, C. A., (2021). Innovative modifications in food processing to reduce the levels of mycotoxins. *Current Opinion in Food Science, 38*, 155–161.
97. Ölmez, H., (2013). Water consumption, reuse, and reduction strategies in food processing. *Sustainable Food Processing*, 401–434.
98. Ottman, N., Ruokolainen, L., Suomalainen, A., Sinkko, H., Karisola, P., Lehtimäki, J., & Fyhrquist, N., (2019). Soil exposure modifies the gut microbiota and supports immune tolerance in a mouse model. *Journal of Allergy and Clinical Immunology, 143*(3), 1198–1206. e1112.
99. Pandiselvam, R., Subhashini, S., Banuu, P. E., Kothakota, A., Ramesh, S., & Shahir, S., (2019). Ozone based food preservation: A promising green technology for enhanced food safety. *Ozone: Science and Engineering, 41*(1), 17–34.
100. Pankaj, S., Shi, H., & Keener, K. M., (2018). A review of novel physical and chemical decontamination technologies for aflatoxin in food. *Trends in Food Science and Technology, 71*, 73–83.
101. Pankaj, S. K., Bueno-Ferrer, C., Misra, N., Milosavljević, V., O'donnell, C., Bourke, P., & Cullen, P., (2014). Applications of cold plasma technology in food packaging. *Trends in Food Science and Technology, 35*(1), 5–17.
102. Paulikienė, S., Venslauskas, K., Raila, A., Žvirdauskienė, R., & Naujokienė, V., (2020). The influence of ozone technology on reduction of carrot loss and environmental IMPACT. *Journal of Cleaner Production, 244*, 118734.
103. Pexara, A., & Govaris, A., (2020). Foodborne viruses and innovative non-thermal food-processing technologies. *Foods, 9*(11), 1520.
104. Phan, K. T. K., Phan, H. T., Boonyawan, D., Intipunya, P., Brennan, C. S., Regenstein, J. M., & Phimolsiripol, Y., (2018). Non-thermal plasma for elimination of pesticide residues in mango. *Innovative Food Science and Emerging Technologies, 48*, 164–171.
105. Piemontese, L., Messia, M. C., Marconi, E., Falasca, L., Zivoli, R., Gambacorta, L., & Solfrizzo, M., (2018). Effect of gaseous ozone treatments on DON, microbial

contaminants and technological parameters of wheat and semolina. *Food Additives and Contaminants: Part A, 35*(4), 761–772.
106. Pinela, J., & Ferreira, I., (2017). Nonthermal physical technologies to decontaminate and extend the shelf-life of fruits and vegetables: Trends aiming at quality and safety. *Critical Reviews in Food Science and Nutrition, 57*(10), 2095–2111.
107. Porto, Y. D., Trombete, F. M., Freitas-Silva, O., De Castro, I. M., Direito, G. M., & Ascheri, J. L. R., (2019). Gaseous ozonation to reduce aflatoxins levels and microbial contamination in corn grits. *Microorganisms, 7*(8), 220.
108. Premi, M., & Khan, K. A., (2018). Irradiation technology: Processing of fruits and vegetables. In: *Technological Interventions in the Processing of Fruits and Vegetables* (pp. 209–226). Apple Academic Press.
109. Priyadarshini, A., Rajauria, G., O'Donnell, C. P., & Tiwari, B. K., (2019). Emerging food processing technologies and factors impacting their industrial adoption. *Critical Reviews in Food Science and Nutrition, 59*(19), 3082–3101.
110. Qi, L., Li, Y., Luo, X., Wang, R., Zheng, R., Wang, L., & Chen, Z., (2016). Detoxification of zearalenone and ochratoxin A by ozone and quality evaluation of ozonized corn. *Food Additives and Contaminants: Part A, 33*(11), 1700–1710.
111. Randeniya, L. K., & De Groot, G. J., (2015). Non-thermal plasma treatment of agricultural seeds for stimulation of germination, removal of surface contamination and other benefits: A review. *Plasma Processes and Polymers, 12*(7), 608–623.
112. Ravindran, R., & Jaiswal, A. K., (2019). Wholesomeness and safety aspects of irradiated foods. *Food Chemistry, 285*, 363–368.
113. Rentschler, H. C., Nagy, R., & Mouromseff, G., (1941). Bactericidal effect of ultraviolet radiation. *Journal of Bacteriology, 41*(6), 745–774.
114. Režek, J. A., Nutrizio, M., Djekić, I., Pleslić, S., & Chemat, F., (2021). Internet of nonthermal food processing technologies (IoNTP): Food industry 4.0 and sustainability. *Applied Sciences, 11*(2), 686.
115. Ribeiro, N. G., Xavier-Santos, D., Campelo, P. H., Guimarães, J. T., Pimentel, T. C., Duarte, M. C. K., & Cruz, A. G., (2022). Dairy foods and novel thermal and non-thermal processing: A bibliometric analysis. *Innovative Food Science and Emerging Technologies*, 102934.
116. Ricciardi, F. E., Plazzotta, S., Conte, A., & Manzocco, L., (2021). Effect of pulsed light on microbial inactivation, sensory properties and protein structure of fresh ricotta cheese. *LWT, 139*, 110556.
117. Rice, R. G., & Graham, D. M., (2001). *US FDA Regulatory Approval of Ozone as an Antimicrobial Agent–What is Allowed and What Needs to be Understood* (Vol. 29, No. 5, pp. 22–31). Ozone News.
118. Rissato, S. R., Galhiane, M. S., Knoll, F. R., & Apon, B. M., (2004). Supercritical fluid extraction for pesticide multiresidue analysis in honey: Determination by gas chromatography with electron-capture and mass spectrometry detection. *Journal of Chromatography A, 1048*(2), 153–159.
119. Rivalain, N., Roquain, J., & Demazeau, G., (2010). Development of high hydrostatic pressure in biosciences: Pressure effect on biological structures and potential applications in biotechnologies. *Biotechnology Advances, 28*(6), 659–672.
120. Rodríguez-Bencomo, J. J., Sanchis, V., Viñas, I., Martín-Belloso, O., & Soliva-Fortuny, R., (2020). Formation of patulin-glutathione conjugates induced by pulsed

light: A tentative strategy for patulin degradation in apple juices. *Food Chemistry, 315*, 126283.

121. Saito-Shida, S., Nemoto, S., & Matsuda, R., (2014). Multiresidue analysis of pesticides in vegetables and fruits by supercritical fluid extraction and liquid chromatography-tandem mass spectrometry. *Food Hygiene and Safety Science, 55*(3), 142–151.
122. Sale, A., & Hamilton, W., (1967). Effects of high electric fields on microorganisms: I. Killing of bacteria and yeasts. *Biochimica et Biophysica Acta – General Subjects, 148*(3), 781–788.
123. Santos, A. A. P., Vela-Paredes, R. S., Santos, A. S., Costa, N. S., Canniatti-Brazaca, S. G., Calori-Domingues, M. A., & Augusto, P. E. D., (2018). Ozone treatment to reduce deoxynivalenol (DON) and zearalenone (ZEN) contamination in wheat bran and its impact on nutritional quality. *Food Additives and Contaminants: Part A, 35*(6), 1189–1199.
124. Sartori, R., Higino, M., Bastos, L., & Mendes, M., (2017). Supercritical extraction of pesticides from banana: Experimental and modeling. *The Journal of Supercritical Fluids, 128*, 149–158.
125. Schottroff, F., Fröhling, A., Zunabovic-Pichler, M., Krottenthaler, A., Schlüter, O., & Jäger, H., (2018). Sublethal injury and viable but non-culturable (VBNC) state in microorganisms during preservation of food and biological materials by non-thermal processes. *Frontiers in Microbiology*, 2773.
126. Shankar, R., Kaushik, U., & Bhat, S. A., (2014). The emerging technology in the sector of food technology-the non-thermal technology. *International Journal of Innovation and Applied Studies, 6*(4), 941.
127. Sharareh, M., Maryam, T., Azadeh, K., & Ali, M. S., (2015). Effect of UV irradiation on quantity and structure of aflatoxins using HPLC, GC/MS and FT-IR techniques. *Ludus Vitalis, 11*(1), 64–69.
128. Shen, Y., Zhu, D., Xi, P., Cai, T., Cao, X., Liu, H., & Li, J., (2021). Effects of temperature-controlled ultrasound treatment on sensory properties, physical characteristics, and antioxidant activity of cloudy apple juice. *LWT, 142*, 111030.
129. Simpson, A. M. A., & Mitch, W. A., (2022). Chlorine and ozone disinfection and disinfection byproducts in postharvest food processing facilities: A review. *Critical Reviews in Environmental Science and Technology, 52*(11), 1825–1867.
130. Singh, H., Bhardwaj, S. K., Khatri, M., Kim, K. H., & Bhardwaj, N., (2021). UVC radiation for food safety: An emerging technology for the microbial disinfection of food products. *Chemical Engineering Journal, 417*, 128084.
131. Singh, P. K., Kumar, S., Kumar, P., & Bhat, Z., (2012). Pulsed light and pulsed electric field-emerging non thermal decontamination of meat. *American Journal of Food Technology, 7*(9), 506–516.
132. Sitzmann, W., & Münch, E., (1988). Das ELCRACK® verfahren: Ein neues verfahren zur verarbeitung tierischer rohstoffe. *Die Fleischmehlindustrie, 40*(2), 22–28.
133. Sivaranjani, S., Prasath, V. A., Pandiselvam, R., Kothakota, A., & Khaneghah, A. M., (2021). Recent advances in applications of ozone in the cereal industry. *LWT, 146*, 111412.
134. Sujayasree, O., Chaitanya, A., Bhoite, R., Pandiselvam, R., Kothakota, A., Gavahian, M., & Mousavi, K. A., (2022). Ozone: An advanced oxidation technology to enhance

sustainable food consumption through mycotoxin degradation. *Ozone: Science and Engineering, 44*(1), 17–37.

135. Tao, Y., Zheng, Z., Yu, Y., Xu, J., Liu, X., Wu, X., & Zheng, Y., (2018). Supercritical fluid chromatography–tandem mass spectrometry-assisted methodology for rapid enantiomeric analysis of fenbuconazole and its chiral metabolites in fruits, vegetables, cereals, and soil. *Food Chemistry, 241*, 32–39.
136. Tappi, S., Tylewicz, U., & Dalla, R. M., (2020). Effect of nonthermal technologies on functional food compounds. In: *Sustainability of the Food System* (pp. 147–165): Elsevier.
137. Ten, B. L., Pfohl, K., Avramidis, G., Wieneke, S., Viöl, W., & Karlovsky, P., (2017). Plasma-based degradation of mycotoxins produced by *Fusarium*, *Aspergillus* and *Alternaria* species. *Toxins, 9*(3), 97.
138. Thacur, B., & Nelson, P., (1998). High pressure processing and preservation of foods. *Food Reviews International, 14*, 427–447.
139. Thirumdas, R., Sarangapani, C., & Annapure, U. S., (2015). Cold plasma: A novel non-thermal technology for food processing. *J. Food Biophysics, 10*(1), 1–11.
140. Tolcha, T., Gemechu, T., Al-Hamimi, S., Megersa, N., & Turner, C., (2020). High density supercritical carbon dioxide for the extraction of pesticide residues in onion with multivariate response surface methodology. *Molecules, 25*(4), 1012.
141. USFDA, (2017). *CFR-Code of Federal Regulations Title 21*. Retrieved from https://www.accessdata.fda.gov/scripts/cdrh/cfdocs/cfcfr/cfrsearch.cfm (accessed on 23 January 2024).
142. Valverde, A., Aguilera, A., Rodriguez, M., & Brotons, M., (2009). Evaluation of a multiresidue method for pesticides in cereals using supercritical fluid extraction and gas chromatographic detection. *Journal of Environmental Science and Health Part B, 44*(3), 204–213.
143. Vega-Gálvez, A., Giovagnoli, C., Pérez-Won, M., Reyes, J. E., Vergara, J., Miranda, M., & Di Scala, K., (2012). Application of high hydrostatic pressure to aloe vera (*Aloe barbadensis Miller*) gel: Microbial inactivation and evaluation of quality parameters. *Innovative Food Science and Emerging Technologies, 13*, 57–63.
144. Vijayalakshmi, S., Nadanasabhapathi, S., Kumar, R., & Sunny, K. S., (2018). Effect of pH and pulsed electric field process parameters on the aflatoxin reduction in model system using response surface methodology. *Journal of Food Science and Technology, 55*(3), 868–878.
145. Vorobiev, E., & Lebovka, N., (2010). Enhanced extraction from solid foods and biosuspensions by pulsed electrical energy. *Food Eng. Rev., 2*, 95–108. In.
146. Vounakis, H., (2020). Food irradiation harmonization in the European Union. In: *Irradiation for Food Safety and Quality* (pp. 26–31). CRC Press.
147. Wang, X., Wang, S., Yan, Y., Wang, W., Zhang, L., & Zong, W., (2020). The degradation of *Alternaria* mycotoxins by dielectric barrier discharge cold plasma. *Food Control, 117*, 107333.
148. Wang, Y., Shang, J., Cai, M., Liu, Y., & Yang, K., (2022). Detoxification of mycotoxins in agricultural products by non-thermal physical technologies: A review of the past five years. *Critical Reviews in Food Science and Nutrition*, 1–12.

149. Watkins, D., (2012). Regulatory and legislative issues for non-thermal technologies: An EU perspective. In: *Novel Thermal and Non-Thermal Technologies for Fluid Foods* (pp. 473–494): Elsevier.
150. Wei, J. N., Zeng, X. A., Tang, T., Jiang, Z., & Liu, Y. Y., (2018). Unfolding and nanotube formation of ovalbumin induced by pulsed electric field. *Innovative Food Science and Emerging Technologies, 45*, 249–254.
151. Wertheim, J. H., (1960). *Radiation Sterilization of Fluid Food Products.* In: Google Patents.
152. Wielogorska, E., Ahmed, Y., Meneely, J., Graham, W. G., Elliott, C. T., & Gilmore, B. F., (2019). A holistic study to understand the detoxification of mycotoxins in maize and impact on its molecular integrity using cold atmospheric plasma treatment. *Food Chemistry, 301*, 125281.
153. Woldemariam, H. W., & Emire, S. A., (2019). High pressure processing of foods for microbial and mycotoxins control: Current trends and future prospects. *Cogent Food and Agriculture, 5*(1), 1622184.
154. Wu, Y., Cheng, J. H., & Sun, D. W., (2021). Blocking and degradation of aflatoxins by cold plasma treatments: Applications and mechanisms. *Trends in Food Science and Technology, 109*, 647–661.
155. Zhang, Y., Hou, Y., Zhang, Y., Chen, J., Chen, F., Liao, X., & Hu, X., (2012). Reduction of diazinon and dimethoate in apple juice by pulsed electric field treatment. *Journal of the Science of Food and Agriculture, 92*(4), 743–750.
156. Zhang, Y., Xiao, Z., Chen, F., Ge, Y., Wu, J., & Hu, X., (2010). Degradation behavior and products of malathion and chlorpyrifos spiked in apple juice by ultrasonic treatment. *Ultrasonics Sonochemistry, 17*(1), 72–77.
157. Zhang, Y., Zhang, W., Liao, X., Zhang, J., Hou, Y., Xiao, Z., & Hu, X., (2010). Degradation of diazinon in apple juice by ultrasonic treatment. *Ultrasonics Sonochemistry, 17*(4), 662–668.
158. Zhang, Z. H., Wang, L. H., Zeng, X. A., Han, Z., & Brennan, C. S., (2019). Non-thermal technologies and its current and future application in the food industry: A review. *International Journal of Food Science and Technology, 54*(1), 1–13.
159. Zhou, D., Li, T., Cong, K., Suo, A., & Wu, C., (2022). Influence of cold plasma on quality attributes and aroma compounds in fresh-cut cantaloupe during low temperature storage. *LWT, 154*, 112893.
160. Zhou, R., Zhou, R., Yu, F., Xi, D., Wang, P., Li, J., & Ostrikov, K. K., (2018). Removal of organophosphorus pesticide residues from *Lyciumbarbarum* by gas phase surface discharge plasma. *Chemical Engineering Journal, 342*, 401–409.
161. Zhou, Y., Watkins, P., Oiseth, S., Cochet-Broch, M., Sikes, A. L., Chen, C., & Buckow, R., (2021). High pressure processing improves the sensory quality of sodium-reduced chicken sausage formulated with three anion types of potassium salt. *Food Control, 126*, 108008.
162. Zhu, F., (2018). Effect of ozone treatment on the quality of grain products. *Food Chemistry, 264*, 358–366.

INDEX

D

H

I

J

K

L

M

N

O

P

Q

R

S

T

Y

Z